Fctrow

An
Introduction
to
Bioinformatics
Algorithms

An Introduction to Bioinformatics Algorithms

Neil C. Jones
Pavel A. Pevzner

A Bradford Book
The MIT Press
Cambridge, Massachusetts
London, England

MIT Press books may be purchased at special quantity discounts for business or sales promotional use. For information, please email special_sales@mitpress.mit.edu or write to Special Sales Department, The MIT Press, 5 Cambridge Center, Cambridge, MA 02142.

Typeset in 10/13 Lucida Bright by the authors using LaTeX 2_ε.
Printed and bound in the United States of America.

Library of Congress Cataloging-in-Publication Data

Jones, Neil C.
    An introduction to bioinformatics algorithms/ by Neil C. Jones and Pavel A. Pevzner.
        p.   cm.—(computational molecular biology series)
    "A Bradford book."
    Includes bibliographical references and index (p.   ).
    ISBN 0-262-10106-8 (hc : alk. paper)
    1. Bioinformatics. 2. Algorithms. I. Pevzner, Pavel. II. Title
QH324.2.J66 2004
570'.285—dc22   2004048289
                                                                        CIP

10 9 8 7 6 5 4 3 2

𝄪 Computational Molecular Biology

Sorin Istrail, Pavel Pevzner, and Michael Waterman, editors

Computational molecular biology is a new discipline, bringing together computational, statistical, experimental, and technological methods, which is energizing and dramatically accelerating the discovery of new technologies and tools for molecular biology. The MIT Press Series on Computational Molecular Biology is intended to provide a unique and effective venue for the rapid publication of monographs, textbooks, edited collections, reference works, and lecture notes of the highest quality.

Computational Molecular Biology: An Algorithmic Approach
Pavel A. Pevzner, 2000

Computational Methods for Modeling Biochemical Networks
James M. Bower and Hamid Bolouri, editors, 2001

Current Topics in Computational Molecular Biology
Tao Jiang, Ying Xu, and Michael Q. Zhang, editors, 2002

Gene Regulation and Metabolism: Postgenomic Computation Approaches
Julio Collado-Vides, editor, 2002

Microarrays for an Integrative Genomics
Isaac S. Kohane, Alvin Kho, and Atul J. Butte, 2002

Kernel Methods in Computational Biology
Bernhard Schölkopf, Koji Tsuda, and Jean-Philippe Vert, 2004

An Introduction to Bioinformatics Algorithms
Neil C. Jones and Pavel A. Pevzner, 2004

For my Grandfather, who lived to write and vice versa.
—NCJ

To Chop-up and Manny.
—PAP

Contents in Brief

Subject	Exhaustive Search	Greedy Algorithms	Dynamic Programming Algorithms	Divide-and-Conquer Algorithms	Graph Algorithms	Combinatorial Pattern Matching	Clustering and Trees	Hidden Markov Models	Randomized Algorithms
	4	5	6	7	8	9	10	11	12
Mapping DNA	○								
Sequencing DNA					○				
Comparing Sequences			○	○		○			
Predicting Genes			○						
Finding Signals	○	○						○	○
Identifying Proteins					○				
Repeat Analysis						○			
DNA Arrays					○		○		
Genome Rearrangements		○							
Molecular Evolution							○		

Featuring historical perspectives from:

Contents

Preface

In the early 1990s when one of us was teaching his first bioinformatics class, he was not sure that there would be enough students to teach. Although the Smith-Waterman and **BLAST** algorithms had already been developed they had not become the household names among biologists that they are today. Even the term "bioinformatics" had not yet been coined. DNA arrays were viewed by most as intellectual toys with dubious practical application, except for a handful of enthusiasts who saw a vast potential in the technology. A few bioinformaticians were developing new algorithmic ideas for nonexistent data sets: David Sankoff laid the foundations of genome rearrangement studies at a time when there was practically no gene order data, Michael Waterman and Gary Stormo were developing motif finding algorithms when there were very few promoter samples available, Gene Myers was developing sophisticated fragment assembly tools when no bacterial genome has been assembled yet, and Webb Miller was dreaming about comparing billion-nucleotide-long DNA sequences when the $172, 282$-nucleotide Epstein-Barr virus was the longest GenBank entry. GenBank itself just recently made a transition from a series of bound (paper!) volumes to an electronic database on magnetic tape that could be sent to scientists worldwide.

One has to go back to the mid-1980s and early 1990s to fully appreciate the revolution in biology that has taken place in the last decade. However, bioinformatics has affected more than just biology—it has also had a profound impact on the computational sciences. Biology has rapidly become a large source of new algorithmic and statistical problems, and has arguably been the target for more algorithms than any of the other fundamental sciences. This link between computer science and biology has important educational implications that change the way we teach computational ideas to biologists, as well as how applied algorithmics is taught to computer scientists.

For many years computer science was taught to only computer scientists, and only rarely to students from other disciplines. A biology student in an algorithms class would be a surprising and unlikely (though entirely welcome) guest in the early 1990s. But these things change; many biology students now take some sort of Algorithms 101. At the same time, curious computer science students often take Genetics 101 and Bioinformatics 101. Although these students are still relatively rare, keep in mind that the number of bioinformatics classes in the early 1990s was so small as to be considered nonexistent. But that number is not so small now. We envision that undergraduate bioinformatics classes will become a permanent component at every major university. This is a feature, not a bug.

This is an introductory textbook on bioinformatics algorithms and the computational ideas that have driven them through the last twenty years. There are many important probabilistic and statistical techniques that we do not cover, nor do we cover many important research questions that bioinformaticians are currently trying to answer. We deliberately do not cover all areas of computational biology; for example, important topics like protein folding are not even discussed. The very first bioinformatics textbooks were Waterman, 1995 (108), which contains excellent coverage of DNA statistics and Gusfield, 1997 (44) which includes an encyclopedia of string algorithms. Durbin et al., 1998 (31) and Baldi and Brunak, 1997 (7) emphasize Hidden Markov Models and machine learning techniques; Baxevanis and Ouellette, 1998 (10) is an excellent practical guide to bioinformatics; Mount, 2001 (76) excels in showing the connections between biological problems and bioinformatics techniques; and Bourne and Weissig, 2002 (15) focuses on protein bioinformatics. There are also excellent web-based lecture notes for many bioinformatics courses and we learned a lot about the pedagogy of bioinformatics from materials on the World Wide Web by Serafim Batzoglou, Dick Karp, Ron Shamir, Martin Tompa, and others.

Website

We have created an extensive website to accompany this book at
                    http://www.bioalgorithms.info
This website contains a number of features that complement the book. For example, though this book does not contain a glossary, we provide this service, a searchable index, and a set of community message boards, at the above web address. Technically savvy students can also download practical

bioinformatics exercises, sample implementations of the algorithms in this book, and sample data to test them with. Instructors and students may find the prepackaged lecture notes on the website to be especially helpful. It is our hope that this website be used as a repository of information that will help introduce students to the diverse world of bioinformatics.

Acknowledgements

We are indebted to those who kindly agreed to be featured in the biographical sketches scattered throughout the book. Their insightful and heartfelt responses definitely made these the most interesting part of this book. Their life stories and views of the challenges that lay ahead will undoubtedly inspire students in the exploration of the unknown. There are many more scientists whose bioboxes we would like to have in this book and it is only the page limit (which turned out to be 200 pages too small) that prevented us from commissioning more of them. Special thanks go to Ethan Bier who inspired us to include biographical sketches in this book.

This book would not have been possible without the diligent teaching assistants in bioinformatics courses taught during the winter and fall of 2003 and 2004: Derren Barken, Bryant Forsgren, Eugene Ke, Coleman Mosley, and Degui Zhi all helped find technical errors, refine practical exercises, and design problems in the book. Helen Wu and John Allison spent many hours making technical figures, which is a thankless task like no other. We are also grateful to Vagisha Sharma who was kind enough to read the book from cover to cover and provide insightful comments and, unfortunately, bugs in the pseudocode. Steve Wasserman provided us with invaluable comments from a biologist's point of view that eventually led to new sections in the book. Alkes Price and Haixu Tang pointed out ambiguities and helped clarify the chapters on graphs and clustering. Ben Raphael and Patricia Jones provided feedback on the early chapters and helped avoid some potential misunderstandings. Dan Gilbert, of Dan Gilbert Art Group, Inc. kindly provided us with Triazzles to illustrate the problems of DNA sequence assembly.

Our special thanks go to Randall Christopher, the artist behind the website www.kleemanandmike.com. Randall illustrated the book and designed many unique graphical representations of some bioinformatics algorithms.

It has been a pleasure to work with Robert Prior of The MIT Press. With sufficient patience and prodding, he managed to keep us on track. We also appreciate the meticulous copyediting of G. W. Helfrich.

Finally, we thank the many students in different undergraduate and graduate bioinformatics classes at UCSD who provided comments on earlier versions of this book.

PAP would like to thank several people who taught him different aspects of computational molecular biology. Andrey Mironov taught him that common sense is perhaps the most important ingredient of any applied research. Mike Waterman was a terrific teacher at the time PAP moved from Moscow to Los Angeles, both in science and in life. PAP also thanks Alexander Karzanov, who taught him combinatorial optimization, which, surprisingly, remains the most useful set of skills in his computational biology research. He especially thanks Mark Borodovsky who convinced him to switch into the field of bioinformatics in 1985, when it was an obscure discipline with an uncertain future.

PAP also thanks his former students, postdocs, and lab members who taught him most of what he knows: Vineet Bafna, Guillaume Bourque, Sridhar Hannenhalli, Steffen Heber, Earl Hubbell, Uri Keich, Zufar Mulyukov, Alkes Price, Ben Raphael, Sing-Hoi Sze, Haixu Tang, and Glenn Tesler.

NCJ would like to thank his mentors during undergraduate school— Toshihiko Takeuchi, Harry Gray, John Baldeschwieler, and Schubert Soares—for patiently but firmly teaching him that persistence is one of the more important ingredients in research. Also, he thanks the admissions committee at the University of California, San Diego who gambled on a chemist-turned-programmer, hopefully for the best.

Neil Jones and Pavel Pevzner
La Jolla, California, 2004

1 *Introduction*

Imagine Alice, Bob, and two piles of ten rocks. Alice and Bob are bored one Saturday afternoon so they play the following game. In each turn a player may either take one rock from a single pile, or one rock from both piles. Once the rocks are taken, they are removed from play; the player that takes the last rock wins the game. Alice moves first.

It is not immediately clear what the winning strategy is, or even if there is one. Does the first player (or the second) always have an advantage? Bob tries to analyze the game and realizes that there are too many variants in the game with two piles of ten rocks (which we will refer to as the *10+10 game*). Using a reductionist approach, he first tries to find a strategy for the simpler 2+2 game. He quickly sees that the second player—himself, in this case—wins any 2+2 game, so he decides to write the "winning recipe":

> If Alice takes one rock from each pile, I will take the remaining rocks and win. If Alice takes one rock, I will take one rock from the same pile. As a result, there will be only one pile and it will have two rocks in it, so Alice's only choice will be to take one of them. I will take the remaining rock to win the game.

Inspired by this analysis, Bob makes a leap of faith: the second player (i.e., himself) wins in any $n+n$ game, for $n \geq 2$. Of course, every hypothesis must be confirmed by experiment, so Bob plays a few rounds with Alice. It turns out that sometimes he wins and sometimes he loses. Bob tries to come up with a simple recipe for the 3+3 game, but there are a large number of different game sequences to consider, and the recipe quickly gets too complicated. There is simply no hope of writing a recipe for the 10+10 game because the number of different strategies that Alice can take is enormous.

Meanwhile, Alice quickly realizes that she will always lose the 2+2 game, but she does not lose hope of finding a winning strategy for the 3+3 game.

Moreover, she took Algorithms 101 and she understands that recipes written in the style that Bob uses will not help very much: recipe-style instructions are not a sufficiently expressive language for describing algorithms. Instead, she begins by drawing the following table filled with the symbols ↑, ←, ↘, and *. The entry in position (i, j) (i.e., the ith row and the jth column) describes the moves that Alice will make in the $i + j$ game, with i and j rocks in piles A and B respectively. A ← indicates that she should take one stone from pile B. A ↑ indicates that she should take one stone from pile A. A ↘ indicates that she should take one stone from each pile, and * indicates that she should not bother playing the game because she will definitely lose against an opponent who has a clue.

	0	1	2	3	4	5	6	7	8	9	10
0	*	←	*	←	*	←	*	←	*	←	*
1	↑	↘	↑	↘	↑	↘	↑	↘	↑	↘	↑
2	*	←	*	←	*	←	*	←	*	←	*
3	↑	↘	↑	↘	↑	↘	↑	↘	↑	↘	↑
4	*	←	*	←	*	←	*	←	*	←	*
5	↑	↘	↑	↘	↑	↘	↑	↘	↑	↘	↑
6	*	←	*	←	*	←	*	←	*	←	*
7	↑	↘	↑	↘	↑	↘	↑	↘	↑	↘	↑
8	*	←	*	←	*	←	*	←	*	←	*
9	↑	↘	↑	↘	↑	↘	↑	↘	↑	↘	↑
10	*	←	*	←	*	←	*	←	*	←	*

For example, if she is faced with the 3+3 game, she finds a ↘ in the third row and third column, indicating that she should take a rock from each pile. This makes Bob take the first move in a 2+2 game, which is marked with a *. No matter what he does, Alice wins. Suppose Bob takes a rock from pile B—this leads to the 2+1 game. Alice again consults the table by reading the entry at (2,1), seeing that she should also take a rock from pile B leaving two rocks in A. However, if Bob had instead taken a rock from pile A, Alice would consult entry (1,2) to find ↑. She again should also take a rock from pile A, leaving two rocks in pile B.

Impressed by the table, Bob learns how to use it to win the 10+10 game. However, Bob does not know how to construct a similar table for the 20+20 game. The problem is not that Bob is stupid, but that he has not studied algorithms. Even if, through sheer luck, Bob figured how to always win the

20+20 game, he could neither say with confidence that it would work no matter what Alice did, nor would he even be able to write down the recipe for the general $n + n$ game. More embarrassing to Bob is that the a general 10+10+10 game with three piles would turn into an impossible conundrum for him.

There are two things Bob could do to remedy his situation. First, he could take a class in algorithms to learn how to solve problems like the rock puzzle. Second, he could memorize a suitably large table that Alice gives him and use that to play the game. Leading questions notwithstanding, what would you do as a biologist?

Of course, the answer we expect to hear from most rational people is "Why in the world do I care about a game with two nerdy people and a bunch of rocks? I'm interested in biology, and this game has nothing to do with me." This is not actually true: the rock game is in fact the ubiquitous *sequence alignment* problem in disguise. Although it is not immediately clear what DNA sequence alignment and the rock game have in common, the computational idea used to solve both problems is the same. The fact that Bob was not able to find the strategy for the game indicates that he does not understand how alignment algorithms work either. He might disagree if he uses alignment algorithms or BLAST[1] on a daily basis, but we argue that since he failed to come up with a strategy for the 10+10 rock game, he will also fail when confronted with a new flavor of alignment problem or a particularly complex similarity analysis. More troubling to Bob, he may find it difficult to compete with the scads of new biologists who think algorithmically about biological problems.[2]

Many biologists are comfortable using algorithms like BLAST without really understanding how the underlying algorithm works. This is not substantially different from a diligent robot following Alice's winning strategy table, but it does have an important consequence. BLAST solves a particular problem only approximately and it has certain systematic weaknesses. We're not picking on BLAST here: the reason that BLAST has these limitations is, in part, because of the particular problem that it solves. Users who do not know how BLAST works might misapply the algorithm or misinterpret the results it returns. Biologists sometimes use bioinformatics tools simply as computational protocols in quite the same way that an uninformed mathematician

---

1. BLAST is a database search tool—a Google for biological sequences—that will be introduced later in this book.
2. These "new biologists" have probably already found another even more elegant solution of the rocks problem that does not require the construction of a table.

might use experimental protocols without any background in biochemistry or molecular biology. In either case, important observations might be missed or incorrect conclusions drawn. Besides, intellectually interesting work can quickly become mere drudgery if one does not really understand it.

Many recent bioinformatics books cater to this sort of protocol-centric practical approach to bioinformatics. They focus on parameter settings, specific features of application, and other details without revealing the ideas behind the algorithms. This trend often follows the tradition of biology books of presenting material as a collection of facts and discoveries. In contrast, introductory books in algorithms usually focus on ideas rather than on the details of computational recipes.

Since bioinformatics is a computational science, a bioinformatics textbook should strive to present the principles that drive an algorithm's design, rather than list a stamp collection of the algorithms themselves. We hope that describing the intellectual content of bioinformatics will help retain your interest in the subject. In this book we attempt to show that a handful of algorithmic ideas can be used to solve a large number of bioinformatics problems. We feel that focusing on ideas has more intellectual value and represents a better long-term investment: protocols change quickly, but the computational ideas don't seem to.

We pursued a goal of presenting both the foundations of algorithms and the important results in bioinformatics under the same cover. A more thorough approach for a student would be to take an Introduction to Algorithms course followed by a Bioinformatics course, but this is often an unrealistic expectation in view of the heavy course load biologists have to take. To make bioinformatics ideas accessible to biologists we appeal to the innate algorithmic intuition of the student and try to avoid tedious proofs. The technical details are hidden unless they are absolutely necessary.[3]

This book covers both new and old areas in computational biology. Some topics, to our knowledge, have never been discussed in a textbook before, while others are relatively old-fashioned and describe some experimental approaches that are rarely used these days. The reason for including older topics is twofold. First, some of them still remain the best examples for introducing algorithmic ideas. Second, our goal is to show the progression of ideas in the field, with the implicit warning that hot areas in bioinformatics seem to come and go with alarming speed.

---

3. In some places we hide important computational and biological details and try to simplify the presentation. We will unavoidably be blamed later for "trivializing" bioinformatics.

One observation gained from teaching bioinformatics classes is that the interest of computer science students, who usually know little of biology, fades quickly when the students are introduced to biology without links to computational issues. The same happens to biologists if they are presented with seemingly unnecessary formalism with no links to real biological problems. To hold a student's interest, it is necessary to introduce biology and algorithms simultaneously. Our rather eclectic table of contents is a demonstration that attempts to reach this goal result in a somewhat interleaved organization of the material. However, we have tried to maintain a consistent algorithmic theme (e.g., graph algorithms) throughout each chapter.

Molecular biology and computer science are complex fields whose terminology and nomenclature can be formidable to the outsider. Bioinformatics merges the two fields, and adds a healthy dose of statistics, combinatorics, and other branches of mathematics. Like modern biologists who have to master the dense language of mathematics and computer science, mathematicians and computer scientists working in bioinformatics have to learn the language of biology. Although the question of who faces the bigger challenge is a topic hotly debated over pints of beer, this is not the first "invasion" of foreigners into biology; seventy years ago a horde of physicists infested biology labs, ultimately to revolutionize the field by deciphering the mystery of DNA.

Two influential scientists are credited with crossing the barrier between physics and biology: Max Delbrück and Erwin Schrödinger. Trained as physicists, their entrances into the field of biology were remarkably different. Delbrück, trained as an atomic physicist by Niels Bohr, quickly became an expert in genetics; in 1945 he was already teaching genetics to other biologists.[4] Schrödinger, on the other hand, never turned into a "certified" geneticist and remained somewhat of a biological dilettante. However, his book *What Is Life?*, published in 1944, was influential to an entire generation of physicists and biologists. Both James Watson (a biology student who wanted to be a naturalist) and Francis Crick (a physicist who worked on magnetic mines) switched careers to DNA science after reading Shrödinger's book. Another Nobel laureate, Sydney Brenner, even admitted to stealing a copy from the public library in Johannesburg, South Africa.

Like Delbrück and Schrödinger, there is great variety in the biological background of today's computer scientists-turned-bioinformaticians. Some of them have become experts in biology—though very few put on lab coats

---

4. Delbrück founded the famous phage genetics courses at Cold Spring Harbor Laboratory.

and perform experiments—while others remain biological dilettantes. Although there exists an opinion that every bioinformatician should be an expert in *both* biology and computer science, we are not sure that this is feasible. First, it takes a lot of work just to master one of the two, so perhaps understanding two in equal amounts is a bit much. Second, it is good to recall that the first pioneers of DNA science were, in fact, self-proclaimed dilettantes. James Watson knew almost no organic or physical chemistry before he started working on the double helix; Francis Crick, being a physicist, knew very little biology. Neither saw any need to know about (let alone memorize) the chemical structure of the four nucleotide bases when they discovered the structure of DNA.[5] When asked by Erwin Chargaff how they could possibly expect to resolve the structure of DNA without knowing the structures of its constituents, they responded that they could always look up the structures in a book if the need arose. Of course, they *understood* the physical principles behind a compound's structure.

The reality is that even the most biologically oriented bioinformaticians are experts only in some specific area of biology. Like Delbrück, who probably would never have passed an exam in biology in the 1930s (when zoology and botany remained the core of the mainstream biological curriculum), a typical modern-day bioinformatician is unlikely to pass the sequence of organic chemistry, biochemistry, and structural biochemistry classes that a "real" biologist has to take. The question of how much biology a good computer scientist–turned–bioinformatician has to know seems to be best answered with "enough to deeply understand the biological problem and to turn it into an adequate computational problem." This book provides a very brief introduction to biology. We do not claim that this is the best approach. Fortunately, an interested reader can use Watson's approach and look up the biological details in the books when the need arises, or read pages 1 through 1294 of Alberts and colleagues' (including Watson) book *Molecular Biology of the Cell* (3).

This book is what we, as computer scientists, believe that a modern biologist ought to know about computer science if he or she would be a successful researcher.

---

5. Accordingly, we do not present anywhere in this book the chemical structures of either nucleotides or amino acids. No algorithm in this book requires knowledge of their structure.

2 *Algorithms and Complexity*

This book is about how to design algorithms that solve biological problems. We will see how popular bioinformatics algorithms work and we will see what principles drove their design. It is important to understand how an algorithm works in order to be confident in its results; it is even more important to understand an algorithm's design methodology in order to identify its potential weaknesses and fix them.

Before considering any algorithms in detail, we need to define loosely what we mean by the word "algorithm" and what might qualify as one. In many places throughout this text we try to avoid tedious mathematical formalisms, yet leave intact the rigor and intuition behind the important concept.

2.1 What Is an Algorithm?

Roughly speaking, an *algorithm* is a sequence of instructions that one must perform in order to solve a well-formulated *problem*. We will specify problems in terms of their *inputs* and their *outputs*, and the algorithm will be the method of translating the inputs into the outputs. A well-formulated problem is unambiguous and precise, leaving no room for misinterpretation.

In order to solve a problem, some entity needs to carry out the steps specified by the algorithm. A human with a pen and paper would be able to do this, but humans are generally slow, make mistakes, and prefer not to perform repetitive work. A computer is less intelligent but can perform simple steps quickly and reliably. A computer cannot understand English, so algorithms must be rephrased in a programming language such as C or Java in order to give specific instructions to the processor. Every detail must be specified to the computer in exactly the right format, making it difficult to describe algorithms; trifling details that a person would naturally understand must be specified. If a computer were to put on shoes, one would need to tell it to find a pair that both matches and fits, to put the left shoe on the left foot, the right shoe on the right, and to tie the laces.[1] In this book, however, we prefer to simply leave it at "Put on a pair of shoes."

However, to understand how an algorithm works, we need some way of listing the steps that the algorithm takes, while being neither too vague nor too formal. We will use *pseudocode*, whose elementary operations are summarized below. Pseudocode is a language computer scientists often use to describe algorithms: it ignores many of the details that are required in a programming language, yet it is more precise and less ambiguous than, say, a recipe in a cookbook. Individually, the operations do not solve any particularly difficult problems, but they can be grouped together into minialgorithms called *subroutines* that do.

In our particular flavor of pseudocode, we use the concepts of *variables*, *arrays*, and *arguments*. A variable, written as x or *total*, contains some numerical value and can be assigned a new numerical value at different points throughout the course of an algorithm. An array of n elements is an ordered collection of n variables $a_1, a_2, \ldots, a_n$. We usually denote arrays by boldface letters like $\mathbf{a} = (a_1, a_2, \ldots, a_n)$ and write the individual elements as a_i where i is between 1 and n. An algorithm in pseudocode is denoted by a name, followed by the list of arguments that it requires, like MAX(a, b) below; this is followed by the statements that describe the algorithm's actions. One can *invoke* an algorithm by passing it the appropriate values for its arguments. For example, MAX$(1, 99)$ would return the larger of 1 and 99. The operation **return** reports the result of the program or simply signals its end. Below are brief descriptions of the elementary commands that we use in the pseudocode throughout this book.[2]

Assignment

Format: $a \leftarrow b$

Effect: Sets the variable a to the value b.

---

1. It is surprisingly difficult to write an unambiguous set of instructions on how to tie a shoelace.
2. An experienced computer programmer might be confused by our not using "**end if**" or "**end for**", which is the conventional practice. We rely on indentation to demarcate blocks of pseudocode.

Example: $b \leftarrow 2$
                    $a \leftarrow b$

Result: The value of a is 2

Arithmetic

Format: $a + b, a - b, a \cdot b, a/b, a^b$

Effect: Addition, subtraction, multiplication, division, and exponentiation of numbers.

Example: $\text{DIST}(x1, y1, x2, y2)$
                    1 $dx \leftarrow (x2 - x1)^2$
                    2 $dy \leftarrow (y2 - y1)^2$
                    3 **return** $\sqrt{(dx + dy)}$

Result: $\text{DIST}(x1, y1, x2, y2)$ computes the Euclidean distance between points with coordinates $(x1, y1)$ and $(x2, y2)$. $\text{DISTANCE}(0, 0, 3, 4)$ returns 5.

Conditional

Format: **if** A is true
                         B
                    else
                         C

Effect: If statement A is true, executes instructions **B**, otherwise executes instructions **C**. Sometimes we will omit "**else C**," in which case this will either execute **B** or not, depending on whether A is true.

Example: $\text{MAX}(a, b)$
                    1 **if** $a < b$
                    2   **return** b
                    3 **else**
                    4   **return** a

Result: $\text{MAX}(a, b)$ computes the maximum of the numbers a and b. For example, $\text{MAX}(1, 99)$ returns 99.

for loops

Format: **for** $i \leftarrow a$ **to** b
           B

Effect: Sets i to a and executes instructions **B**. Sets i to $a + 1$ and executes instructions **B** again. Repeats for $i = a + 2, a + 3, \ldots, b - 1, b$.[3]

Example: SUMINTEGERS(n)
      1 $sum \leftarrow 0$
      2 **for** $i \leftarrow 1$ **to** n
      3      $sum \leftarrow sum + i$
      4 **return** sum

Result: SUMINTEGERS(n) computes the sum of integers from 1 to n. SUM-INTEGERS(10) returns $1 + 2 + \cdots + 10 = 55$.

while loops

Format: **while** A is true
           B

Effect: Checks the condition A. If it is true, then executes instructions **B**. Checks A again; if it's true, it executes **B** again. Repeats until A is not true.

Example: ADDUNTIL(b)
      1 $i \leftarrow 1$
      2 $total \leftarrow i$
      3 **while** $total \leq b$
      4      $i \leftarrow i + 1$
      5      $total \leftarrow total + i$
      6 **return** i

Result: ADDUNTIL(b) computes the smallest integer i such that $1 + 2 + \cdots + i$ is larger than b. For example, ADDUNTIL(25) returns 7, since

---

3. If a is larger than b, this loop operates in the reverse order: it sets i to a and executes instructions **B**, then repeats for $i = a - 1, a - 2, \ldots, b + 1, b$.

$1 + 2 + \cdots + 7 = 28$, which is larger than 25, but $1 + 2 + \cdots + 6 = 21$, which is smaller than 25.

Array access

Format: a_i

Effect: The ith number of array $\mathbf{a} = (a_1, \ldots a_i, \ldots a_n)$. For example, if $\mathbf{F} = (1, 1, 2, 3, 5, 8, 13)$, then $F_3 = 2$, and $F_4 = 3$.

Example: FIBONACCI(n)
            1 $F_1 \leftarrow 1$
            2 $F_2 \leftarrow 1$
            3 **for** $i \leftarrow 3$ **to** n
            4      $F_i \leftarrow F_{i-1} + F_{i-2}$
            5 **return** F_n

Result: FIBONACCI(n) computes the nth Fibonacci number. FIBONACCI(8) returns 21.

While computer scientists are accustomed to the pseudocode jargon above, we fear that some biologists reading it might decide that this book is too cryptic and therefore useless. Although modern biologists deal with algorithms on a daily basis, the language they use to describe an algorithm might be closer to the language used in a cookbook, like the pumpkin pie recipe in figure 2.1. Accordingly, some bioinformatics books are written in this familiar lingo as an effort to make biologists feel at home with different bioinformatics concepts. Unfortunately, the cookbook language is insufficient to describe more complex algorithmic ideas that are necessary for even the simplest tools in bioinformatics. The problem is that natural languages are not suitable for communicating algorithmic ideas more complex than the pumpkin pie. Computer scientists have yet to invent anything better than pseudocode for this purpose, so we use it in this book.

To illustrate more concretely the distinction between pseudocode and an informal language, we can write an "algorithm" to create a pumpkin pie that mimics the recipe shown in figure 2.1. The admittedly contrived pseudocode below, MAKEPUMPKINPIE, is quite a bit more explicit.

$1\frac{1}{2}$ cups canned or cooked pumpkin
1 cup brown sugar, firmly packed
$\frac{1}{2}$ teaspoon salt
2 teaspoons cinnamon
1 teaspoon ginger
2 tablespoons molasses
3 eggs, slightly beaten
12 ounce can of evaporated milk
1 unbaked pie crust

Combine pumpkin, sugar, salt, ginger, cinnamon, and molasses. Add eggs and milk and mix thoroughly. Pour into unbaked pie crust and bake in hot oven (425 degrees Fahrenheit) for 40 to 45 minutes, or until knife inserted comes out clean.

Figure 2.1   A recipe for pumpkin pie.

MAKEPUMPKINPIE(*pumpkin, sugar, salt, spices, eggs, milk, crust*)
1   PREHEATOVEN(425)
2   *filling* ← MIXFILLING(*pumpkin, sugar, salt, spices, eggs, milk*)
3   *pie* ← ASSEMBLE(*crust, filling*)
4   **while**   knife inserted does not come out clean
5   BAKE(*pie*)
6   **output** "Pumpkin pie is complete"
7   **return** *pie*

MIXFILLING(*pumpkin, sugar, salt, spices, eggs, milk*)
1   *bowl* ← Get a bowl from cupboard
2   PUT(*pumpkin, bowl*)
3   PUT(*sugar, bowl*)
4   PUT(*salt, bowl*)
5   PUT(*spices, bowl*)
6   STIR(*bowl*)
7   PUT(*eggs, bowl*)
8   PUT(*milk, bowl*)
9   STIR(*bowl*)
10   *filling* ← Contents of *bowl*
11   **return** *filling*

MAKEPUMPKINPIE *calls* (i.e., activates) the subroutine MIXFILLING, which uses **return** to return the pie filling. The operation **return** terminates the execution of the subroutine and returns a result to the routine that called it, in this case MAKEPUMPKINPIE. When the pie is complete, MAKEPUMPKINPIE notifies and returns the pie to whomever requested it. The entity *pie* in MAKEPUMPKINPIE is a *variable* that represents the pie in the various stages of cooking.

A subroutine, such as MIXFILLING, will normally need to return the result of some important calculation. However, in some cases the inputs to the subroutine might be invalid (e.g., if you gave the algorithm watermelon instead of pumpkin). In these cases, a subroutine may return no value at all and **output** a suitable error message. When an algorithm is finished calculating a result, it naturally needs to output that result and stop executing. The operation **output** displays information to an interested user.[4]

A subtle observation is that MAKEPUMPKINPIE does not in fact make a pumpkin pie, but only tells you *how* to make a pumpkin pie at a fairly abstract level. If you were to build a machine that follows these instructions, you would need to make it specific to a particular kitchen and be tirelessly explicit in all the steps (e.g., how many times and how hard to stir the filling, with what kind of spoon, in what kind of bowl, etc.) This is exactly the difference between pseudocode (the abstract sequence of steps to solve a well-formulated computational problem) and computer code (a set of detailed instructions that one particular computer will be able to perform). We reiterate that the function of pseudocode in this book is only to communicate the idea behind an algorithm, and that to actually use an algorithm in this book you would need to turn the pseudocode into computer code, which is not always easy.

We will often avoid tedious details in the specification of an algorithm by specifying parts of it in English (e.g., "Get a bowl from cupboard"), using operations that are not listed in our description of pseudocode, or by omitting certain details that are unimportant. We assume that, in the case of confusion, the reader will fill in the details using pseudocode operations in a sensible way.

---

4. Exactly how this is done remains beyond the scope of pseudocode and really does not matter.

2.2   Biological Algorithms versus Computer Algorithms

Nature uses algorithm-like procedures to solve biological problems, for example, in the process of *DNA replication*. Before a cell can divide, it must first make a complete copy of all its genetic material.

DNA replication proceeds in phases, each of which requires an elaborate cooperation between different types of molecules. For the sake of simplicity, we describe the replication process as it occurs in bacteria, rather than the replication process in humans or other mammals, which is quite a bit more involved. The basic mechanism was proposed by James Watson and Francis Crick in the early 1950s, but could only be verified through the ingenious Meselson-Stahl experiment of 1957. The replication process starts from a pair of complementary[5] strands of DNA and ends up with two pairs of complementary strands.[6]

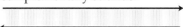

1. A molecular machine (in other words, a protein complex) called a *DNA helicase*, binds to the DNA at certain positions called *replication origins*.

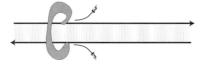

2. Helicase wrenches apart the two strands of DNA, creating a so-called *replication fork*. The two strands are complementary and run in opposite directions (one strand is denoted $3' \rightarrow 5'$, the other $5' \rightarrow 3'$). Two other molecular machines, *topoisomerase* and *single-strand binding protein* (not shown) bind to the single strands to help relieve the instability of single-stranded DNA.

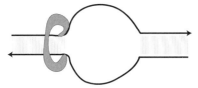

---

5. Complementarity is described in chapter 3.

6. It is possible that computer scientists will spontaneously abort due to the complexity of this system. While biologists feel at home with a description of DNA replication, computer scientists may find it too overloaded with unfamiliar terms. This example only illustrates what biologists use as "pseudocode;" the terms here are not crucial for understanding the rest of the book.

3. *Primers*, which are short single strands of RNA, are synthesized by a protein complex called *primase* and latch on to specific positions in the newly
   opened strands, providing an anchor for the next step. Without primers,
   the next step cannot begin.

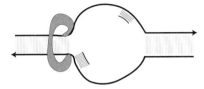

4. A *DNA polymerase* (yet another molecular machine) binds to each freshly
   separated *template* strand of the DNA; the DNA polymerase traverses the
   parent strands only in the $3' \rightarrow 5'$ direction. Therefore, the DNA poly-
   merases attached to the two DNA strands move in opposite directions.

5. At each nucleotide, DNA polymerase matches the template strand's nucleotide with the complementary base, and adds it to the growing synthesized chain. Prior to moving to the next nucleotide, DNA polymerase
   checks to ensure that the correct base has been paired at the current position; if not, it removes the incorrect base and retries.

   Since DNA polymerase can only traverse DNA in the $3' \rightarrow 5'$ direction,
   and since the two strands of DNA run in opposite directions, only one
   strand of the template DNA can be used by polymerase to continuously
   synthesize its complement; the other strand requires occasional stopping
   and restarting. This results in short segments called *Okazaki fragments*.

6. Another molecular machine, *DNA ligase*, repairs the gaps in the newly
   synthesized DNA's backbone, effectively linking together all Okazaki fragments into a single molecule and cleaning any breaks in the primary strand.

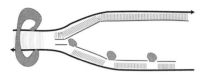

7. When all the DNA has been copied in such a manner, the original strands separate, so that two pairs of DNA strands are formed, each pair consisting of one old and one newly synthesized strand.

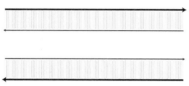

Obviously, an astounding amount of molecular logistics is required to ensure completely accurate DNA replication: DNA helicase separates strands, DNA polymerase ensures proper complementarity, and so on. However, in terms of the logic of the process, none of this complicated molecular machinery actually matters—to mimic this process in an algorithm we simply need to take a string which represents the DNA and return a copy of it.

String Duplication Problem:
Given a string of letters, return a copy.

   Input: A string $\mathbf{s} = (s_1, s_2, \ldots, s_n)$ of length n, as an array of characters.

   Output: A string representing a copy of $\mathbf{s}$.

Of course, this is a particularly easy problem to solve and yields absolutely no interesting algorithmic intuition. However it is still illustrative to write the pseudocode. The STRINGCOPY program below uses the string $\mathbf{t}$ to hold a copy of the input string $\mathbf{s}$, and returns the result $\mathbf{t}$.

STRINGCOPY$(\mathbf{s}, n)$
1   **for** $i \leftarrow 1$ **to** n
2       $t_i \leftarrow s_i$
3   **return** t

While STRINGCOPY is a trivial algorithm, the number of operations that a real computer performs to copy a string is surprisingly large. For one particular computer architecture, we may end up issuing thousands of instructions to a computer processor. Computer scientists distance themselves from this complexity by inventing programming languages that allow one to ignore many of these details. Biologists have not yet invented a similar "language" to describe biological algorithms working in the cell.

The amount of "intelligence" that the simplest organism, such as a bacterium, exhibits to perform any routine task—including replication—is amazing. Unlike STRINGCOPY, which only performs abstract operations, the bacterium really *builds* new DNA using materials that are floating near the replication fork. What would happen if it ran out? To prevent this, a bacterium examines the surroundings, imports new materials from outside, or moves off to forage for food. Moreover, it waits to begin copying its DNA until sufficient materials are available. These observations, let alone the coordination between the individual molecules, lead us to wonder whether even the most sophisticated computer programs can match the complicated behavior displayed by even a single-celled organism.

2.3   The Change Problem

The first—and often the most difficult—step in solving a computational problem is to identify precisely what the problem is. By using the techniques described in this book, you can then devise an algorithm that solves it. However, you cannot stop there. Two important questions to ask are: "Does it work correctly?" and "How long will it take?" Certainly you would not be satisfied with an algorithm that only returned correct results half the time, or took 600 years to arrive at an answer. Establishing reasonable expectations for an algorithm is an important step in understanding how well the algorithm solves the problem, and whether or not you trust its answer.

A problem describes a class of computational tasks. A problem *instance* is one particular input from that class. To illustrate the difference between a problem and an instance of a problem, consider the following example. You find yourself in a bookstore buying a fairly expensive pen for $4.23 which you pay for with a $5 bill (fig. 2.2). You would be due 77 cents in change, and the cashier now makes a decision as to exactly how you get it.[7] You would be annoyed at a fistful of 77 pennies or 15 nickels and 2 pennies, which raises the question of how to make change in the least annoying way. Most cashiers try

---

7. A penny is the smallest denomination in U.S. currency. A dollar is 100 pennies, a quarter is 25 pennies, a dime is 10, and a nickel is 5.

Figure 2.2   The subtle difference between a problem (top) and an instance of a problem (bottom).

to minimize the number of coins returned for a particular quantity of change. The example of 77 cents represents an instance of the United States Change problem, which we can formulate as follows.[8]

---

8. Though this problem is not at particularly relevant to biology, it serves as a useful tool to illustrate a number of different algorithmic approaches.

United States Change Problem:
Convert some amount of money into the fewest number of coins.

   Input: An amount of money, M, in cents.

   Output: The smallest number of quarters q, dimes d, nickels
   n, and pennies p whose values add to M (i.e., $25q + 10d +$
   $5n + p = M$ and $q + d + n + p$ is as small as possible).

The algorithm that is used by cashiers all over the United States to solve
this problem is simple:

USCHANGE(M)
1   **while** $M > 0$
2         $c \leftarrow$ Largest coin that is smaller than (or equal to) M
3   Give coin with denomination c to customer
4         $M \leftarrow M - c$

A slightly more detailed description of this algorithm is:

USCHANGE(M)
1   Give the integer part of $M/25$ quarters to customer.
2   Let *remainder* be the remaining amount due the customer.
3   Give the integer part of *remainder*$/10$ dimes to customer.
4   Let *remainder* be the remaining amount due the customer.
5   Give the integer part of *remainder*$/5$ nickels to customer.
6   Let *remainder* be the remaining amount due the customer.
7   Give *remainder* pennies to customer.

A pseudocode version of the above algorithm is:

USCHANGE(M)
1   $r \leftarrow M$
2   $q \leftarrow r/25$
3   $r \leftarrow r - 25 \cdot q$
4   $d \leftarrow r/10$
5   $r \leftarrow r - 10 \cdot d$
6   $n \leftarrow r/5$
7   $r \leftarrow r - 5 \cdot n$
8   $p \leftarrow r$
9   **return** (q, d, n, p)

When $r/25$ is not a whole number, we take the *floor* of $r/25$, that is, the integer part[9] of $r/25$. When the cashier runs USCHANGE(77), it returns three quarters, no dimes or nickels, and two pennies, which is the desired result (there is no other combination that has fewer coins and adds to 77 cents). First, the variable r is set to 77. Then q, the number of quarters, is set to the value 3, since $\lfloor 77/25 \rfloor = 3$. The variable r is then updated in line 3 to be equal to 2, which is the difference between the amount of money the cashier is changing (77 cents) and the three quarters he has chosen to return. The variables d and n—dimes and nickels, respectively—are subsequently set to 0 in lines 4 and 6, since $\lfloor 2/10 \rfloor = 0$ and $\lfloor 2/5 \rfloor = 0$; r remains unchanged on lines 5 and 7 since d and n are 0. Finally, the variable p, which stands for "pennies," is set to 2, which is the amount in variable r. The values of four variables—q, d, n, and p—are returned as the solution to the problem.[10]

2.4   Correct versus Incorrect Algorithms

As presented, USCHANGE lacks elegance and generality. Inherent in the algorithm is the assumption that you are changing United States currency, and that the cashier has an unlimited supply of each denomination—generally quarters are harder to come by than dimes. We would like to generalize the algorithm to accommodate different denominations without requiring a completely new algorithm for each one. To accomplish this, however, we must first generalize the problem to provide the algorithm with the denominations that it can change M into. The new Change problem below assumes

---

9. The floor of $77/25$, denoted $\lfloor 3.08 \rfloor$, is 3.
10. Inevitably, an experienced computer programmer will wring his or her hands at returning multiple, rather than single, answers from a subroutine. This is not actually a problem, but how this really works inside a computer is irrelevant to our discussion of algorithms.

that there are d denominations, rather than the four of the previous problem. These denominations are represented by an array $\mathbf{c} = (c_1, \ldots, c_d)$. For simplicity, we assume that the denominations are given in decreasing order of value. For example, in the case of the United States Change problem, $\mathbf{c} = (25, 10, 5, 1)$, whereas in the European Union Change problem, $\mathbf{c} = (20, 10, 5, 2, 1)$.

Change Problem:

Convert some amount of money M into given denominations, using the smallest possible number of coins.

> **Input:** An amount of money M, and an array of d denominations $\mathbf{c} = (c_1, c_2, \ldots, c_d)$, in decreasing order of value $(c_1 > c_2 > \cdots > c_d)$.
>
> **Output:** A list of d integers $i_1, i_2, \ldots, i_d$ such that $c_1 i_1 + c_2 i_2 + \cdots + c_d i_d = M$, and $i_1 + i_2 + \cdots + i_d$ is as small as possible.

We can solve this problem with an even simpler five line pseudocode than the previous USCHANGE algorithm.[11]

BETTERCHANGE($M, \mathbf{c}, d$)
1 $r \leftarrow M$
2 **for** $k \leftarrow 1$ **to** d
3      $i_k \leftarrow r / c_k$
4      $r \leftarrow r - c_k \cdot i_k$
5 **return** $(i_1, i_2, \ldots, i_d)$

We say that an algorithm is *correct* when it can translate every input instance into the correct output. An algorithm is *incorrect* when there is at least one input instance for which the algorithm does not produce the correct output. At first this seems unbalanced: if an algorithm fails on even a single input instance, then the whole algorithm is judged incorrect. This reflects a critical, yet healthy, pessimism that you should maintain when designing an algorithm: unless you can justify that an algorithm always returns correct results, you should consider it to be wrong.[12]

---

11. This is a trap! Try to figure out why this is wrong. That is, find some set of inputs for which this new algorithm does not return the correct answer.
12. Some problems are so difficult, however, that no practical algorithm that is correct has been found. Often, researchers rely on *approximation* algorithms (described in chapter 5) to produce

BETTERCHANGE is not a correct algorithm. Suppose we were changing 40 cents into coins with denominations of $c_1 = 25$, $c_2 = 20$, $c_3 = 10$, $c_4 = 5$, and $c_5 = 1$. BETTERCHANGE would incorrectly return 1 quarter, 1 dime, and 1 nickel, instead of 2 twenty-cent pieces. As contrived as this may seem, in 1875 a twenty-cent coin existed in the United States. Between 1865 and 1889, the U.S. Treasury even produced three-cent coins. How sure can we be that BETTERCHANGE returns the minimal number of coins for our modern currency, or for foreign countries? Determining the conditions under which BETTERCHANGE is a correct algorithm is left as a problem at the end of this chapter.

To correct the BETTERCHANGE algorithm, we could consider every possible combination of coins with denominations $c_1, c_2, \ldots, c_d$ that adds to M, and return the combination with the fewest. We do not need to consider any combinations with $i_1 > M/c_1$, or $i_2 > M/c_2$ (in general, i_k should not exceed M/c_k), because we would otherwise be returning an amount of money strictly larger than M. The pseudocode below uses the symbol $\sum$, meaning summation: $\sum_{i=1}^{m} a_i = a_1 + a_2 + a_3 + \cdots + a_m$. The pseudocode also uses the notion of "infinity" (∞) as an initial value for $smallestNumberOfCoins$; there are a number of ways to carry this out in a real computer, but the details are not important here.

BRUTEFORCECHANGE($M, \mathbf{c}, d$)
1    $smallestNumberOfCoins \leftarrow \infty$
2   **for each** $(i_1, \ldots, i_d)$ **from** $(0, \ldots, 0)$ **to** $(M/c_1, \ldots, M/c_d)$
3        $valueOfCoins \leftarrow \sum_{k=1}^{d} i_k c_k$
4   **if** $valueOfCoins = M$
5            $numberOfCoins \leftarrow \sum_{k=1}^{d} i_k$
6   **if** $numberOfCoins < smallestNumberOfCoins$
7                $smallestNumberOfCoins \leftarrow numberOfCoins$
8                $\mathbf{bestChange} \leftarrow (i_1, i_2, \ldots, i_d)$
9   **return** ($\mathbf{bestChange}$)

Line 2 iterates over every combination $(i_1, i_2, \ldots, i_d)$ of the d indices,[13] and

---

results. The implicit acknowledgment that we make in using those types of algorithms is that some better solution probably exists, but we were unable to find it.

13. An *array index* points to an element in an array. For example, if $\mathbf{c} = \{1, 1, 2, 3, 5, 8, 13, 21, 34\}$, then the index of element 8 is 6, while the index of element 34 is 9.

stops when it has reached $(M/c_1, M/c_2, \ldots, M/c_{d-1}, M/c_d)$:

$$
\begin{array}{ccccc}
(\quad 0, & 0, & \ldots, & 0, & 0 \quad) \\
(\quad 0, & 0, & \ldots, & 0, & 1 \quad) \\
(\quad 0, & 0, & \ldots, & 0, & 2 \quad) \\
& & \vdots & & \\
(\quad 0, & 0, & \ldots, & 0, & \frac{M}{c_d} \quad) \\
(\quad 0, & 0, & \ldots, & 1, & 0 \quad) \\
(\quad 0, & 0, & \ldots, & 1, & 1 \quad) \\
(\quad 0, & 0, & \ldots, & 1, & 2 \quad) \\
& & \vdots & & \\
(\quad 0, & 0, & \ldots, & 1, & \frac{M}{c_d} \quad) \\
& & \vdots & & \\
(\quad \frac{M}{c_1}, & \frac{M}{c_2}, & \ldots, & \frac{M}{c_{d-1}} - 1, & 0 \quad) \\
(\quad \frac{M}{c_1}, & \frac{M}{c_2}, & \ldots, & \frac{M}{c_{d-1}} - 1, & 1 \quad) \\
(\quad \frac{M}{c_1}, & \frac{M}{c_2}, & \ldots, & \frac{M}{c_{d-1}} - 1, & 2 \quad) \\
& & \vdots & & \\
(\quad \frac{M}{c_1}, & \frac{M}{c_2}, & \ldots, & \frac{M}{c_{d-1}} - 1, & \frac{M}{c_d} \quad) \\
(\quad \frac{M}{c_1}, & \frac{M}{c_2}, & \ldots, & \frac{M}{c_{d-1}}, & 0 \quad) \\
(\quad \frac{M}{c_1}, & \frac{M}{c_2}, & \ldots, & \frac{M}{c_{d-1}}, & 1 \quad) \\
(\quad \frac{M}{c_1}, & \frac{M}{c_2}, & \ldots, & \frac{M}{c_{d-1}}, & 2 \quad) \\
& & \vdots & & \\
(\quad \frac{M}{c_1}, & \frac{M}{c_2}, & \ldots, & \frac{M}{c_{d-1}}, & \frac{M}{c_d} \quad) \\
\end{array}
$$

We have omitted some details from the BRUTEFORCECHANGE algorithm. For example, there is no pseudocode operation that performs summation of d integers at one time, nor does it include any way to iterate over every combination of d indices. These subroutines are left as problems at the end of this chapter because they are instructive to work out in detail. We have made the hidden assumption that given any set of denominations we can change any amount of money M. This may not be true, for example in the (unlikely) case that the monetary system has no pennies (that is, $c_d > 1$).

How do we know that BRUTEFORCECHANGE does not suffer from the same problem as BETTERCHANGE did, namely that some input instance returns an incorrect result? Since BRUTEFORCECHANGE explores *all possible* combinations of denominations, it will eventually come across an optimal solution and record it as such in the **bestChange** array. Any combination of coins that adds to M must have at least as many coins as the optimal combination, so BRUTEFORCECHANGE will never overwrite **bestChange** with a suboptimal solution.

We revisit the Change problem in future chapters to improve on this solution. So far we have answered only one of the two important algorithmic questions ("Does it work?", but not "How fast is it?"). We shall see that BruteForceChange is not particularly speedy.

2.5   Recursive Algorithms

Recursion is one of the most ubiquitous algorithmic concepts. Simply, an algorithm is recursive if it calls itself.

The *Towers of Hanoi* puzzle, introduced in 1883 by a French mathematician, consists of three pegs, which we label from left to right as 1, 2, and 3, and a number of disks of decreasing radius, each with a hole in the center. The disks are initially stacked on the left peg (peg 1) so that smaller disks are on top of larger ones. The game is played by moving one disk at a time between pegs. You are only allowed to place smaller disks on top of larger ones, and any disk may go onto an empty peg. The puzzle is solved when all of the disks have been moved from peg 1 to peg 3.

Towers of Hanoi Problem:
Output a list of moves that solves the Towers of Hanoi.

    Input: An integer n.

    Output: A sequence of moves that will solve the n-disk Towers of Hanoi puzzle.

Solving the puzzle with one disk is easy: move the disk to the right peg. The two-disk puzzle is not much harder: move the small disk to the middle peg, then the large disk to the right peg, then the small disk to the right peg to rest on top of the large disk. The three-disk puzzle is somewhat harder, but the following sequence of seven moves solves it:

- Move disk from peg 1 to peg 3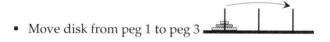

- Move disk from peg 1 to peg 2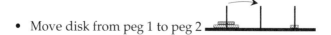

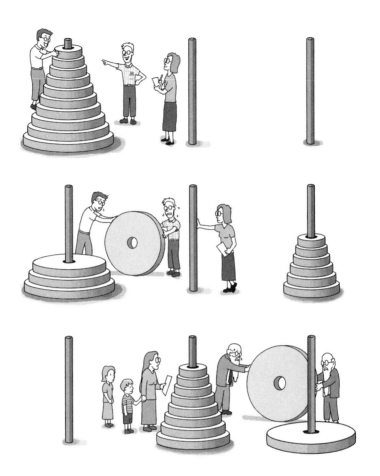

- Move disk from peg 3 to peg 2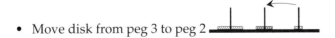

- Move disk from peg 1 to peg 3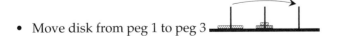

- Move disk from peg 2 to peg 1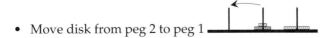

- Move disk from peg 2 to peg 3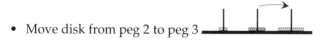

- Move disk from peg 1 to peg 3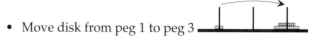

Now we will figure out how many steps are required to solve a four-disk puzzle. You cannot complete this game without moving the largest disk. However, in order to move the largest disk, we first had to move all the smaller disks to an empty peg. If we had four disks instead of three, then we would first have to move the top three to an empty peg (7 moves), then move the largest disk (1 move), then again move the three disks from their temporary peg to rest on top of the largest disk (another 7 moves). The whole procedure will take $7 + 1 + 7 = 15$ moves. More generally, to move a stack of size n from the left to the right peg, you first need to move a stack of size $n - 1$ from the left to the middle peg, and then from the middle peg to the right peg once you have moved the nth disk to the right peg. To move a stack of size $n - 1$ from the middle to the right, you first need to move a stack of size $n - 2$ from the middle to the left, then move the $(n - 1)$th disk to the right, and then move the stack of $n - 2$ from the left to the right peg, and so on.

At first glance, the Towers of Hanoi problem looks difficult. However, the following recursive algorithm solves the Towers of Hanoi problem with n disks. The iterative version of this algorithm is more difficult to write and analyze, so we do not present it here.

HANOITOWERS($n, fromPeg, toPeg$)
1   **if** $n = 1$
2   **output** "Move disk from peg $fromPeg$ to peg $toPeg$"
3   **return**
4   $unusedPeg \leftarrow 6 - fromPeg - toPeg$
5   HANOITOWERS($n - 1, fromPeg, unusedPeg$)
6   **output** "Move disk from peg $fromPeg$ to peg $toPeg$"
7   HANOITOWERS($n - 1, unusedPeg, toPeg$)
8   **return**

The variables $fromPeg$, $toPeg$, and $unusedPeg$ refer to the three different pegs so that HANOITOWERS($n, 1, 3$) moves n disks from the first peg to the third peg. The variable $unusedPeg$ represents which of the three pegs can

Table 2.1   The result of $6 - fromPeg - toPeg$ for all possible values of $fromPeg$ and $toPeg$.

$fromPeg$	$toPeg$	$unusedPeg$
1	2	3
1	3	2
2	1	3
2	3	1
3	1	2
3	2	1

serve as a temporary destination for the first $n-1$ disks. Note that $fromPeg + toPeg + unusedPeg$ is always equal to $1+2+3 = 6$, so the value of the variable $unusedPeg$ can be computed as $6 - fromPeg - toPeg$ which is determined in line 4 (see table 2.1). The subsequent statements (lines 5–7) then solve the smaller problem of moving the stack of size $n-1$ first to the temporary space, moving the largest disk, and then moving the $n-1$ small disks to the final destination. Note that we do not have to specify *which* disk the player should move from $fromPeg$ to $toPeg$: it is always the top disk currently residing on $fromPeg$ that gets moved.

Although the solution can be expressed in 8 lines of pseudocode, it requires a surprisingly long time to run. To solve a five-disk tower requires 31 moves, but to solve a hundred-disk tower would require more moves than there are atoms in the universe. The fast growth of the number of moves that HANOITOWERS requires is easy to see by noticing that every time HANOITOWERS$(n, 1, 3)$ is called, it calls itself twice for $n-1$, which in turn triggers four calls for $n-2$, and so on. We can illustrate this situation in a *recursion tree*, which is shown in figure 2.3. A call to HANOITOWERS$(4, 1, 3)$ results in calls HANOITOWERS$(3, 1, 2)$ and HANOITOWERS$(3, 2, 3)$; each of these results in calls to HANOITOWERS$(2, 1, 3)$, HANOITOWERS$(2, 3, 2)$ and HANOITOWERS$(2, 2, 1)$, HANOITOWERS$(2, 1, 3)$, and so on. Each call to the subroutine HANOITOWERS requires some amount of time, so we would like to know how much time the algorithm will take. This is determined in section 2.7.

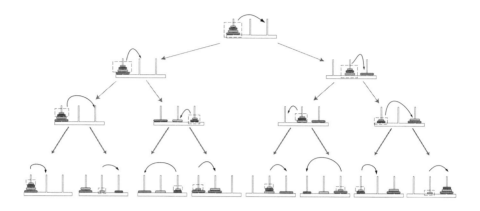

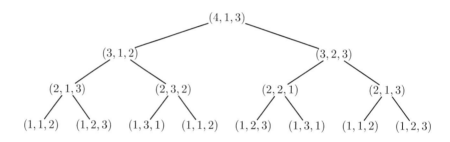

Figure 2.3   The recursion tree for a call to HANOITOWERS $(4, 1, 3)$, which solves the Towers of Hanoi problem of size 4. At each point in the tree, (i, j, k) stands for HANOITOWERS (i, j, k).

2.6   Iterative versus Recursive Algorithms

Recursive algorithms can often be rewritten to use iterative loops instead, and vice versa; it is a matter of elegance and clarity that dictates which technique is easier to use. Consider the problem of sorting a list of integers into ascending order.

Sorting Problem:
Sort a list of integers.

> **Input:** A list of n distinct integers $\mathbf{a} = (a_1, a_2, \ldots, a_n)$.

> **Output:** Sorted list of integers, that is, a reordering $\mathbf{b} = (b_1, b_2, \ldots, b_n)$ of integers from $\mathbf{a}$ such that $b_1 < b_2 < \cdots < b_n$.

The following algorithm, called SELECTIONSORT, is a naive but simple iterative method to solve the Sorting problem. First, SELECTIONSORT finds the smallest element in $\mathbf{a}$, and moves it to the first position by swapping it with whatever happens to be in the first position (i.e., a_1). Next, SELECTIONSORT finds the second smallest element in $\mathbf{a}$, and moves it to the second position, again by swapping with a_2. At the ith iteration, SELECTIONSORT finds the ith smallest element in $\mathbf{a}$, and moves it to the ith position. This is an intuitive approach at sorting, but is not the best-known one. If $\mathbf{a} = (7, 92, 87, 1, 4, 3, 2, 6)$, SELECTIONSORT$(\mathbf{a}, 8)$ takes the following seven steps:

$$(\overset{\frown}{7}, 92, 87, 1, 4, 3, 2, 6)$$

$$(1, \overset{\frown}{92}, 87, 7, 4, 3, 2, 6)$$

$$(1, 2, \overset{\frown}{87}, 7, 4, 3, 92, 6)$$

$$(1, 2, 3, \overset{\frown}{7}, 4, 87, 92, 6)$$

$$(1, 2, 3, 4, \overset{\frown}{7}, 87, 92, 6)$$

$$(1, 2, 3, 4, 6, \overset{\frown}{87}, 92, 7)$$

$$(1, 2, 3, 4, 6, 7, \overset{\frown}{92}, 87)$$

$$(1, 2, 3, 4, 6, 7, 87, 92)$$

SELECTIONSORT$(\mathbf{a}, n)$
1   **for** $i \leftarrow 1$ **to** $n - 1$
2       $a_j \leftarrow$ Smallest element among $a_i, a_{i+1}, \ldots a_n$.
3   Swap a_i and a_j
4   **return a**

Line 2 of SELECTIONSORT finds the smallest element over all elements of $\mathbf{a}$ that come after i, and fits nicely into a subroutine as follows. The subroutine

INDEXOFMIN(**array**, $first, last$) works with **array** and returns the index of the smallest element between positions $first$ and $last$ by examining each element from $array_{first}$ to $array_{last}$.

INDEXOFMIN(**array**, $first, last$)
1   $index \leftarrow first$
2   **for** $k \leftarrow first + 1$ **to** $last$
3   **if** $array_k < array_{index}$
4         $index \leftarrow k$
5   **return** $index$

For example, if $\mathbf{a} = (7, 92, 87, 1, 4, 3, 2, 6)$, then INDEXOFMIN($\mathbf{a}, 1, 8$) would be 4, since $a_4 = 1$ is smaller than any other element in $(a_1, a_2, \ldots, a_8)$. Similarly, INDEXOFMIN($\mathbf{a}, 5, 8$) would be 7, since $a_7 = 2$ is smaller than any other element in (a_5, a_6, a_7, a_8). We can now write SELECTIONSORT using this subroutine.

SELECTIONSORT($\mathbf{a}, n$)
1   **for** $i \leftarrow 1$ **to** $n - 1$
2      $j \leftarrow$ INDEXOFMIN($\mathbf{a}, i, n$)
3   Swap elements a_i and a_j
4   **return a**

To illustrate the similarity between recursion and iteration, we could instead have written SELECTIONSORT recursively (reusing INDEXOFMIN from above):

RECURSIVESELECTIONSORT($\mathbf{a}, first, last$)
1   **if** $first < last$
2      $index \leftarrow$ INDEXOFMIN($\mathbf{a}, first, last$)
3   Swap a_{first} with a_{index}
4      $\mathbf{a} \leftarrow$ RECURSIVESELECTIONSORT($\mathbf{a}, first + 1, last$)
5   **return a**

In this case, RECURSIVESELECTIONSORT($\mathbf{a}, 1, n$) performs exactly the same operations as SELECTIONSORT($\mathbf{a}, n$).

It may seem contradictory at first that RECURSIVESELECTIONSORT calls itself to get an answer, but the key to understanding this algorithm is to realize that each time it is called, it works on a smaller set of elements from the list until it reaches the end of the list; at the end, it no longer needs to recurse.

The reason that the recursion does not continue indefinitely is because the algorithm works toward a point at which it "bottoms out" and no longer needs to recurse—in this case, when $first = last$.

As convoluted as it may seem at first, recursion is often the most natural way to solve many computational problems as it was in the Towers of Hanoi problem, and we will see many recursive algorithms in the coming chapters. However, recursion can often lead to very inefficient algorithms, as this next example shows.

The Fibonacci sequence is a mathematically important, yet very simple, progression of numbers. The series was first studied in the thirteenth century by the early Italian mathematician Leonardo Pisano Fibonacci, who tried to compute the number of offspring of a pair of rabbits over the course of a year (fig. 2.4). Fibonacci reasoned that one pair of adult rabbits could create a new pair of rabbits in about the same time that it takes bunnies to grow into adults. Thus, in any given period, each pair of adult rabbits produces a new pair of baby rabbits, and all baby rabbits grow into adult rabbits.[14] If we let F_n represent the number of rabbits in period n, then we can determine the value of F_n in terms of F_{n-1} and F_{n-2}. The number of adult rabbits at time period n is equal to the number of rabbits (adult and baby) in the previous time period, or F_{n-1}. The number of baby rabbits at time period n is equal to the number of adult rabbits in F_{n-1}, which is F_{n-2}. Thus, the total number of rabbits at time period n is the number of adults plus the number of babies, that is, $F_n = F_{n-1} + F_{n-2}$, with $F_1 = F_2 = 1$. Consider the following problem:

Fibonacci Problem:
Calculate the nth Fibonacci number.

   Input: An integer n.

   Output: The nth Fibonacci number $F_n = F_{n-1} + F_{n-2}$ (with $F_1 = F_2 = 1$).

The simplest recursive algorithm, shown below, calculates F_n by calling itself to compute F_{n-1} and F_{n-2}. As figure 2.5 shows, this approach results in a large amount of duplicated effort: in calculating F_{n-1} we find the value

---

14. Fibonacci faced the challenge of adequately formulating the problem he was studying, one of the more difficult parts of bioinformatics research. The Fibonacci view of rabbit life is overly simplistic and inadequate: in particular, rabbits never die in his model. As a result, after just a few generations, the number of rabbits will be larger than the number of atoms in the universe.

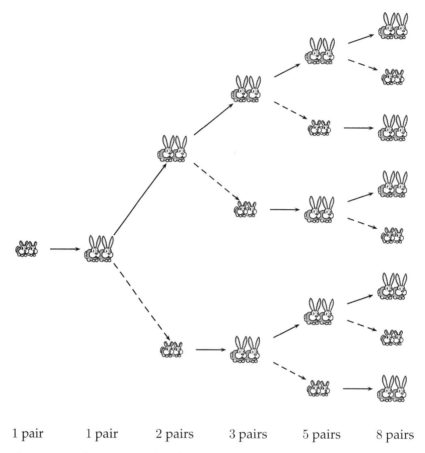

1 pair   1 pair   2 pairs   3 pairs   5 pairs   8 pairs

Figure 2.4   Fibonacci's model of rabbit expansion. A dashed line from a pair of big rabbits to a pair of little rabbits means that the pair of adult rabbits had bunnies.

of F_{n-2}, but we calculate it again from scratch in order to determine F_n. Therefore, most of the effort in this algorithm is wasted recomputing values that are already known.

RECURSIVEFIBONACCI(n)
1 **if** $n = 1$ **or** $n = 2$
2   **return** 1
3 **else**
4      $a \leftarrow$ RECURSIVEFIBONACCI($n - 1$)
5      $b \leftarrow$ RECURSIVEFIBONACCI($n - 2$)
6   **return** $a + b$

However, by using an array to save previously computed Fibonacci numbers, we can calculate the nth Fibonacci number without repeating work.

FIBONACCI(n)
1 $F_1 \leftarrow 1$
2 $F_2 \leftarrow 1$
3 **for** $i \leftarrow 3$ **to** n
4      $F_i \leftarrow F_{i-1} + F_{i-2}$
5 **return** F_n

In the language of the next section, FIBONACCI is a linear-time algorithm, while RECURSIVEFIBONACCI is an exponential-time algorithm. What this example has shown is not that an iterative algorithm is superior to a recursive algorithm, but that the two methods may lead to algorithms that require different amounts of time to solve the same problem instance.

2.7 Fast versus Slow Algorithms

Real computers require a certain amount of time to perform an operation such as addition, subtraction, or testing the conditions in a **while** loop. A supercomputer might take 10^{-9} second to perform an addition, while a hand calculator might take 10^{-5} second. Suppose that you had a computer that took 10^{-9} second to perform an elementary operation such as addition, and that you knew how many operations a particular algorithm would perform. You could estimate the running time of the algorithm simply by taking the product of the number of operations and the time per operation. However, computing devices are constantly improving, leading to a decreasing time per operation, so your notion of the running time would soon be outdated. Rather than computing an algorithm's running time on every computer, we rely on the total number of operations that the algorithm performs to de-

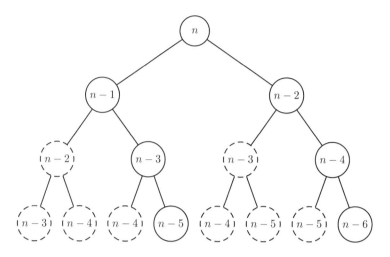

Figure 2.5 The recursion tree for RECURSIVEFIBONACCI(n). Vertices enclosed in dashed circles represent duplicated effort—the same value had been calculated in another vertex in the tree at a higher level. As the tree grows larger, the number of dashed vertices increases exponentially ($2^i - 2$ at level i), while the number of regular vertices increases linearly (2 per level).

scribe its running time, since this is an attribute of the algorithm, and not an attribute of the computer you happen to be using.

Unfortunately, determining how many operations an algorithm will perform is not always easy. We can see that USCHANGE will always perform 17 operations (one for each assignment, subtraction, multiplication, and division), but this is a very simple algorithm. An algorithm like SELECTIONSORT, on the other hand, will perform a different number of operations depending on what it receives as input: it will take less time to sort a 5-element list than it will to sort a 5000-element list. You might be tempted to think that SELECTIONSORT will take 1000 times longer to sort a 5000-element array than it will to sort a 5-element array. But you would be wrong. As we will see, it actually takes on the order of $1000^2 = 1,000,000$ times longer, no matter what kind of computer you use. It is typically the case that the larger the input is, the longer the algorithm takes to process it.

If we know how to compute the number of basic operations that an algorithm performs, then we have a basis to compare it against a different algorithm that solves the same problem. Rather than tediously count every multiplication and addition, we can perform this comparison by gaining a high-level understanding of the growth of each algorithm's operation count as the size of the input increases. To illustrate this, suppose an algorithm $\mathcal{A}$ performs $11n^3$ operations on an input of size n, and a different algorithm, $\mathcal{B}$, solves the same problem in $99n^2 + 7$ operations. Which algorithm, $\mathcal{A}$ or $\mathcal{B}$, is faster? Although, $\mathcal{A}$ may be faster than $\mathcal{B}$ for some small n (e.g., for n between 0 and 9), $\mathcal{B}$ will become faster with large n (e.g., for all $n \geq 10$). Since n^3 is, in some sense, a "faster-growing" function than n^2 with respect to n, the constants 11, 99, and 7 do not affect the competition between the two algorithms for large n (see figure 2.6). We refer to $\mathcal{A}$ as a cubic algorithm and to $\mathcal{B}$ as a quadratic algorithm, and say that $\mathcal{A}$ is less efficient than $\mathcal{B}$ because it performs more operations to solve the same problem when n is large. Thus, we will often be somewhat imprecise when we count operations in algorithms—the behavior of algorithms on small inputs does not matter.

Let us estimate how long BRUTEFORCECHANGE will take on an input instance of M cents, and denominations $(c_1, c_2, \ldots, c_d)$. To calculate the total number of operations in the **for** loop, we can take the approximate number of operations performed in each iteration and multiply this by the total number of iterations. Since there are roughly $\frac{M}{c_1} \cdot \frac{M}{c_2} \cdots \frac{M}{c_d}$ iterations, the **for** loop performs on the order of $d \cdot \frac{M^d}{c_1 \cdot c_2 \cdots c_d}$ operations, which dwarfs the other operations in the algorithm.

This type of algorithm is often referred to as an *exponential* algorithm in contrast to quadratic, cubic, or other *polynomial* algorithms. The expression for the running time of exponential algorithms includes a term like M^d, where d is a *parameter* of the problem (i.e., d may deliberately be made arbitrarily large by changing the input to the algorithm), while the running time of a polynomial algorithm is bounded by a term like M^k where k is a constant not related to the size of any parameters. For example, an algorithm with running time M^1 (linear), M^2 (quadratic), M^3 (cubic), or even M^{2005} is polynomial. Of course, an algorithm with running time M^{2005} is not very practical, perhaps less so than some exponential algorithms, and much effort in computer science goes into designing faster and faster polynomial algorithms. Since d may be large when the algorithm is called with a long list of denominations [e.g., $\mathbf{c} = (1, 2, 3, 4, 5, \ldots, 100)$], we see that BRUTE-FORCECHANGE can take a very long time to execute.

We have seen that the running time of an algorithm is often related to the *size* of its input. However, the running time of an algorithm can also vary among inputs of the *same* size. For example, suppose SELECTIONSORT first

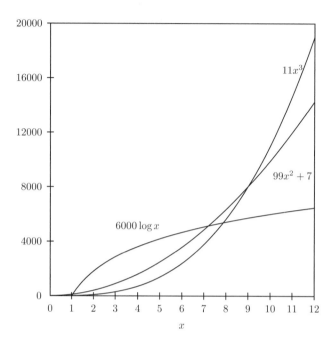

Figure 2.6 A comparison of a logarithmic ($h(x) = 6000 \log x$), a quadratic ($f(x) = 99x^2 + 7$), and a cubic ($g(x) = 11x^3$) function. After $x = 8$, both $f(x)$ and $g(x)$ are larger than $h(x)$. After $x = 9$, $g(x)$ is larger than $f(x)$, even though for values 0 through 9, $f(x)$ is larger than $g(x)$. The functions that we chose here are irrelevant and arbitrary: any three (positive-valued) functions with leading terms of $\log x$, x^2, and x^3 respectively would exhibit the same basic behavior, though the crossover points might be different.

checked to see if its input were already sorted. It would take this modified SELECTIONSORT less time to sort an ordered list of 5000 elements than it would to sort an unordered list of 5000 elements. As we see in the next section, when we speak of the running time of an algorithm as a function of input size, we refer to that one input—or set of inputs—of a particular size that the algorithm will take the longest to process. In the modified SELECTIONSORT, that input would be any not-already-sorted list.

2.8   Big-O Notation

Computer scientists use the *Big-O* notation to describe concisely the running time of an algorithm. If we say that the running time of an algorithm is quadratic, or $O(n^2)$, it means that the running time of the algorithm on an input of size n is limited by a quadratic function of n. That limit may be $99.7n^2$ or $0.001n^2$ or $5n^2 + 3.2n + 99993$; the main factor that describes the growth rate of the running time is the term that grows the fastest with respect to n, for example n^2 when compared to terms like $3.2n$, or 99993. All functions with a leading term of n^2 have more or less the same rate of growth, so we lump them into one class which we call $O(n^2)$. The difference in behavior between two quadratic functions in that class, say $99.7n^2$ and $5n^2 + 3.2n + 99993$, is negligible when compared to the difference in behavior between two functions in different classes, say $5n^2 + 3.2n$ and $1.2n^3$. Of course, $99.7n^2$ and $5n^2$ are different functions and we would prefer an algorithm that takes $5n^2$ operations to an algorithm that takes $99.7n^2$. However, computer scientists typically ignore the leading constant and pay attention only to the fastest-growing term.

When we write $f(n) = O(n^2)$, we mean that the function $f(n)$ does not grow faster than a function with a leading term of cn^2, for a suitable choice of the constant c. A formal definition of Big-O notation, which is helpful in analyzing an algorithm's running time, is given in figure 2.7.

The relationship $f(n) = O(n^2)$ tells us that $f(n)$ does not grow faster than some quadratic function, but it does not tell us whether $f(n)$ grows slower than any quadratic function. In other words, $2n = O(n^2)$, but this valid statement is not as informative as it could be; $2n = O(n)$ is more precise. We say that the Big-O relationship establishes an *upper bound* on the growth of a function: if $f(n) = O(g(n))$, then the function f grows no faster than the function g. A similar concept exists for *lower bounds*, and we use the notation $f(n) = \Omega(g(n))$ to indicate that f grows no slower than g. If, for some function g, an algorithm's time grows no faster than g *and* no slower than g, then we say that g is a *tight bound* for the algorithm's running time. For example, if an algorithm requires $2n \log n$ time, then technically, it is an $O(n^2)$ algorithm even though this is a misleadingly loose bound. A tight bound on the algorithm's running time is actually $O(n \log n)$. Unfortunately, it is often easier to prove a loose bound than a tight one.

In keeping with the healthy dose of pessimism toward an algorithm's correctness, we measure an algorithm's efficiency as its *worst case* efficiency, which is the largest amount of time an algorithm can take given the worst

A function $f(x)$ is "Big-O of $g(x)$", or $O(g(x))$, when $f(x)$ is less than or equal to $g(x)$ to within some constant multiple c. If there are a few points x such that $f(x)$ is not less than $c \cdot g(x)$, this does not affect our overall understanding of f's growth. Mathematically speaking, the Big-O notation deals with *asymptotic* behavior of a function as its input grows arbitrarily large, beyond some (arbitrary) value x_0.

Definition 2.1 *A function $f(x)$ is $O\left(g(x)\right)$ if there are positive real constants c and x_0 such that $f(x) \leq cg(x)$ for all values of $x \geq x_0$.*

For example, the function $3x = O(.2x^2)$, but at $x = 1$, $3x > .2x^2$. However, for all $x > 15$, $.2x^2 > 3x$. Here, $x_0 = 15$ represents the point at which $3x$ is bounded above by $.2x^2$. Notice that this definition blurs the advantage gained by mere constants: $5x^2 = O(x^2)$, even though it would be wrong to say that $5x^2 \leq x^2$.

Like Big-O notation, which governs an upper bound on the growth of a function, we can define a relationship that reflects a lower bound on the growth of a function.

Definition 2.2 *A function $f(x)$ is $\Omega\left(g(x)\right)$ if there are positive real constants c and x_0 such that $f(x) \geq cg(x)$ for all values of $x \geq x_0$.*

If $f(x) = \Omega(g(x))$, then f is said to grow "faster" than g.

Now, if $f(x) = O(g(x))$ and $f(x) = \Omega(g(x))$ then we know very precisely how $f(x)$ grows with respect to $g(x)$. We call this the Θ relationship.

Definition 2.3 *A function $f(x)$ is $\Theta\left(g(x)\right)$ if $f(x) = O\left(g(x)\right)$ and $f(x) = \Omega\left(g(x)\right)$.*

Figure 2.7   Definitions of the Big-O, Ω, and Θ notations.

possible input of a given size. The advantage to considering the worst case efficiency of an algorithm is that we are guaranteed that our algorithm will never behave worse than our worst case estimate, so we are never surprised or disappointed. Thus, when we derive a Big-O bound, it is a bound on the worst case efficiency.

We illustrate the above notion of efficiency by analyzing the two sorting algorithms, SELECTIONSORT and RECURSIVESELECTIONSORT. The parameter that describes the input size is n, the number of integers in the input list, so we wish to determine the efficiency of the algorithms as a function of n.

The SELECTIONSORT algorithm makes $n - 1$ iterations in the **for** loop and analyzes $n - i + 1$ elements $a_i, \ldots, a_n$ in iteration i. In the first iteration, it analyzes all n elements, at the next one it analyzes $n - 1$ elements, and so on. Therefore, the approximate number of operations performed in SELECTION-SORT is: $n + (n-1) + (n-2) + \cdots + 2 + 1 = 1 + 2 + \cdots + n = n(n+1)/2$.[15] At each iteration, the same swapping of array elements occurs, so SELECTIONSORT requires roughly $n(n + 1)/2 + 3n$ operations, which is $O(n^2)$ operations.[16] Again, because we can safely ignore multiplicative constants and terms that are smaller than the fastest-growing term, our calculations are somewhat imprecise but yield an overall picture of the function's growth.

We will now consider RECURSIVESELECTIONSORT. Let $T(n)$ denote the amount of time that RECURSIVESELECTIONSORT takes on an n-element array. Calling RECURSIVESELECTIONSORT on an n-element array involves finding the smallest element (roughly n operations), followed by a recursive call on a list with $n - 1$ elements, which performs $T(n - 1)$ operations. Calling RECURSIVESELECTIONSORT on a 1-element list requires 1 operation (one for the **if** statement), so the following equations hold.

$$
\begin{aligned}
T(n) &= n + T(n - 1) \\
T(1) &= 1
\end{aligned}
$$

Therefore,

$$
\begin{aligned}
T(n) &= n + T(n - 1) \\
&= n + (n - 1) + T(n - 2) \\
&= n + (n - 1) + (n - 2) + \cdots + 3 + 2 + T(1) \\
&= O(n^2).
\end{aligned}
$$

Thus, calling RECURSIVESELECTIONSORT on an n element array will require roughly the same $O(n^2)$ time as calling SELECTIONSORT. Since RECURSIVESELECTIONSORT always performs the same operations on a list of size n, we can be certain that this is a tight analysis of the running time of the algorithm. This is why using SELECTIONSORT to sort a 5000-element array takes $1,000,000$ times longer than it does to sort a 5-element array: $5,000^2 = 1,000,000 \cdot 5^2$.

---

15. Here we rely on the fact that $1 + 2 + \cdots + n = n(n + 1)/2$.
16. Each swapping requires three (rather than two) operations.

Of course, this does not show that the Sorting problem requires $O(n^2)$ time to solve. All we have shown so far is that two particular algorithms, RECURSIVESELECTIONSORT and SELECTIONSORT, require $O(n^2)$ time; in fact, we will see a different sorting algorithm in chapter 7 that runs in $O(n \log n)$ time.

We can use the same technique to calculate the running time of HANOITOWERS called on a tower of size n. Let $T(n)$ denote the number of disk moves that HANOITOWERS(n) performs. The following equations hold.

$$
\begin{aligned}
T(n) &= 2 \cdot T(n-1) + 1 \\
T(1) &= 1
\end{aligned}
$$

This recurrence relation produces the following sequence: $1, 3, 7, 15, 31, 63$, and so on. We can solve it by adding 1 to both sides and noticing

$$T(n) + 1 = 2 \cdot T(n-1) + 1 + 1 = 2(T(n-1) + 1).$$

If we introduce a new variable, $U(n) = T(n) + 1$, then $U(n) = 2 \cdot U(n-1)$. Thus, we have changed the problem to the following recurrence relation.

$$
\begin{aligned}
U(n) &= 2 \cdot U(n-1) \\
U(1) &= 2
\end{aligned}
$$

This gives rise to the sequence $2, 4, 8, 16, 32, 64, \ldots$ and it is easy to see that $U(n) = 2^n$. Since $T(n) = U(n) - 1$, we see that $T(n) = 2^n - 1$. Thus, HANOITOWERS is an exponential algorithm, which we hinted at in section 2.5.

2.9 Algorithm Design Techniques

Over the last forty years, computer scientists have discovered that many algorithms share similar ideas, even though they solve very different problems. There appear to be relatively few basic techniques that can be applied when designing an algorithm, and we cover some of them later in this book in varying degrees of detail. For now we will mention the most common algorithm design techniques, so that future examples can be categorized in terms of the algorithm's design methodology.

To illustrate the design techniques, we will consider a very simple problem that plagues nearly everyone with a cordless phone. Suppose your cordless phone rings, but you have misplaced the handset somewhere in your home.

How do you find it? To complicate matters, you have just walked into your home with an armful of groceries, and it is dark out, so you cannot rely solely on eyesight.

2.9.1 Exhaustive Search

An *exhaustive search*, or *brute force*, algorithm examines every possible alternative to find one particular solution.

For example, if you used the brute force algorithm to find the ringing telephone, you would ignore the ringing of the phone, as if you could not hear it, and simply walk over every square inch of your home checking to see if the phone was present. You probably would not be able to answer the phone before it stopped ringing, unless you were very lucky, but you would be guaranteed to eventually find the phone no matter where it was.

BRUTEFORCECHANGE is a brute force algorithm, and chapter 4 introduces some additional examples of such algorithms—these are the easiest algorithms to design and understand, and sometimes they work acceptably for certain practical problems in biology. In general, though, brute force algorithms are too slow to be practical for anything but the smallest instances

and we will spend most of this book demonstrating how to avoid the brute force algorithms or how to finesse them into faster versions.

2.9.2   Branch-and-Bound Algorithms

In certain cases, as we explore the various alternatives in a brute force algorithm, we discover that we can omit a large number of alternatives, a technique that is often called *branch-and-bound*, or *pruning*.

Suppose you were exhaustively searching the first floor and heard the phone ringing above your head. You could immediately rule out the need to search the basement or the first floor. What may have taken three hours may now only take one, depending on the amount of space that you can rule out.

2.9.3 Greedy Algorithms

Many algorithms are iterative procedures that choose among a number of alternatives at each iteration. For example, a cashier can view the Change problem as a series of decisions he or she has to make: which coin (among d denominations) to return first, which to return second, and so on. Some of these alternatives may lead to correct solutions while others may not. Greedy algorithms choose the "most attractive" alternative at each iteration, for example, the largest denomination possible. USCHANGE used quarters, then dimes, then nickels, and finally pennies (in that order) to make change for M. By greedily choosing the largest denomination first, the algorithm avoided any combination of coins that included fewer than three quarters to make change for an amount larger than or equal to 75 cents. Of course, we showed that the generalization of this greedy strategy, BETTERCHANGE, produced incorrect results when certain new denominations were included.

In the telephone example, the corresponding greedy algorithm would simply be to walk in the direction of the telephone's ringing until you found it. The problem here is that there may be a wall (or an expensive and fragile vase) between you and the phone, preventing you from finding it. Unfortunately, these sorts of difficulties frequently occur in most realistic problems. In many cases, a greedy approach will seem "obvious" and natural, but will be subtly wrong.

2.9.4 Dynamic Programming

Some algorithms break a problem into smaller subproblems and use the solutions of the subproblems to construct the solution of the larger one. During this process, the number of subproblems may become very large, and some algorithms solve the same subproblem repeatedly, needlessly increasing the running time. Dynamic programming organizes computations to avoid re-computing values that you already know, which can often save a great deal of time. The Ringing Telephone problem does not lend itself to a dynamic programming solution, so we consider a different problem to illustrate the technique.

Suppose that instead of answering the phone you decide to play the "Rocks" game from the previous chapter with two piles of rocks, say ten in each. We remind the reader that in each turn, one player may take either one rock (from either pile) or two rocks (one from each pile). Once the rocks are taken, they are removed from play. The player that takes the last rock wins the game. You make the first move.

To find the winning strategy for the $10 + 10$ game, we can construct a table, which we can call **R**, shown below. Instead of solving a problem with 10 rocks in each pile, we will solve a more general problem with n rocks in one pile and m rocks in another (the $n + m$ game) where n and m are arbitrary. If Player 1 can always win the game of $5 + 6$, then we would say $R_{5,6} = W$, but if Player 1 has no winning strategy against a player that always makes the right moves, we would write $R_{5,6} = L$. Computing $R_{n,m}$ for an arbitrary n and m seems difficult, but we can build on smaller values. Some games, notably $R_{0,1}$, $R_{1,0}$, and $R_{1,1}$, are clearly winning propositions for Player 1 since in the first move, Player 1 can win. Thus, we fill in entries $(1, 1)$, $(0, 1)$ and $(1, 0)$ as W.

	0	1	2	3	4	5	6	7	8	9	10
0		W									
1	W	W									
2											
3											
4											
5											
6											
7											
8											
9											
10											

After the entries $(0, 1)$, $(1, 0)$, and $(1, 1)$ are filled, one can try to fill other entries. For example, in the $(2, 0)$ case, the only move that Player 1 can make leads to the $(1, 0)$ case that, as we already know, is a winning position for his opponent. A similar analysis applies to the $(0, 2)$ case, leading to the following result:

	0	1	2	3	4	5	6	7	8	9	10
0		W	L								
1	W	W									
2	L										
3											
4											
5											
6											
7											
8											
9											
10											

In the $(2, 1)$ case, Player 1 can make three different moves that lead respectively to the games of $(1, 1)$, $(2, 0)$, or $(1, 0)$. One of these cases, $(2, 0)$, leads to a losing position for his opponent and therefore $(2, 1)$ is a winning position. The case $(1, 2)$ is symmetric to $(2, 1)$, so we have the following table:

	0	1	2	3	4	5	6	7	8	9	10
0		W	L								
1	W	W	W								
2	L	W									
3											
4											
5											
6											
7											
8											
9											
10											

Now we can fill in $R_{2,2}$. In the $(2, 2)$ case, Player 1 can make three different moves that lead to entries $(2, 1)$, $(1, 2)$, and $(1, 1)$. All of these entries are winning positions for his opponent and therefore $R_{2,2} = L$

	0	1	2	3	4	5	6	7	8	9	10
0		W	L								
1	W	W	W								
2	L	W	L								
3											
4											
5											
6											
7											
8											
9											
10											

We can proceed filling in **R** in this way by noticing that for the entry (i, j) to be L, the entries above, diagonally to the left and directly to the left, must be W. These entries $((i - 1, j), (i - 1, j - 1)$, and $(i, j - 1))$ correspond to the three possible moves that player 1 can make.

	0	1	2	3	4	5	6	7	8	9	10
0		W	L	W	L	W	L	W	L	W	L
1	W	W	W	W	W	W	W	W	W	W	W
2	L	W	L	W	L	W	L	W	L	W	L
3	W	W	W	W	W	W	W	W	W	W	W
4	L	W	L	W	L	W	L	W	L	W	L
5	W	W	W	W	W	W	W	W	W	W	W
6	L	W	L	W	L	W	L	W	L	W	L
7	W	W	W	W	W	W	W	W	W	W	W
8	L	W	L	W	L	W	L	W	L	W	L
9	W	W	W	W	W	W	W	W	W	W	W
10	L	W	L	W	L	W	L	W	L	W	L

The ROCKS algorithm determines if Player 1 wins or loses. If Player 1 wins in an $n+m$ game, ROCKS returns W. If Player 1 loses, ROCKS returns L. The ROCKS algorithm introduces an artificial initial condition, $R_{0,0} = L$ to simplify the pseudocode.

ROCKS(n, m)
1   $R_{0,0} = L$
2   **for** $i \leftarrow 1$ **to** n
3   **if** $R_{i-1,0} = W$
4           $R_{i,0} \leftarrow L$
5   **else**
6           $R_{i,0} \leftarrow W$
7   **for** $j \leftarrow 1$ **to** m
8   **if** $R_{0,j-1} = W$
9           $R_{0,j} \leftarrow L$
10   **else**
11          $R_{0,j} \leftarrow W$
12 **for** $i \leftarrow 1$ **to** n
13   **for** $j \leftarrow 1$ **to** m
14   **if** $R_{i-1,j-1} = W$ **and** $R_{i,j-1} = W$ **and** $R_{i-1,j} = W$
15              $R_{i,j} \leftarrow L$
16   **else**
17              $R_{i,j} \leftarrow W$
18 **return** $R_{n,m}$

In point of fact, a faster algorithm to solve the Rocks puzzle relies on the simple pattern in **R**, and checks to see if n and m are both even, in which case the player loses.

FASTROCKS(n, m)
1   **if** n and m are both even
2   **return** L
3   **else**
4   **return** W

However, though FASTROCKS is more efficient than ROCKS, it may be difficult to modify it for other games, for example a game in which each player can move up to three rocks at a time from the piles. This is one example where the slower algorithm is more instructive than a faster one. But obviously, it is usually better to use the faster one when you really need to solve the problem.

2.9.5   Divide-and-Conquer Algorithms

One big problem may be hard to solve, but two problems that are half the size may be significantly easier. In these cases, *divide-and-conquer* algorithms fare well by doing just that: splitting the problem into smaller subproblems, solving the subproblems independently, and combining the solutions of sub-problems into a solution of the original problem. The situation is usually more complicated than this and after splitting one problem into subprob-lems, a divide-and-conquer algorithm usually splits these subproblems into even smaller sub-subproblems, and so on, until it reaches a point at which it no longer needs to recurse. A critical step in many divide-and-conquer algorithms is the recombining of solutions to subproblems into a solution for a larger problem. Often, this merging step can consume a considerable amount of time. We will see examples of this technique in chapter 7.

2.9.6   Machine Learning

Another approach to the phone search problem is to collect statistics over the course of a year about where you leave the phone, learning where the phone tends to end up most of the time. If the phone was left in the bathroom 80% of the time, in the bedroom 15% of the time, and in the kitchen 5% of the time, then a sensible time-saving strategy would be to start the search in the bathroom, continue to the bedroom, and finish in the kitchen. Machine learning algorithms often base their strategies on the computational analysis of previously collected data.

2.9.7   Randomized Algorithms

If you happen to have a coin, then before even starting to search for the phone, you could toss it to decide whether you want to start your search on the first floor if the coin comes up heads, or on the second floor if the coin comes up tails. If you also happen to have a die, then after deciding on the second floor, you could roll it to decide in which of the six rooms on the sec-ond floor to start your search.[17] Although tossing coins and rolling dice may be a fun way to search for the phone, it is certainly not the intuitive thing to do, nor is it at all clear whether it gives you any algorithmic advantage over a deterministic algorithm. We will learn how randomized algorithms

---

17. Assuming that you have a large house, of course.

help solve practical problems, and why some of them have a competitive advantage over deterministic algorithms.

2.10   Tractable versus Intractable Problems

We have described a correct algorithm that solves the Change problem, but requires exponential time to do so. This does not mean that all algorithms that solve the Change problem will require exponential time. Showing that a particular algorithm requires exponential time is much different than showing that a *problem* cannot be solved by *any* algorithm in less than exponential time. For example, we showed that RECURSIVEFIBONACCI required exponential time to compute the *n*th Fibonacci number, while FIBONACCI solves the same problem in linear $O(n)$ time.[18]

We have seen that algorithms can be categorized according to their complexity. In the early 1970s, computer scientists discovered that *problems* could also be categorized according to their inherent complexity. It turns out that some problems, such as listing every subset of an *n*-element set, require exponential time—no algorithm, no matter how clever, could possibly solve the problem in less than exponential time. Other problems, such as sorting a list of integers, require only polynomial time. Somewhere between the polynomial problems and the exponential problems lies one particularly important category of problems called the $\mathcal{NP}$-*complete* problems. These are problems

---

18. There exists an algorithm even faster than $O(n)$ to compute the *n*-th Fibonacci number; it does not calculate all of the Fibonacci numbers in the sequence up to n.

that appear to be quite difficult, in that no polynomial-time algorithm for any of these problems has yet been found. However, nobody can seem to prove that polynomial-time algorithms for these problems are impossible, so nobody can rule out the possibility that these problems are actually efficiently solvable. One particularly famous example of an $\mathcal{NP}$-complete problem is the Traveling Salesman problem, which has a wide variety of practical applications in biology.

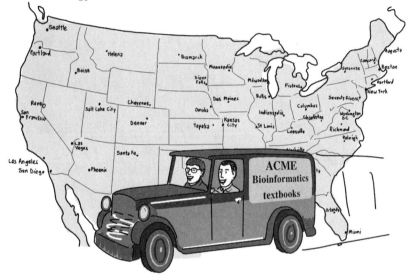

Traveling Salesman Problem:
Find the shortest path through a set of cities, visiting each city only one time.

> **Input:** A map of cities, roads between the cities, and distances along the roads.

> **Output:** A sequence of roads that will take a salesman through every city on the map, such that he will visit each city exactly once, and will travel the shortest total distance.

The critical property of $\mathcal{NP}$-complete problems is that, if one $\mathcal{NP}$-complete problem is solvable by a polynomial-time algorithm, then *all* $\mathcal{NP}$-complete problems can be solved by minor modifications of the same algorithm. The fact that nobody has yet found that magic algorithm, after half a century of research, suggests that it may not exist. However, this has not yet been proven mathematically. It turns out that thousands of algorithmic problems are actually instances of the Traveling Salesman problems in disguise.[19] Taken together, these problems form the class of $\mathcal{NP}$-complete problems.

Despite the lack of mathematical proof, attempting to find a polynomial algorithm for an $\mathcal{NP}$-complete problem will likely result in failure and a whole lot of wasted time. Unfortunately, proving that a problem really is $\mathcal{NP}$-complete, and not just superficially difficult, is somewhat of an undertaking. In fact, it is not yet known whether or not some important bioinformatics problems are $\mathcal{NP}$-complete.

2.11 Notes

The word "algorithm" derived from the name of the ninth-century Arab mathematician, al-Khwarizmi, of the court of Caliph Mamun in Baghdad. al-Khwarizmi was a scholar in the House of Wisdom, where Greek scientific works were translated; much of the mathematical knowledge in medieval Europe was derived from Latin translations of his books. More recent books on the general topic of algorithms are Knuth, 1998 (57); Aho, Hopcroft, and Ullman, 1983 (1); and Cormen et al., 2001 (24).

The notion of $\mathcal{NP}$-completeness was proposed in the early 1970s by Stephen Cook (23), and Leonid Levin (65), and was further analyzed by Richard Karp in 1972 (53) who demonstrated a rich variety of $\mathcal{NP}$-complete problems. Garey and Johnson (39) authored an encyclopedic reference of $\mathcal{NP}$-complete problems.

---

19. Although these problems may have nothing in common with the Traveling Salesman problem—no cities, no roads, no distances—the Traveling Salesman problem can be converted into each one, and vice versa.

Richard Karp, born 1935 in Boston, is a Professor at the University of California at Berkeley, with a principal appointment in computer science and additional appointments in mathematics, bioengineering, and operations research. He attended Boston Latin School and Harvard University, where he received a PhD in Applied Mathematics in 1959. From 1959 to 1968 he was a member of the Mathematical Sciences Department of the IBM Research Center in Yorktown Heights, NY. He has been a faculty member at the University of California at Berkeley since 1968 (with the exception of the period 1995–99, when he was a professor at the University of Washington). Since 1988 he has also been a research scientist at the International Computer Science Institute, a non-profit research company in Berkeley. Karp says:

> Ever since my undergraduate days I have had a fascination with combinatorial algorithms. These puzzle-like problems involve searching through a finite but vast set of possibilities for a pattern or structure that meets certain requirements or is of minimum cost. Examples in bioinformatics include sequence assembly, multiple alignment of sequences, phylogeny construction, the analysis of genomic rearrangements, and the modeling of gene regulation. For some combinatorial problems, there are elegant and efficient algorithms that proceed like clockwork to find the required solution, but most are less tractable and require either a very long computation or a compromise on a solution that may not be optimal.

In 1972, Karp developed an approach to showing that many seemingly hard combinatorial problems are equivalent in the sense that either all of them or none of them are efficiently solvable (a problem is considered efficiently solvable if it can be solved by a polynomial algorithm). These problems are the "$\mathcal{NP}$-Complete" problems. Over the years, thousands of examples have been added to his original list of twenty-one $\mathcal{NP}$-complete problems, yet despite intensive effort none of these problems has been shown to be efficiently solvable. Many computer scientists (including Karp) believe that none of them ever will be.

Karp began working in bioinformatics circa 1991, attracted by the belief that computational methods might reveal the secret inner workings of living organisms. He says:

> [I hoped] that my experience in studying combinatorial algorithms could be useful in cracking those secrets. I have indeed been able to apply my skills in this new area, but only after coming to understand that solving biological problems requires far more than clever algorithms: it involves a creative partnership between biologists and mathematical scientists to arrive at an appropriate mathematical model, the acquisition and use of diverse sources of data, and statistical methods to show that the biological patterns and regularities that we discover could not be due to chance. My recent work is concerned with analyzing the transcriptional regulation of genes, discovering conserved regulatory pathways, and analyzing genetic variation in humans.
>
> There have been spectacular advances in biology since 1991, most notable being the sequencing of genomes. I believe that we are now poised to understand—and possibly even reprogram— the gene regulatory networks and the metabolic networks that control cellular processes. By comparing many related organisms, we hope to understand how these networks evolved. Effectively, we are trying to find the genetic basis of complex diseases so that we can develop more effective modes of treatment.

2.12   Problems

Problem 2.1

Write an algorithm that, given a list of n numbers, returns the largest and smallest numbers in the list. Estimate the running time of the algorithm. Can you design an algorithm that performs only $3n/2$ comparisons to find the smallest and largest numbers in the list?

Problem 2.2

Write two algorithms that iterate over every index from $(0, 0, \ldots, 0)$ to $(n_1, n_2, \ldots, n_d)$. Make one algorithm recursive, and the other iterative.

Problem 2.3

Is $\log n = O(n)$? Is $\log n = \Omega(n)$? Is $\log n = \Theta(n)$?

Problem 2.4

You are given an unsorted list of $n - 1$ distinct integers from the range 1 to n. Write a linear-time algorithm to find the missing integer.

Problem 2.5

Though FIBONACCI(n) is fast, it uses a fair amount of space to store the array $\mathbf{F}$. How much storage will it require to calculate the nth Fibonacci number? Modify the algorithm to require a constant amount of storage, regardless of the value of n.

Problem 2.6

Prove that

$$F_n = \frac{1}{\sqrt{5}}(\phi^n - \overline{\phi}^n)$$

where F_n is the nth Fibonacci number, $\phi = \frac{1+\sqrt{5}}{2}$ and $\overline{\phi} = \frac{1-\sqrt{5}}{2}$.

Problem 2.7

Design an algorithm for computing the n-th Fibonacci number that requires less than $O(n)$ time. *Hint:* You probably want to use the result from problem 2.6. However, computing ϕ^n naively still requires $O(n)$ time because each multiplication is a single operation.

Problem 2.8

Propose a more realistic model of rabbit life (and death) that limits the life span of rabbits by k years. For example, if $k = 2.5$, then the corresponding sequence $1, 1, 2, 3, 4$ grows more slowly than the Fibonacci seqence. Write a recurrence relation and pseudocode to compute the number of rabbits under this model. Will the number of rabbits ever exceed the number of atoms in the universe (under these assumptions)?

Problem 2.9

Write an iterative (i.e., nonrecursive) algorithm to solve the Hanoi Tower problem.

Problem 2.10

Prove that $\sum_{i=1}^{n} i = \frac{n(n+1)}{2}$.

Problem 2.11

Prove that $\sum_{i=1}^{n} 2^i = 2^{n+1} - 2$ and that $\sum_{i=1}^{n} 2^{-i} = 1 - 2^{-n}$.

Problem 2.12

We saw that BETTERCHANGE is an incorrect algorithm for the set of denominations $(25, 20, 10, 5, 1)$. Add a new denomination to this set such that BETTERCHANGE will return the correct change combination for any value of M.

Problem 2.13

Design an algorithm that computes the average number of coins returned by the program USCHANGE(M) as M varies from 1 to 100.

Problem 2.14

Given a set of arbitrary denominations $\mathbf{c} = (c_1, c_2, \ldots, c_d)$, write an algorithm that can decide whether BETTERCHANGE is a correct algorithm when run on $\mathbf{c}$.

Problem 2.15

A king stands on the upper left square of the chessboard. Two players make turns moving the king either one square to the right or one square downward or one square along a diagonal in the southeast direction. The player who can place the king on the lower right square of the chessboard wins. Who will win? Describe the winning strategy.

Problem 2.16

Bob and Alice are bored one Saturday afternoon so they invent the following game. Initially, there are n rocks in a single pile. At each turn, one of the two players can split any pile of rocks that has more than 1 rock into two piles of arbitrary size such that the size of each of the two new piles must add up to the size of the original big pile. No player can split a pile that has only a single rock, and the last person to move wins. Does one of the two players, first or second, have an advantage? Explain which player will win for each value of n.

Problem 2.17

There are n bacteria and 1 virus in a Petri dish. Within the first minute, the virus kills one bacterium and produces another copy of itself, and all of the remaining bacteria reproduce, making 2 viruses and $2 \cdot (n - 1)$ bacteria. In the second minute, each of the viruses kills a bacterium and produces a new copy of itself (resulting in 4 viruses and $2(2(n - 1) - 2) = 4n - 8$ bacteria; again, the remaining bacteria reproduce. This process continues every minute. Will the viruses eventually kill all the bacteria? If so, design an algorithm that computes how many steps it will take. How does the running time of your algorithm depend on n?

Problem 2.18

A very large bioinformatics department at a prominent university has a mix of 100 professors: some are honest and hard-working, while others are deceitful and do not like students. The honest professors always tell the truth, but the deceitful ones sometimes tell the truth and sometimes lie. You can ask any professors the following question about any other professor: "Professor Y, is Professor X honest?" Professor Y will answer with either "yes" or "no." Design an algorithm that, with no more than 198 questions, would allow you to figure out which of the 100 professors are honest (thus identifying possible research advisors). It is known that there are more honest than dishonest professors.

Problem 2.19

You are given an 8×8 table of natural numbers. In any one step, you can either double each of the numbers in any one row, or subtract 1 from each of the numbers in any one column. Devise an algorithm that transforms the original table into a table of all zeros. What is the running time of your algorithm?

Problem 2.20

There are n black, m green, and k brown chameleons on a deserted island. When two chameleons of different colors meet they both change their color to the third one (e.g., black and green chameleons become brown). For each choice of n, m, and k decide whether it is possible that after some time all the chameleons on the island are the same color (if you think that it is always possible, check the case $n = 1$, $m = 3$, and $k = 5$).

3 *Molecular Biology Primer*

To understand bioinformatics in any meaningful way, it is necessary for a computer scientist to understand some basic biology, just as it is necessary for a biologist to understand some basic computer science. This chapter provides a short and informal introduction to those biological fundamentals. We scanned existing bioinformatics books to find out how much biological material was "relevant" to those books and we were surprised how little biological knowledge was actually presented. It would be safe to say that the minimum biological background one needs in order to digest a typical bioinformatics book could fit into ten pages.[1] In this chapter we give a brief introduction to biology that covers most of the computational concepts discussed in bioinformatics books. Some of the sections in this chapter are not directly related to the rest of the book, but we present them to convey the fascinating story of molecular biology in the twentieth century.

3.1 What Is Life Made Of?

Biology at the microscopic level began in 1665 when a maverick and virtuoso performer of public animal dissections, Robert Hooke, discovered that organisms are composed of individual compartments called *cells*. Cell theory, further advanced by Matthias Schleiden and Theodor Schwann in the 1830s, marked an important milestone: it turned biology into a science beyond the reach of the naked eye. In many ways, the study of life became the study of cells.

---

1. This is not to say that computer scientists should limit themselves to these ten pages. More detailed discussions can be found in introductory biology textbooks like Brown (17), Lewin (66), or Alberts (3).

A great diversity of cells exist in nature, but they all have some common features. All cells have a life cycle: they are born, eat, replicate, and die. During the life cycle, a cell has to make many important decisions. For example, if a cell were to attempt to replicate before it had collected all of the necessary nutrients to do so, the result would be a disaster. However, cells do not have brains. Instead, these decisions are manifested in complex networks of chemical reactions, called *pathways*, that synthesize new materials, break other materials down for spare parts, or signal that the time has come to eat or die. The amazingly reliable and complex algorithm that controls the life of the cell is still beyond our comprehension.

One can envision a cell as a complex mechanical system with many moving parts. Not only does it store all of the *information* necessary to make a complete replica of itself, it also contains all the *machinery* required to collect and manufacture its components, carry out the copying process, and kick-start its new offspring. In macroscopic terms, a cell would be roughly analogous to a car factory that could mine for ore, fabricate girders and concrete pillars, and assemble an exact working copy of itself, all the while building family sedans with no human intervention.

Despite the complexity of a cell, there seems to be a few organizing principles that are conserved across all organisms. All life on this planet depends on three types of molecule: DNA, RNA, and proteins.[2] Roughly speaking, a cell's DNA holds a vast library describing how the cell works. RNA acts to transfer certain short pieces of this library to different places in the cell, at which point those smaller volumes of information are used as templates to synthesize proteins. Proteins form enzymes that perform biochemical reactions, send signals to other cells, form the body's major components (like the keratin in our skin), and otherwise perform the actual work of the cell. DNA, RNA, and proteins are examples of *strings* written in either the four-letter alphabet of DNA and RNA or the twenty-letter alphabet of proteins. This meshes well with Schrödinger's visionary idea about an "instruction book" of life scribbled in a secret code. It took a long time to figure out that DNA, RNA, and proteins are the main players in the cells. Below we give a brief summary of how this was discovered.

---

2. To be sure, other types of molecules, like lipids, play a critical role in maintaining the cell's structure, but DNA, RNA, and proteins are the three primary types of molecules that biologists study.

3.2   What Is the Genetic Material?

Schleiden's and Schwann's studies of cells were further advanced by the discovery of threadlike chromosomes in the cell nucleii. Different organisms have different numbers of chromosomes, suggesting that they might carry information specific for each species. This fit well with the work of the Augustinian monk Gregor Mendel in the 1860s, whose experiments with garden peas suggested the existence of *genes* that were responsible for inheritance. Evidence that traits (more precisely, genes) are located on chromosomes came in the 1920s through the work of Thomas Morgan. Unlike Mendel, Morgan worked in New York City and lacked the garden space to cultivate peas, so he instead used fruit flies for his experiments: they have a short life span and produce numerous offspring. One of these offspring turned out to have white eyes, whereas wild flies had red eyes. This one white-eyed male fly born in Morgan's "fly room" in New York City became the cornerstone of modern genetics.

The white-eyed male fly was mated with its red-eyed sisters and the offspring were followed closely for a few generations. The analysis of offspring revealed that white eyes appeared predominantly in males, suggesting that a gene for eye color resides on the X chromosome (which partly determines the gender of a fruit fly). Thus, Morgan suspected that genes were located on chromosomes. Of course, Morgan had no idea what chromosomes were themselves made of.

Morgan and his students proceeded to identify other mutations in flies and used ever more sophisticated techniques to assign these mutations to certain locations on chromosomes. Morgan postulated that the genes somehow responsible for these mutations were also positioned at these locations. His group showed that certain genes are inherited together, as if they were a single unit. For example, Morgan identified mutants with a black body color (normal flies are gray) and mutants with vestigial wings. He proceeded to cross black flies with vestigial wings with gray flies with normal wings, expecting to see a number of gray flies with vestigial wings, gray flies with normal wings, black flies with vestigial wings, and black flies with normal wings. However, the experiment produced a surprisingly large number of normal flies (gray body, normal wings) and a surprisingly large number of double mutants (black body, vestigial wings). Morgan immediately proposed a hypothesis that such *linked* genes reside close together on a chromosome. Moreover, he theorized, the more tightly two genes are linked (i.e., the more often they are inherited together), the closer they are on a chromosome.

Morgan's student Alfred Sturtevant pursued Morgan's chromosome theory and constructed the first genetic map of a chromosome that showed the order of genes. Sturtevant studied three genes: *cn*, which determines eye color; *b*, which determines body color; and *vg*, which determines wing size. Sturtevant crossed double-mutant *b* and *vg* flies with normal flies and saw that about 17% of the offspring had only a single mutation. However, when Sturtevant crossed double-mutant *b* and *cn* flies he found that 9% of the offspring had only a single mutation. This implied that *b* and *cn* reside closer together than *b* and *vg*; a further experiment with *cn* and *vg* mutants demonstrated an 8% single mutation rate. Combined together, these three observations showed that *b* lies on one side of *cn* and *vg* on the other. By studying many genes in this way, it is possible to determine the ordering of genes. However, the nature of genes remained an elusive and abstract concept for many years, since it was not clear how genes encoded information and how they passed that information to the organism's progeny.

3.3   What Do Genes Do?

By the early 1940s, biologists understood that a cell's traits were inherent in its genetic information, that the genetic information was passed to its offspring, and that the genetic information was organized into genes that resided on chromosomes. They did not know what the chromosomes were made of or what the genes actually did to give rise to a cell's traits. George Beadle and Edward Tatum were the first to identify the job of the gene, without actually revealing the true nature of genetic information. They worked with the bread mold *Neurospora*, which can survive by consuming very simple nutrients like sucrose and salt. To be able to live on such a limited diet, *Neurospora* must have some proteins (enzymes) that are able to convert these simple nutrients into "real food" like amino acids and the other molecules necessary for life. It was known that proteins performed this type of chemical "work" in the cell.

In 1941 Beadle and Tatum irradiated *Neurospora* with x-rays and examined its growth on the usual "spartan" medium. Not surprisingly, some irradiated *Neurospora* spores failed to grow on this diet. Beadle and Tatum conjectured that x-rays introduced some mutations that possibly "destroyed" one of the genes responsible for processing *Neurospora*'s diet into real food. Which particular gene was destroyed remained unclear, but one of the experiments revealed that the irradiated *Neurospora* survived and even flourished when

Beadle and Tatum supplemented its spartan diet with vitamin B_6. An immediate conclusion was that x-rays damaged a gene that produces a protein (enzyme) responsible for the synthesis of B_6. The simplest explanation for this observation was that the role of a gene was to produce proteins. The rule of "one gene, one protein" remained the dominant thinking for the next half-century until biologists learned that one gene may produce a multitude of proteins.

3.4   What Molecule Codes for Genes?

DNA was discovered in 1869 by Johann Friedrich Miescher when he isolated a substance he called "nuclein" from the nuclei of white blood cells. By the early 1900s it was known that DNA (nuclein) was a long molecule consisting of four types of bases: adenine (A), thymine (T), guanine (G), and cytosine (C). Originally, biologists discovered five types of bases, the fifth being uracil (U), which is chemically similar to thymine. By the 1920s, nucleic acids were grouped into two classes called DNA and RNA, that differ slightly in their base composition: DNA uses T while RNA uses U.

DNA, or *deoxyribonucleic acid*, is a simple molecule consisting of a sugar (a common type of organic compound), a phosphate group (containing the element phosphorus), and one of four nitrogenous bases (A, T, G, or C). The chemical bonds linking together nucleotides in DNA are always the same such that the backbone of a DNA molecule is very regular. It is the A, T, C, and G bases that give "individuality" to each DNA molecule.

Ironically, for a long time biologists paid little attention to DNA since it was thought to be a repetitive molecule incapable of encoding genetic information. They thought that each nucleotide in DNA followed another in an unchanging long pattern like ATGCATGCATGCATGCATGC, like synthetic polymers. Such a simple sequence could not serve as Schrödinger's codescript, so biologists remained largely uninterested in DNA. This changed in 1944 when Oswald Avery and colleagues proved that genes indeed reside on DNA.

3.5   What Is the Structure of DNA?

The modern DNA era began in 1953 when James Watson and Francis Crick (fig. 3.1) determined the double helical structure of a DNA molecule. Just 3 years earlier, Erwin Chargaff discovered a surprising one-to-one ratio of

Figure 3.1   Watson and Crick puzzling about the structure of DNA. (Photo courtesy of Photo Researchers, Inc.)

the adenine-to-thymine and guanine-to-cytosine content in DNA (known as the *Chargaff rule*). In 1951, Maurice Wilkins and Rosalind Franklin obtained sharp x-ray images of DNA that suggested that DNA is a helical molecule.

Watson and Crick were facing a three-dimensional jigsaw puzzle: find a helical structure made out of DNA subunits that explains the Chargaff rule. When they learned of the Chargaff rule, Watson and Crick wondered whether A might be chemically attracted to T (and G to C) during DNA replication. If this was the case, then the "parental" strand of DNA would be complementary to the "child" strand, in the sense that ATGACC is complementary to TACTGG. After manipulating paper and metal Tinkertoy representations of bases[3] Watson and Crick arrived at the very simple and elegant double-stranded helical structure of DNA. The two strands were held together by hydrogen bonds between specific base pairings: A-T and C-G. The key ingredient in their discovery was the chemical logic begind the complementary relationship between nucleotides in each strand—it explained the

---

3. Computers were not common at that time, so they built a six-foot tall metal model of DNA. Amusingly, they ran out of the metal pieces and ended up cutting out cardboard ones to take their place.

Chargaff rule, since **A** was predicted to pair with **T**, and **C** with **G**. Thus, the nucleotide string of one strand completely defined the nucleotide string of the other. This is, in fact, the key to DNA replication, and the missing link between the DNA molecule and heredity. As Watson and Crick gently put it in their one-page paper on April 25, 1953: "It has not escaped our notice that the specific pairing we have postulated immediately suggests a possible copying mechanism for the genetic material."

3.6 What Carries Information between DNA and Proteins?

The double helix provided the key to DNA replication, but the question remained as to how DNA (a long but simple molecule) generates an enormous variety of different proteins. The DNA content of a cell does not change over time, but the concentrations of different proteins do. DNA is written in a four-letter alphabet while proteins are written in a twenty-letter alphabet. The key insight was that different pieces of a long DNA molecule coded for different proteins. But what was the code that translated texts written in a four-letter alphabet into texts written in a twenty-letter alphabet? How was this code read and executed?

First, we must realize that there are two types of cells: those that encapsulate their DNA in a *nucleus* and those that do not. The former are referred to as *eukaryotic* cells and the latter are *prokaryotic* cells. All multicellular organisms (like flies or humans) are eukaryotic, while most unicellular organisms (like bacteria) are prokaryotic. For our purposes, the major difference between prokaryotes and eukaryotes is that prokaryotic genes are continuous strings, while they are broken into pieces (called *exons*) in eukaryotes. Human genes may be broken into as many as 50 exons, separated by seemingly meaningless pieces called *introns*, whose function researchers are still trying to determine.

Understanding the connection between DNA and proteins began with the realization that proteins could not be made directly from DNA, since in eukaryotes DNA resides within the nucleus, whereas protein synthesis had been observed to happen outside the nucleus, in the *cytoplasm*. Therefore, some unknown agent had to somehow transport the genetic information from the DNA in the nucleus to the cytoplasm. In the mid 1950s Paul Zamecnik discovered that protein synthesis in the cytoplasm happens with the help of certain large molecules called *ribosomes* that contain RNA. This led to the suspicion that RNA could be the intermediary agent between DNA and proteins. Finally, in 1960 Benjamin Hall and and Sol Spiegelman demonstrated that RNA forms duplexes with single-stranded DNA, proving that the RNA (responsible for the synthesis of a particular protein) is complementary to the DNA segment (i.e., the gene) that codes for the protein. Thus, DNA served as a template used to copy a particular gene into *messenger RNA (mRNA)* that carries the gene's genetic information to the ribosome to make a particular protein.[4]

Chemically speaking, RNA, or *ribonucleic acid*, is almost the same as DNA. There are two main differences between RNA and DNA: there is no T base in RNA—the similar base U takes its place—and an oxygen atom is added to the sugar component. These two seemingly minor differences have a major impact on the biological roles of the two molecules. DNA is mostly inert and almost always double-stranded, helping it to serve as a static repository for information. RNA, on the other hand, is more chemically active and it usually lives in a single-stranded form. The effect is that RNA can carry short messages from the DNA to the cellular machinery that builds protein, and it can actively participate in important chemical reactions.

In 1960 Jerard Hurwitz and Samuel Weiss identified a molecular machine (composed of many proteins) that uses DNA as a template and adds ribonucleotide by ribonucleotide to make RNA. This process is called *transcription* and the molecular machine responsible for this process got the name *RNA polymerase*. Despite the advances in our understanding of the copying of DNA into RNA, how RNA polymerase knows where to start and stop transcribing DNA remains one of the many unsolved bioinformatics problems. Furthermore, the transcription of a gene into mRNA is tightly controlled, so that not all genes produce proteins at all times. Though some basic mechanisms of how gene transcription is controlled are known, a comprehensive understanding for all genes is still beyond our grasp.

In eukaryotes, a gene is typically broken into many pieces but it still produces a coherent protein. To do so, these cells have to cut the introns out of the RNA transcript and concatenate all the exons together prior to the mRNA entering the ribosome. This process of cutting and pasting the "raw" RNA version of the gene into the mRNA version that enters the ribosome is called *splicing* and is quite complicated at a molecular level.

---

4. Later biologists discovered that not all RNAs are destined to serve as templates for building proteins. Some RNAs (like transfer RNA described below) play a different role.

```
DNA:     TAC  CGC  GGC  TAT  TAC  TGC  CAG  GAA  GGA  ACT
RNA:     AUG  GCG  CCG  AUA  AUG  ACG  GUC  CUU  CCU  UGA
Protein: Met  Ala  Pro  Ile  Met  Thr  Val  Leu  Pro  Stop
```

Figure 3.2 The transcription of DNA into RNA, and the translation of RNA into a protein. Every amino acid is denoted with three letters, for example Met stands for the amino acid Methionine.

3.7 How Are Proteins Made?

In 1820 Henry Braconnot identified the first amino acid, glycine. By the early 1900s all twenty amino acids had been discovered and their chemical structure identified. Since the early 1900s when Emil Hermann Fischer showed that amino acids were linked together into linear chains to form proteins, proteins became the focus of biochemistry and molecular biology. It was postulated that the properties of proteins were defined by the composition and arrangement of their amino acids, which we now accept as true.

To uncover the code responsible for the transformation of DNA into protein, biologists conjectured that triplets of consecutive letters in DNA (called *codons*) were responsible for the amino acid sequence in a protein. Thus, a particular 30-base pair gene in DNA will make a protein of a specific 10 amino acids in a specific order, as in figure 3.2. There are $4^3 = 64$ different codons, which is more than three times as large as the number of amino acids. To explain this redundancy biologists conjectured that the *genetic code* responsible for transforming DNA into protein is degenerate: different triplets of nucleotides may code for the same amino acid. Biologists raced to find out which triplets code for which amino acids and by the late 1960s discovered the *genetic code* (table 3.1).[5] The *triplet rule* was therefore confirmed and is now accepted as fact.

Unlike the regular double-helical structure of DNA, the three-dimensional structure of proteins is highly variable. Researchers invest a large amount of effort into finding the structure of each protein; it is this structure that determines what role a protein plays in the cell—does it participate in the DNA replication process, or does it take part in some pathway that helps the cell metabolize sugar faster? Proteins perform most of the chemical work

---

5. The exact genetic code and the set of start and stop codons may vary by species from the standard genetic code presented in table 3.1. For example, mitochondrial DNA or single-cell protozoan ciliates use a slightly different table.

Table 3.1 The genetic code, from the perspective of mRNA. The codon for methionine, or AUG, also acts as a "start" codon that initiates transcription. This code is translated as in figure 3.2.

		U		C		A		G	
U	UUU	Phe	UCU	Ser	UAU	Tyr	UGU	Cys
	UUC	Phe	UCC	Ser	UAC	Tyr	UGC	Cys
	UUA	Leu	UCA	Ser	UAA	Stop	UGA	Stop
	UUG	Leu	UCG	Ser	UAG	Stop	UGG	Trp
C	CUU	Leu	CCU	Pro	CAU	His	CGU	Arg
	CUC	Leu	CCC	Pro	CAC	His	CGC	Arg
	CUA	Leu	CCA	Pro	CAA	Gln	CGA	Arg
	CUG	Leu	CCG	Pro	CAG	Gln	CGG	Arg
A	AUU	Ile	ACU	Thr	AAU	Asn	AGU	Ser
	AUC	Ile	ACC	Thr	AAC	Asn	AGC	Ser
	AUA	Ile	ACA	Thr	AAA	Lys	AGA	Arg
	AUG	Met	ACG	Thr	AAG	Lys	AGG	Arg
G	GUU	Val	GCU	Ala	GAU	Asp	GGU	Gly
	GUC	Val	GCC	Ala	GAC	Asp	GGC	Gly
	GUA	Val	GCA	Ala	GAA	Glu	GGA	Gly
	GUG	Val	GCG	Ala	GAG	Glu	GGG	Gly

in the cell, including copying DNA, moving materials inside the cell, and communicating with nearby cells. Biologists used to believe that one gene coded for one protein, but a more complex picture emerged recently with the discovery of *alternative splicing*, allowing one gene to code for many proteins.

Many chemical systems in the cell require *protein complexes*, which are groups of proteins that clump together into a large structure. A protein complex, known as RNA polymerase, begins *transcribing* a gene by copying its DNA base sequence into a short RNA base sequence (pairing a DNA T with an RNA A, a DNA A with an RNA U, and so on) called *messenger RNA*,[6] or mRNA. This short molecule is then attacked by large molecular complexes known as *ribosomes*, which read consecutive codons and locate the corresponding amino acid for inclusion in the growing polypeptide chain. Ribosomes are, in effect, molecular factories where proteins are assembled.

To help with the location of the proper amino acid for a given codon, a spe-

---

6. More precisely, this is the case in prokaryotes. In eukaryotes, this RNA template undergoes the splicing process above to form mRNA.

cial type of RNA, called *transfer RNA* (tRNA), performs a specific and elegant function. There are twenty types of tRNAs, and twenty types of amino acids. Each type of amino acid binds to a different tRNA, and the tRNA molecules have a three-base segment (called an *anticodon*) that is complementary to the codon in the mRNA. As in DNA base-pairing, the anticodon on the tRNA sticks to the codon on the RNA, which makes the amino acid available to the ribosome to add to the polypeptide chain. When one amino acid has been added, the ribosome shifts one codon to the right, and the process repeats. The process of turning an mRNA into a protein is called *translation*, since it translates information from the RNA (written in a four-letter alphabet) into the protein (written in 20-letter alphabet). All proteins, including the ones necessary for this process, are produced *by* this process.

This flow of information,

$$DNA \rightarrow transcription \rightarrow RNA \rightarrow translation \rightarrow protein,$$

is emphatically referred to as *the central dogma in molecular biology*.

3.8 How Can We Analyze DNA?

Over the years, biologists have learned how to analyze DNA. Below we describe some important techniques for copying, cutting, pasting, measuring, and probing DNA.

3.8.1 Copying DNA

Why does one need to copy DNA, that is, to obtain a large number of identical DNA fragments? From a computer science perspective, having the same string in 10^9 copies does not mean much since it does not increase the total amount of information. However, most experimental techniques (like gel electrophoresis, used for measuring DNA length) require many copies of the same DNA fragment. Since it is difficult to detect a single molecule or even a hundred molecules with modern instrumentation, amplifying DNA to yield millions or billions of identical copies is often a prerequisite of further analysis.

One method, *polymerase chain reaction* or *PCR*, is the Gutenberg printing press for DNA and is illustrated in figure 3.3. PCR amplifies a short (100- to 500-nucleotide) DNA fragment and produces a large number of identical DNA strings. To use PCR, one must know a pair of short (20- to 30-letter) strings in the DNA flanking the area of interest and design two *PCR primers,* synthetic DNA fragments identical to these strings.

Suppose we want to generate a billion copies of a DNA fragment of 500 nucleotides, that we know happens to be flanked by the 20-mer nucleotide sequence X on the left and the 20-mer nucleotide sequence Y on the right. PCR repeats a cycle of three operations: *denaturation, priming,* and *extension* to double the number of DNA fragments in every iteration. Therefore, after thirty iterations of PCR we will have on the order of 2^{30} DNA fragments, which is more than a billion copies. To start PCR, we only need a single copy of the target DNA, some artificially synthesized 20-nucleotide long DNA fragment $\overline{X}$ (many copies), some 20-nucleotide long DNA fragment Y (many copies), and billions of "spare" nucleotides (A,T,G,C).[7] We also need a molecular machine that will copy an existing DNA strand to produce a new DNA strand, and for this purpose we hijack DNA polymerase. DNA polymerase has an ability to add a complementary copy to a single-stranded DNA as long as there is a primer (i.e., $\overline{X}$ and Y) attached to the DNA strand and a sufficient supply of spare nucleotides

The denaturation step simply amounts to heating double-stranded DNA to separate it into two single strands (fig. 3.3 (top)). Priming is cooling down the solution to allow primers $\overline{X}$ and Y to hybridize to their complementary positions in DNA (fig. 3.3 (middle)). In the extension step, DNA polymerase extends the primer to produce two double-stranded DNA copies from single-stranded DNA (fig. 3.3 (bottom)). By repeatedly performing these three steps, one achieves an exponential increase in the amount of DNA, as shown in figure 3.4.

Another way to copy DNA is to *clone* it. In contrast to PCR, cloning does not require any prior information about flanking primers. However, biologists usually have no control over *which* fragment of DNA gets amplified. The process usually starts with breaking DNA into small pieces; to study an individual piece, biologists obtain many identical copies of each piece by cloning the pieces, and then try to select the individual piece of interest. Cloning incorporates a fragment of DNA into a *cloning vector,* which is a DNA molecule originating from a virus or bacterium. In this operation, the cloning vector does not lose its ability for self-replication, but carries the additional incorporated *insert* that the biologist plans to study. Vectors introduce foreign DNA into host cells (such as bacteria) which reproduce in large quantities. The self-replication process creates a large number of copies of the

---

7. $\overline{X}$ stands for the Watson-Crick complement of the 20-mer X.

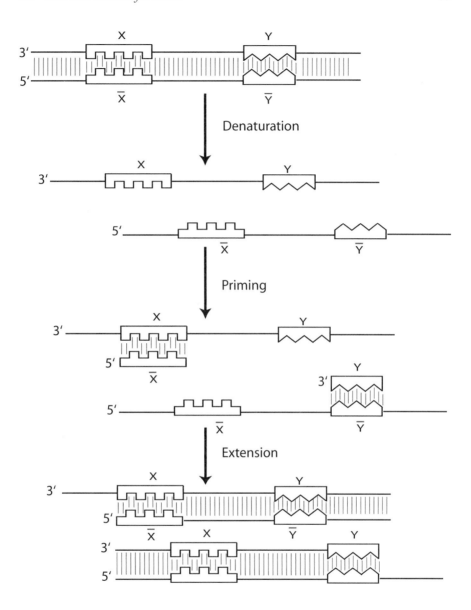

Figure 3.3 The three main operations in the polymerase chain reaction. Denaturation (top) is performed by heating the solution of DNA until the strands separate
(which happens around 70 C). Priming (middle) occurs when an excess amount of
primers $\overline{X}$ and Y are added to the denatured solution and the whole soup is allowed
to cool. Finally, extension (bottom) occurs when DNA polymerase and excess free
nucleotides (more precisely, nucleotide triphosphates) are added.

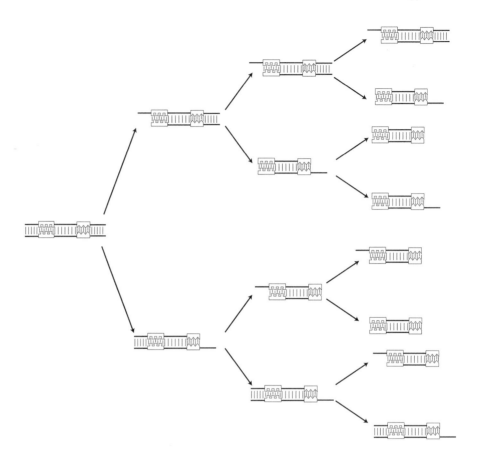

Figure 3.4   The first few iterations of PCR. Within three iterations we can go from one copy of the target DNA to eight copies.

fragment, thus enabling its properties to be studied. A fragment reproduced in this way is called a *clone*. Biologists can make *clone libraries* consisting of thousands of clones (each representing a short, randomly chosen DNA fragment) from the same DNA molecule. For example, the entire human genome can be represented as a library of 30,000 clones, each clone carrying a 100- to 200-kilobase (1000 base pairs) insert.

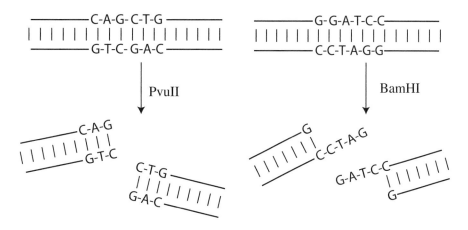

Figure 3.5 Sticky and blunt ends after cutting DNA with restriction enzymes. *Bam*HI and *Pvu*II cut at GGATCC and CAGCTG, respectively, both of which are palindromes. However, the result of *Bam*HI leaves four unmatched nucleotides on each of the strands that are cut (these unmatched nucleotides are called *sticky ends*); if a gene is cut out of one organism with *Bam*HI, it can be inserted into a different sequence that has also been cut with *Bam*HI because the sticky ends act as glue.

3.8.2 Cutting and Pasting DNA

In order to study a gene (more generally, a genomic region) of interest, it is sometimes necessary to cut it out of an organism's genome and reintroduce it into some host organism that is easy to grow, like a bacterium. Fortunately, there exist "scissors" that do just this task: certain proteins destroy the internal bonds in DNA molecules, effectively cutting it into pieces. *Restriction enzymes* are proteins that act as molecular scissors that cut DNA at every occurrence of a certain string (recognition site). For example, the *Bam*HI restriction enzyme cuts DNA into *restriction fragments* at every occurrence of the string GGATCC. Restriction enzymes first bind to the recognition site in the double-stranded DNA and then cut the DNA. The cut may produce blunt or sticky ends, as shown in figure 3.5.

Biologists have many ways to fuse two pieces of DNA together by adding the required chemical bonds. This is usually done by mimicking the processes that happen in the cell all the time: hybridization (based on complementary base-pairing) and ligation (fixing bonds within single strands), shown in figure 3.6.

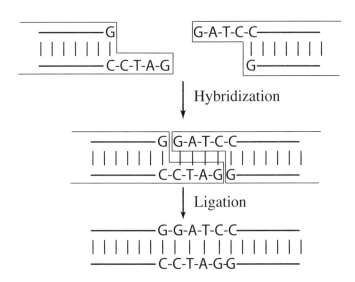

Figure 3.6   Cutting and pasting two fragments that have sticky ends (created by the restriction enzyme *Bam*HI). After hybridization, the bonds in the same DNA strands remain unfixed. The ligation step patches these bonds.

3.8.3   Measuring DNA Length

Gel electrophoresis is a technique that allows a biologist to measure the size of a DNA fragment without actually finding its exact sequence. DNA is a negatively charged molecule that migrates toward the positive pole of an electric field. The gel acts as a molecular "brake" so that long molecules move slower than short ones. The speed of migration of a fragment is related to the fragment's size, so the measurement of the migration distance for a given amount of time allows one to estimate the size of a DNA fragment. But, of course, you cannot actually see DNA molecules, so "molecular light bulbs," which are fluorescent compounds, are attached by a chemical reaction to the ends of the DNA fragments. With these bulbs, biologists can see how far different DNA fragments in a mixture migrate in the gel and thus estimate their respective lengths.

3.8.4   Probing DNA

A common task in biology is to test whether a particular DNA fragment is present in a given DNA solution. This is often done using *hybridization*: the process of joining two complementary DNA strands into a single double-stranded molecule. Biologists often use *probes*, which are single-stranded DNA fragments 20 to 30 nucleotides long that have a known sequence and a fluorescent tag. Hybridization of the probe to some unknown DNA fragment of interest can show a biologist the presence of the probe's complementary sequence in the larger DNA fragment.[8]

We can also probe RNA using a *DNA array* to see if a gene is on or off. A DNA array is essentially composed of "spots" bound to a solid support, such as a glass slide. On each spot are many copies of the complement of one gene's mRNA transcript. If the mRNA content of a cell is poured onto this slide, the mRNA will bind to the single-stranded spots and can be detected with the light-bulb technique described earler. As a result, biologists can find out which genes are producing mRNA in a particular tissue under fixed conditions.

3.9 How Do Individuals of a Species Differ?

The genetic makeup of an individual manifests itself in *traits*, such as hair color, eye color, or susceptibility to malaria. Traits are caused by variations in genes. A surprising observation is that, despite the near similarity of genomes among all humans, no two individuals are quite the same. In fact, the variations among the same gene across different individuals are limited to a handful of different base pairs (if any). Roughly only 0.1% of the 3 billion nucleotide human genome (or 3 million bases) are different between any two individuals. Still, this leaves room for roughly $4^{3,000,000}$ different genomes, and is for all intents and purposes an endless diversity.

In other words, when we speak of "the" genome of a species, we are referring to some sort of "master" genome that is fairly representative of all the possible genomes that an individual of that species could have. While specific individuals of the species may differ in some bases, the basic long DNA sequence is roughly the same in all members of the species. Of course, this handful of differences is critically important, and the large Human Diversity Project is underway to understand how various individuals differ. This will hopefully identify the mutations reponsible for a number of genetic diseases.

---

8. This is, essentially, a constant-time search of a database performed by molecular machines, something computer scientists only fantasize about!

3.10   How Do Different Species Differ?

The genomes of different organisms may be vastly different and amazingly similar.[9] The human genome consists of about 3 billion bases, while the fly genome has a scant 140 million bases. However, an analysis of the genomic sequences for two vastly different organisms (fruit flies and humans) has revealed that many genes in humans and flies are similar. Moreover, as many as 99% of all human genes are conserved across all mammals! Some human genes show strong similarity across not only mammals and flies but also across worms, plants, and (worse yet) deadly bacteria. A species, then, is a collection of individuals whose genomes are "compatible," in the sense of mating.

The likelihood that all currently living species could spontaneously develop the same gene with the same function is quite low, so it seems reasonable to assume that some process must exist that generates new species from old ones. This process is called *evolution*. The theory that all living things have evolved through a process of incremental change over millions of years has been at the heart of biology since the publication in 1859 of Charles Darwin's *On the Origin of Species*. However, only with the discovery of genomic sequences were biologists able to see how these changes are reflected in the genetic texts of existing species.

There are a number of sources for genetic variation across individuals in a species. Errors in the replication of DNA, and bizarre biological processes such as reverse transcription all cause the genomes of any two individuals in a species to be subtly different. However, genetic differences are not entirely spurious; many organisms have inherent processes that enforce genetic variation, so that no two individuals could be the same.[10] Occasionally, a variation in an individual's genome can produce a new trait, perhaps slightly stronger teeth or longer fins. If the mutations in an individual are beneficial in that individual's environment, then that individual will be more likely to be reproductively successful, passing along the mutation to its progeny. If the mutations are harmful, then that individual will be less likely to reproduce and the mutation will die out. This filtering of mutations is called *natural selection*. Over many generations the more successful individuals will become an increasingly large part of the population, to the end that the ben-

---

9. There are some genetic similarities between species that are rather surprising; we will examine some of these similarities in later chapters.
10. For example, chromosomes randomly crossover before they can make offspring. This occurs in *meiosis*, a cell replication mechanism required for most multicellular organisms to reproduce.

eficial mutation gradually takes root in all the living members of a species. As the species, as a whole, takes on the new trait, we say that it *adapts* to its environment.

If a species is divided into two isolated groups and placed into different environments, then the groups will adapt differently.[11] After many more generations, the two groups become so different that their individuals can no longer reproduce with each other, and they have become different species. This process is called *speciation*. Adaptation and speciation together form the basis of the process of evolution by natural selection, and explains the apparent paradox that there is such a diversity of life on the planet, yet so many of the genes seem similar at a sequence level. The recent abundance of genomic sequence data has enabled bioinformaticians to carry out studies that try to unravel the evolutionary relationships among different species. Since evolution by natural selection is the direct effect of adjustments to a species' genomic sequence, it stands to reason that studying the genomic sequences in different species can yield insight into their evolutionary history.

3.11 Why Bioinformatics?

As James Watson and Francis Crick worked to decipher the DNA puzzle, the 30 year-old English architect Michael Ventris tried to decipher an ancient language known as *Linear B*. At the beginning of the twentieth century, archaeologists excavated the ancient city of Knossos located on the island of Crete and found what might have been the palace of King Minos, complete with labyrinth. The archaeologists also found clay tablets with an unfamiliar form of writing. These were letters of an unknown language and there was nothing to compare them to.

The script that the ancient Cretans used (nicknamed "Linear B") remained a mystery for the next fifty years. Linguists at that time thought that Linear B was used to write in some hypothetical Minoan language (i.e., after King Minos) and cut off any investigation into the possibility that the language on

11. For example, a small group of birds might fly to an isolated part of a continent, or a few lizards might float to an island on a log.

the tablets was Greek.[12] In 1936, a fourteen-year-old boy, Michael Ventris, went on a school trip to the Minoan exhibit in London and was fascinated with the legend of the Minotaur and the unsolved puzzle of the Minoan language. After seventeen years of code-breaking, Ventris decoded the Minoan language at about the same time Watson and Crick deciphered the structure of DNA.

Some Linear B tablets had been discovered on the Greek mainland. Noting that certain strings of symbols appeared in the Cretan texts but did not appear in Greek texts, Ventris made the inspired guess that those strings applied to cities on the island. Armed with these new symbols that he could decipher, he soon unlocked much more text, and determined that the underlying language of Linear B was, in fact, just Greek written in a different alphabet. This showed that the Cretan civilization of the Linear B tablets had been part of Greek civilization.

There were two types of clay tablets found at Crete: some written in Linear B and others written in a different script named *Linear A*. Linear A appears to be older than Linear B and linguists think that Linear A is the oldest written language of Europe, a precursor of Greek. Linear A has resisted all attempts at decoding. Its underlying language is still unknown and probably will remain undecoded since it does not seem to relate to any other surviving language in the world. Linear A and Linear B texts are written in alphabets consisting of roughly ninety symbols.

Bioinformatics was born after biologists discovered how to sequence DNA and soon generated many texts in the four-letter alphabet of DNA. DNA is more like Linear A than Linear B when it comes to decoding—we still know very little about the language of DNA. Like Michael Ventris, who mobilized the mathematics of code-breaking to decipher Linear B, bioinformaticians use algorithms, statistics, and other mathematical techniques to decipher the language of DNA.

For example, suppose we have the genomic sequences of two insects that we suspect are somewhat related, evolutionarily speaking—perhaps a fruit fly (*Drosophila melanogaster*) and a malaria mosquito (*Anopheles gamibae*). Taking the Michael Ventris approach, we would like to know what parts of the fruit fly genomic sequence are dissimilar and what parts are similar to the mosquito genomic sequence. Though the means to find this out may not be immediately obvious at this point, the alignment algorithms described later

---

12. For many years biologists thought that proteins rather than DNA represent the language of the cell, which was another mistaken assumption.

in this book allow one to compare any two genes and to detect similarities between them. Unfortunately, it will take an unbearably long time to do so if we want to compare the entire fruit fly genome with the entire mosquito genome. Rather than giving up on the question altogether, biologists combined their efforts with algorithmists and mathematicians to come up with an algorithm (**BLAST**) that solves the problem very quickly and evaluates the statistical significance of any similarities that it finds.

Comparing related DNA sequences is often a key to understanding each of them, which is why recent efforts to sequence many related genomes (e.g., human, chimpanzee, mouse, rat) provide the best hope for understanding the language of DNA. This approach is often referred to as *comparative genomics*. A similar approach was used by the nineteenth century French linguist Jean-François Champollion who decoded the ancient Egyptian language.

The ancient Egyptians used hieroglyphs, but when the Egyptian religion was banned in the fourth century as a pagan cult, knowledge of hieroglyphics was lost. Even worse, the spoken language of Egyptian and its script (known as *demotic*) was lost soon afterward and completely forgotten by the tenth century when Arabic became the language of Egypt. As a result, a script that had been in use since the beginning of the third millennium BC turned into a forgotten language that nobody remembered.

During Napoleon's Egyptian campaign, French soldiers near the city of Rosetta found a stone (now known as the *Rosetta stone*) that was inscribed in three different scripts. Many of Napoleon's officers happened to be classically educated and one of them, a Lieutenant Bouchard, identified the three bands of scripts as hieroglyphic, demotic, and ancient Greek. The last sentence of the Greek inscription read: "This decree shall be inscribed on stelae of hard rock, in sacred characters, both native and Greek." The Rosetta stone thus presented a comparative linguistics problem not unlike the comparative genomics problems bionformaticians face today.

In recent decades biology has raised fascinating mathematical problems and has enabled important biological discoveries. Biologists that reduce bioinformatics to simply "the application of computers in biology" sometimes fail to recognize the rich intellectual content of bioinformatics. Bioinformatics has become a part of modern biology and often dictates new fashions, enables new approaches, and drives further biological developments. Simply using bioinformatics as a tool kit without a reasonable understanding of the main computational ideas is not very different from using a PCR kit without knowing how PCR works.

Bioinformatics is a large branch of biology (or of computer science) and this book presents neither a complete cross section nor a detailed look at any one part of it. Our intent is to describe those algorithmic principles that underlie the solution to several important biological problems to make it possible to understand any other part of the field.

Russell F. Doolittle, born 1931 in Connecticut, is currently a research professor at the Center for Molecular Genetics, University of California, San Diego. His principal research interests center around the evolution of protein structure and function. He has a PhD in biochemistry from Harvard (1962) and did postdoctoral work in Sweden. He was an early advocate of using computers as an aid to characterizing proteins.

For some it may be difficult to envision a time when the World Wide Web did not exist and every academician did not have a computer terminal on his or her desk. It may be even harder to imagine the primitive state of computer hardware and software at the time of the recombinant DNA revolution, which dates back to about 1978. It was in this period that Russell Doolittle, using a DEC PDP11 computer and a suite of home-grown programs, began systematically searching sequences in an effort to find evolutionary and other biological relationships. In 1983 he stunned cancer biologists when he reported that a newly reported sequence for platelet derived growth factor (PDGF) was virtually identical to a previously reported sequence for the oncogene known as ν-sis.[13] This was big news, and the finding served as a wake-up call to molecular biologists: searching all new sequences against up-to-date databases is your first order of business.

Doolittle had actually begun his computer studies on protein sequences much earlier. Fascinated by the idea that the history of all life might be traceable by sequence analysis, he had begun determining and aligning sequences in the early 1960s. When he landed a job at UCSD in 1964, he tried to interest consultants at the university computer center in the problem, but it was clear that the language and cultural divide between them was too great. Because computer people were not interested in learning molecular biology, he would have to learn about computing. He took an elementary course in FORTRAN

---

13. *Oncogenes* are genes in viruses that cause a cancer-like transformation of infected cells. Oncogene ν-sis in the *simian sarcoma virus* causes uncontrolled cell growth and leads to cancer in monkeys. The seemingly unrelated *growth factor* PDGF is a protein that stimulates cell growth.

programming, and, with the help of his older son, developed some simple programs for comparing sequences. These were the days when one used a keypunch machine to enter data on eighty-column cards, packs of which were dropped off at the computer center with the hope that the output could be collected the next day.

In the mid-1960s, Richard Eck and Margaret Dayhoff had begun the Atlas of Protein Sequence and Structure, the forerunner of the Protein Identification Resource (PIR) database. Their original intention was to publish an annual volume of "all the sequences that could fit between two covers." Clearly, no one foresaw the deluge of sequences that was to come once methods had been developed for directly sequencing DNA. In 1978, for example, the entire holding of the atlas, which could be purchased on magnetic tape, amounted to 1081 entries. Realizing that this was a very biased collection of protein sequences, Doolittle began his own database, which, because it followed the format of the atlas, he called NEWAT ("new atlas"). At about the same time he acquired a PDP11 computer, the maximum capacity of which was only 100 kilobytes, much of that occupied by a mini-UNIX operating system. With the help of his secretary and his younger son (eleven years old at the time), Doolittle began typing in every new sequence he could get his hands on, searching each against every other sequence in the collection as they went. This was in keeping with his view that all new proteins come from old proteins, mostly by way of gene duplications. In the first few years of their small enterprise, Doolittle & Son established a number of unexpected connections.

Doolittle admits that in 1978 he knew hardly anything about cancer viruses, but a number of chance happenings put him in touch with the field. For one, Ted Friedmann and Gernot Walter (who was then at the Salk Institute), had sought Doolittle's aid in comparing the sequences of two DNA tumor viruses, simian virus 40 (SV40) and the polyoma virus. This led indirectly to contacts with Inder Verma's group at Salk, which was studying retroviruses and had sequenced an "oncogene" called v-mos in a retrovirus that caused sarcomas in mice. They asked Doolittle to search it for them, but no significant matches were found. Not long afterward (in 1980), Doolittle read an article reporting the nucleotide sequence of an oncogene from an avian sarcoma virus—the famous *Rous sarcoma virus*. It was noted in that article that the Salk team had provided the authors with a copy of their still unpublished mouse sarcoma gene sequence, but no resemblances had been detected. In line with his own project, Doolittle promptly typed the new avian sequence into his computer to see if it might match anything else. He was astonished to find that in fact a match quickly appeared with the still unpublished Salk sequence for the mouse retrovirus oncogene. He immediately telephoned Inder Verma; "Hey, these two sequences are in fact homologous. These proteins must be doing the same thing." Verma, who had just packaged up a manuscript describing the new sequence, promptly unwrapped it and added the new feature. He was so pleased with the outcome that he added Doolittle's name as one of the coauthors.

How was it that the group studying the Rous sarcoma virus had missed this match? It's a reflection on how people were thinking at the time. They had compared the DNA sequences of the two genes without translating them into the corresponding amino acid sequences, losing most of the information as a result. It was another simple but urgent message to the community about how to think about sequence comparisons.

In May of 1983, an article appeared in *Science* describing the characterization of a growth factor isolated from human blood platelets. Harry Antoniades and Michael Hunkapiller had determined 28 amino acid residues from the N-terminal end of PDGF. (It had taken almost 100,000 units of human blood to obtain enough of the growth factor material to get this much sequence.) The article noted that the authors had conducted a limited search of known sequences and hadn't found any similar proteins.

By this time, Doolittle had modem access to a department VAX computer where he now stored his database. He typed in the PDGF partial sequence and set it searching. Twenty minutes later he had the results of the search; human PDGF had a sequence that was virtually identical to that of an oncogene isolated from a woolly monkey. Doolittle describes it as an electrifying moment, enriched greatly by his prior experiences with the other oncogenes. He remembers remarking to his then fifteen-year old son, "Will, this experiment took us five years and twenty minutes." As it happened, he was not alone in enjoying the thrill of this discovery. Workers at the Imperial Cancer Laboratory in London were also sequencing PDGF, and in the spring of 1983 had written to Doolittle asking for a tape of his sequence collection. He had sent them his newest version, fortuitously containing the *v*-sis sequence from the woolly monkey. Just a few weeks before the *Science* article appeared, the group at the ICL replied to Doolittle with an effusive letter of thanks, without mentioning just why the tape had been so valuable. Meanwhile, Doolittle had written to both the PDGF workers and the *v*-sis team, suggesting that they compare notes. As a result, the news of the match was quickly made known, and a spirited race to publication occurred, the report from the Americans appearing in *Science* only a week ahead of the British effort in *Nature*. Doolittle went on to make many other matches during the mid-

1980s, including several more involving oncogenes. For example, he found a relationship between the oncogene ν-jun and the gene regulator GCN4. He describes those days as unusual in that an amateur could still occasionally compete with the professionals. Although he continued with his interests in protein evolution, he increasingly retreated to the laboratory and left bioinformatics to those more formally trained in the field.

4 *Exhaustive Search*

Exhaustive search algorithms require little effort to design but for many problems of interest cannot process inputs of any reasonable size within your lifetime. Despite this problem, exhaustive search, or *brute force* algorithms are often the first step in designing more efficient algorithms.

We introduce two biological problems: *DNA restriction mapping* and *regulatory motif finding*, whose brute force solutions are not practical. We further describe the *branch-and-bound* technique to transform an inefficient brute force algorithm into a practical one. In chapter 5 we will see how to improve our motif finding algorithm to arrive at an algorithm very similar to the popular CONSENSUS motif finding tool. In chapter 12 we describe two randomized algorithms, GibbsSampler and RandomProjections, that use coin-tossing to find motifs.

4.1 Restriction Mapping

Hamilton Smith discovered in 1970 that the *restriction enzyme Hind*II cleaves DNA molecules at every occurrence, or *site*, of the sequences GTGCAC or GTTAAC, breaking a long molecule into a set of *restriction fragments*. Shortly thereafter, maps of restriction sites in DNA molecules, or *restriction maps*, became powerful research tools in molecular biology by helping to narrow the location of certain genetic markers.

If the genomic DNA sequence of an organism is known, then construction of a restriction map for *Hind*II amounts to finding all occurrences of GTGCAC and GTTAAC in the genome. Because the first bacterial genome was sequenced twenty-five years after the discovery of restriction enzymes, for many years biologists were forced to build restriction maps for genomes without prior knowledge of the genomes' DNA sequence.[1]

Several experimental approaches to restriction mapping exist, each with advantages and disadvantages. The distance between two individual restriction sites corresponds to the length of the restriction fragment between those two sites and can be measured by the *gel electrophoresis* technique described in chapter 2. This requires no knowledge of the DNA sequence. Biologists can vary experimental conditions to produce either a *complete* digest [fig. 4.1 (a)] or a *partial* digest [fig. 4.1 (b)] of DNA.[2] The restriction mapping problem can be formulated in terms of recovering positions of points when only pairwise distances between those points are known.

To formulate the restriction mapping problem, we will introduce some notation. A *multiset* is a set that allows duplicate elements (e.g., $\{2, 2, 2, 3, 3, 4, 5\}$ is a multiset with duplicate elements 2 and 3). If $X = \{x_1 = 0, x_2, \ldots, x_n\}$ is a set of n points on a line segment in increasing order, then ΔX denotes the multiset of *all* $\binom{n}{2}$ pairwise distances[3] between points in X:

$$\Delta X = \{x_j - x_i : \ 1 \le i < j \le n\}.$$

For example, if $X=\{0, 2, 4, 7, 10\}$, then $\Delta X=\{2, 2, 3, 3, 4, 5, 6, 7, 8, 10\}$, which are the ten pairwise distances between these points (table 4.1). In restriction mapping, we are given ΔX, the experimental data about fragment lengths. The problem is to reconstruct X from ΔX. For example, could you infer

---

1. While restriction maps were popular research tools in the late 1980s, their role has been somewhat reduced in the last ten to fifteen years with the development of efficient DNA sequencing technologies. Though biologists rarely have to solve the DNA mapping problems in current research, we present them here as illustrations of branch-and-bound techniques for algorithm development.

2. Biologists typically work with billions of identical DNA molecules in solution. A complete digest corresponds to experimental conditions under which *every* DNA molecule at *every* restriction site is cut (i.e., the probability of cut at every restriction site is 1). Every linear DNA molecule with n restriction sites is cut into $n+1$ fragments that are recorded by gel electrophoresis as in figure 4.1 (a). A partial digest corresponds to experimental conditions that cut every DNA molecule at a given restriction site with probability less than 1. As a result, with some probability, the interval between any two (not necessarily consecutive) sites remains uncut, thus generating all fragments shown in figure 4.1 (b).

3. The notation $\binom{n}{k}$, read "n choose k," means "the number of distinct subsets of k elements taken from a (larger) set of n elements," and is given by the expression $\frac{n!}{(n-k)!k!}$. In particular, $\binom{n}{2} = \frac{n(n-1)}{2}$ is the number of different pairs of elements from an n-element set. For example, if $n = 5$, the set $\{1, 2, 3, 4, 5\}$ has $\binom{5}{2} = 10$ subsets formed by two elements: $\{1, 2\}$, $\{1, 3\}$, $\{1, 4\}$, $\{1, 5\}$, $\{2, 3\}$, $\{2, 4\}$, $\{2, 5\}$, $\{3, 4\}$, $\{3, 5\}$, and $\{4, 5\}$. These ten subsets give rise to the elements $x_2 - x_1, x_3 - x_1, x_4 - x_1, x_5 - x_1, x_3 - x_2, x_4 - x_2, x_5 - x_2, x_4 - x_3, x_5 - x_3$, and $x_5 - x_4$.

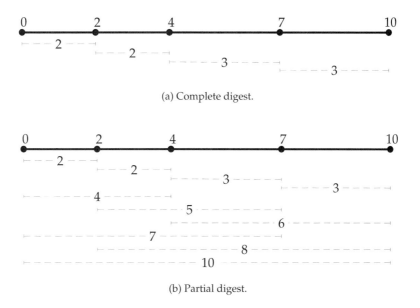

(a) Complete digest.

(b) Partial digest.

Figure 4.1 Different methods of digesting a DNA molecule. A complete digest produces only fragments between consecutive restriction sites, while a partial digest yields fragments between any two restriction sites. Each of the dots represents a restriction site.

that the set $\Delta X = \{2, 2, 3, 3, 4, 5, 6, 7, 8, 10\}$ was derived from $\{0, 2, 4, 7, 10\}$? Though gel electrophoresis allows one to determine the lengths of DNA fragments easily, it is often difficult to judge their multiplicity. That is, the number of different fragments of a given length can be difficult to determine. However, it is experimentally possible to do so with a lot of work, and we assume for the sake of simplifying the problem that this information is given to us as an input to the problem.

Table 4.1 Representation of $\Delta X = \{2, 2, 3, 3, 4, 5, 6, 7, 8, 10\}$ as a two-dimensional table, with the elements of $X = \{0, 2, 4, 7, 10\}$ along both the top and left side. The element at (i, j) in the table is the value $x_j - x_i$ for $1 \le i < j \le n$.

	0	2	4	7	10
0		2	4	7	10
2			2	5	8
4				3	6
7					3
10					

Partial Digest Problem:

Given all pairwise distances between points on a line, reconstruct the positions of those points.

Input: The multiset of pairwise distances L, containing $\binom{n}{2}$ integers.

Output: A set X, of n integers, such that $\Delta X = L$

This *Partial Digest problem*, or *PDP*, is sometimes called the *Turnpike problem* in computer science. Suppose you knew the set of distances between every (not necessarily consecutive) pair of exits on a highway leading from one town to another. Could you reconstruct the geography of the highway from this information? That is, could you find the distance from the first town to each exit? Here, the "highway exits" are the restrictions sites in DNA; the lengths of the resulting DNA restriction fragments correspond to distances between highway exits. Computationally, the only difference between the Turnpike problem and the PDP is that the distances between exits are given in miles in the Turnpike problem, while the distances between restriction sites are given in nucleotides in the PDP.

We remark that it is not always possible to uniquely reconstruct a set X based only on ΔX. For example, for any integer v and set A, one can see that ΔA is equal to $\Delta(A \oplus \{v\})$, where $A \oplus \{v\}$ is defined to be $\{a + v : a \in A\}$, a *shift* of every element in A by v. Also $\Delta A = \Delta(-A)$, where $-A = \{-a : a \in A\}$ is the *reflection* of A. For example, sets $A = \{0, 2, 4, 7, 10\}$, $\Delta(A \oplus \{100\}) = \{100, 102, 104, 107, 110\}$, and $-A = \{-10, -7, -4, -2, 0\}$ all produce the same partial digest. The sets $\{0, 1, 3, 8, 9, 11, 12, 13, 15\}$ and $\{0, 1, 3, 4, 5, 7, 12, 13, 15\}$

present a less trivial example of this problem of nonuniqueness. The partial digests of these two sets is the same multiset of 36 elements[4]:
$\{1_4, 2_4, 3_4, 4_3, 5_2, 6_2, 7_2, 8_3, 9_2, 10_2, 11_2, 12_3, 13, 14, 15\}.$

	0	1	3	4	5	7	12	13	15
0		1	3	4	5	7	12	13	15
1			2	3	4	6	11	12	14
3				1	2	4	9	10	12
4					1	3	8	9	11
5						2	7	8	10
7							5	6	8
12								1	3
13									2
15									

	0	1	3	8	9	11	12	13	15
0		1	3	8	9	11	12	13	15
1			2	7	8	10	11	12	14
3				5	6	8	9	10	12
8					1	3	4	5	7
9						2	3	4	6
11							1	2	4
12								1	3
13									2
15									

In general, sets A and B are said to be *homometric* if $\Delta A = \Delta B$. Let U and V be two sets of numbers. One can verify that the multisets

$$U \oplus V = \{u + v : u \in U, v \in V\}$$

and

$$U \ominus V = \{u - v : u \in U, v \in V\}$$

are homometric (a problem at the end of this chapter). The "nontrivial" nine-point example above came from $U = \{6, 7, 9\}$ and $V = \{-6, 2, 6\}$. Indeed $U \oplus V = \{0, 1, 3, 8, 9, 11, 12, 13, 15\}$ while $U \ominus V = \{0, 1, 3, 4, 5, 7, 12, 13, 15\}$ as illustrated below:

$U \oplus V$	-6	2	6
6	0	8	12
7	1	9	13
9	3	11	15

$U \ominus V$	-6	2	6
6	12	4	0
7	13	5	1
9	15	7	3

While the PDP is to find one set X such that $\Delta X = L$, biologists are often interested in *all* homometric sets.

4.2   Impractical Restriction Mapping Algorithms

The algorithm below, BRUTEFORCEPDP, takes the list L of $\binom{n}{2}$ integers as an input, and returns the set X of n integers such that $\Delta X = L$. We remind the reader that we do not always provide complete details for all of the operations in pseudocode. In particular, we have not provided any subroutine to calculate ΔX; problem 4.1 asks you to do fill in the details for this step.

---

4. The notation 1_4 means that element 1 is repeated four times in this multiset.

BRUTEFORCEPDP(L, n)
1   $M \leftarrow$ maximum element in L
2   **for** every set of $n - 2$ integers $0 < x_2 < \cdots < x_{n-1} < M$
3            $X \leftarrow \{0, x_2, \ldots, x_{n-1}, M\}$
4   Form ΔX from X
5   **if** $\Delta X = L$
6   **return** X
7   **output** "No Solution"

BRUTEFORCEPDP is slow since it examines $\binom{M-1}{n-2}$ different sets of positions, which requires about $O(M^{n-2})$ time.[5]

One might question the wisdom of selecting $n - 2$ *arbitrary* integers from the interval 0 to M. For example, if L does not contain the number 5, there is really no point in choosing any $x_i = 5$, though the above algorithm will do so. Indeed, observing that all points in X have to correspond to *some* distance in ΔX, we can select $n - 2$ distinct elements from L rather than the less constrained selection from the interval $(0, M)$. Since M may be large, even with a small number of points, building a new algorithm that makes choices of x_i based only on elements in L yields an improvement in efficiency.

ANOTHERBRUTEFORCEPDP(L, n)
1   $M \leftarrow$ maximum element in L
2   **for** every set of $n - 2$ integers $0 < x_2 < \cdots < x_{n-1} < M$ from L
3            $X \leftarrow \{0, x_2, \ldots, x_{n-1}, M\}$
4   Form ΔX from X
5   **if** $\Delta X = L$
6   **return** X
7   **output** "No Solution"

This algorithm examines $\binom{|L|}{n-2}$ different sets of integers, but $|L| = \frac{n(n-1)}{2}$, so ANOTHERBRUTEFORCEPDP takes roughly $O(n^{2n-4})$ time. This is still not practical, but since M can be arbitrarily large compared to n, this is actually a more efficient algorithm than BRUTEFORCEPDP. For example, BRUTE-FORCEPDP takes a very long time to execute when called on an input of $L = \{2, 998, 1000\}$, but ANOTHERBRUTEFORCEPDP takes very little time. Here, n is 3 while M is 1000.

---

5. Although a careful mathematical analysis of the running time leads to a somewhat smaller number, it does not help much in practice.

4.3 A Practical Restriction Mapping Algorithm

In 1990, Steven Skiena described a different brute force algorithm for the PDP that works well in practice.[6] First, find the largest distance in L; this must determine the two outermost points of X, and we can delete this distance from L. Now that we have fixed the two outermost points, we select the largest remaining distance in L, call it δ. One of the points that generated δ must be one of the two outermost points, since δ is the *largest* remaining distance; thus, we have one of two choices to place a point: δ from the leftmost point, or δ from the rightmost point. Suppose we decide to include the point that is δ from the leftmost point in the set. We can calculate the pairwise distances between this new position and all the other positions that we have chosen, and ask if these distances are in L. If so, then we remove those distances from L and repeat by selecting the next largest remaining distance in L and so on. If these pairwise distances are not in L, then we choose the position that is δ from the rightmost point, and perform the same query. However, if these pairwise distances are not in L either, then we must have made a bad choice at some earlier step and we need to *backtrack* a few steps, reverse a decision, and try again. If we ever get to the point where L is empty, then we have found a valid solution.[7]

For example, suppose $L = \{2, 2, 3, 3, 4, 5, 6, 7, 8, 10\}$. The size of L is $\binom{n}{2} = \frac{n(n-1)}{2} = 10$, where n is the number of points in the solution. In this case, n must be 5, and we will refer to the positions in X as $x_1 = 0$, x_2, x_3, x_4 and x_5, from left to right, on the line.

Since 10 is the largest distance in L, x_5 must be at position 10, so we remove this distance $x_5 - x_1 = 10$ from L to obtain

$$X = \{0, 10\} \qquad L = \{2, 2, 3, 3, 4, 5, 6, 7, 8\} \ .$$

The largest remaining distance is 8. We have two choices: either $x_4 = 8$ or $x_2 = 2$. Since those two cases are mirror images of each other, without loss of generality, we can assume $x_2 = 2$. After removal of distances $x_5 - x_2 = 8$ and $x_2 - x_1 = 2$ from L, we obtain

$$X = \{0, 2, 10\} \qquad L = \{2, 3, 3, 4, 5, 6, 7\} \ .$$

---

6. To be more accurate, this algorithm works well for high-quality PDP data. However, the bottleneck in the PDP technique is the difficulty in acquiring accurate data.
7. This is an example of a cookbook description; the pseudocode of this algorithm is described below.

Now 7 is the largest remaining distance, so either $x_4 = 7$ or $x_3 = 3$. If $x_3 = 3$, then $x_3 - x_2 = 1$ must be in L, but it is not, so x_4 must be at 7. After removing distances $x_5 - x_4 = 3$, $x_4 - x_2 = 5$, and $x_4 - x_1 = 7$ from L, we obtain

$$X = \{0, 2, 7, 10\} \qquad L = \{2, 3, 4, 6\} \ .$$

Now 6 is the largest remaining distance, and we once again have only two choices: either $x_3 = 4$ or $x_3 = 6$. If $x_3 = 6$, the distance $x_4 - x_3 = 1$ must be in L, but it is not. We are left with only one choice, $x_3 = 4$, and this provides a solution $X = \{0, 2, 4, 7, 10\}$ to the PDP.

The PARTIALDIGEST algorithm shown below works with the list of pairwise distances, L, and uses the function $\text{DELETE}(y, L)$ which removes the value y from L. We use the notation $\Delta(y, X)$ to denote the multiset of distances between a point y and all points in a set X. For example,
$$\Delta(2, \{1, 3, 4, 5\}) = \{1, 1, 2, 3\} \ .$$

PARTIALDIGEST(L)
1   $width \leftarrow$ Maximum element in L
2   DELETE($width, L$)
3   $X \leftarrow \{0, width\}$
4   PLACE(L, X)

PLACE(L, X)
  1   **if** L is empty
  2   **output** X
  3   **return**
  4   $y \leftarrow$ Maximum element in L
  5   **if** $\Delta(y, X) \subseteq L$
  6   Add y to X and remove lengths $\Delta(y, X)$ from L
  7   PLACE(L, X)
  8   Remove y from X and add lengths $\Delta(y, X)$ to L
  9   **if** $\Delta(width - y, X) \subseteq L$
10   Add $width - y$ to X and remove lengths $\Delta(width - y, X)$ from L
11   PLACE(L, X)
12   Remove $width - y$ from X and add lengths $\Delta(width - y, X)$ to L
13   **return**

After each recursive call in PLACE, we undo our modifications to the sets X and L in order to restore them for the next recursive call. It is important to note that this algorithm will list *all* sets X with $\Delta X = L$.

At first glance, this algorithm looks efficient—at each point we examine two alternatives ("left" or "right"), ruling out the obviously incorrect decisions that lead to inconsistent distances. Indeed, this algorithm is very fast for most instances of the PDP since usually only one of the two alternatives, "left" or "right," is viable at any step. It was not clear for a number of years whether or not this algorithm is polynomial in the worst case—sometimes both alternatives are viable. If both "left" and "right" alternatives hold and if this continues to happen in future steps of the algorithm, then the performance of the algorithm starts growing as 2^k where k is the number of such "ambiguous" steps.[8]

Let $T(n)$ be the maximum time PARTIALDIGEST takes to find the solution for an n-point instance of the PDP. If there is only one viable alternative at every step, then PARTIALDIGEST steadily reduces the size of the problem by one and calls itself recursively, so

$$T(n) = T(n-1) + O(n),$$

where $O(n)$ is the work spent adjusting the sets X and L. However, if there are two alternatives, then

$$T(n) = 2T(n-1) + O(n).$$

While the expressions $T(n) = T(n-1) + O(n)$ and $T(n) = 2T(n-1) + O(n)$ bear a superficial similarity in form, they each lead to very different expressions for the algorithm's running time. One is quadratic, as we saw when analyzing SELECTIONSORT, and the other exponential, as we saw with HANOITOWERS. In fact, polynomial algorithms for the PDP were unknown until 2002 when Maurice Nivat and colleagues designed the first one.

4.4   Regulatory Motifs in DNA Sequences

Fruit flies, like humans, are susceptible to infections from bacteria and other pathogens. Although fruit flies do not have as sophisticated an immune system as humans do, they have a small set of *immunity genes* that are usually dormant in the fly genome, but somehow get switched on when the organism gets infected. When these genes are turned on, they produce proteins that destroy the pathogen, usually curing the infection.

One could design an experiment that is rather unpleasant to the flies, but very informative to biologists: infect flies with a bacterium, then grind up the flies and measure (perhaps with a DNA array) which genes are switched on

---

8. There exist pathological examples forcing the algorithm to explore *both* "left" and "right" alternatives at nearly every step.

as an immune response. From this set of genes, we would like to determine
what triggers their activation.   It turns out that many immunity genes in
the fruit fly genome have strings that are reminiscent of TCGGGGATTTCC,
located upstream of the genes' start. These short strings, called NF-κB *binding sites*, are important examples of *regulatory motifs* that turn on immunity
and other genes. Proteins known as *transcription factors* bind to these motifs,
encouraging RNA polymerase to transcribe the downstream genes. Motif
finding is the problem of discovering such motifs without any prior knowledge of how the motifs look.

Ideally, the fly infection experiment would result in a set of upstream regions from genes in the genome, each region containing at least one NF-κB
binding site. Suppose we do not know what the NF-κB pattern looks like,
nor do we know where it is located in the experimental sample. The fly infection experiment requires an algorithm that, given a set of sequences from
a genome, can find short substrings that seem to occur surprisingly often.

"The Gold Bug", by Edgar Allan Poe, helps to illustrate the spirit, if not the
mechanics, of finding motifs in DNA sequences. When the character William
Legrand finds a parchment written by the pirate Captain Kidd, Legrand's
friend says, "Were all the jewels of Golconda awaiting me upon my solution
of this enigma, I am quite sure that I should be unable to earn them." Written
on the parchment in question was

```
53++!305))6*;4826)4+.)4+);806*;48!8'60))85;]8*:+*8!
83(88)5*!;46(;88*96*?;8)*+(;485);5*!2:*+(;4956*2(5*
-4)8'8*; 4069285);)6!8)4++;1(+9;48081;8:8+1;48!85;4
)485!528806*81(+9;48;(88;4(+?34;48)4+;161;:188;+?;
```

Mr. Legrand responds, "It may well be doubted whether human ingenuity can construct an enigma of the kind which human ingenuity may not, by
proper application, resolve." He notices that a combination of three symbols—
; 4 8—appears very frequently in the text. He also knows that Captain
Kidd's pirates speak English and that the most frequent English word is
"the." Proceeding under the assumption that ; 4 8 encodes "the," Mr.
Legrand deciphers the parchment note and finds the pirate treasure. After
making this substitution, Mr. Legrand has a slightly easier text to decipher:

```
53++!305))6*THE26)H+.)H+)TE06*THE!E'60))E5T]E*:+*E!
E3(EE)5*!TH6(TEE*96*?TE)*+(THE5)T5*!2:*+(TH956*2(5*
-H)E'E*T H0692E5)T)6!E)H++T1(+9THE0E1TE:E+1THE!E5TH
)HE5!52EE06*E1(+9THET(EETH(+?3HTHE)H+T161T:1EET+?T
```

You might try to figure out what the symbol ")" might code for in order to complete the puzzle.

Unfortunately, DNA texts are not that easy to decipher, and there is little doubt that nature has constructed an enigma that human ingenuity cannot entirely solve. However, bioinformaticians borrowed Mr. Legrand's method, and a popular approach to motif finding is based on the assumption that frequent or rare words may correspond to regulatory motifs in DNA. It stands to reason that if a word occurs considerably more frequently than expected, then it is more likely to be some sort of "signal," and it is crucially important to figure out the biological meaning of the signal.

This "DNA linguistics" approach is at the heart of the *pattern-driven* approach to signal finding, which is based on enumerating all possible patterns and choosing the most frequent (or the most statistically surprising) among them.

4.5 Profiles

Figure 4.2 (a) presents seven 32-nucleotide DNA sequences generated randomly. Also shown [fig. 4.2 (b)] are the same sequences with the "secret" pattern $P = $ **ATGCAACT** of length $l = 8$ implanted at random positions. Suppose you do not know what the pattern P is, or where in each sequence it has been implanted [fig. 4.2 (c)]. Can you reconstruct P by analyzing the DNA sequences?

We could simply count the number of times each l-mer, or string of length l, occurs in the sample. Since there are only $7 \cdot (32 + 8) = 280$ nucleotides in the sample, it is unlikely that any 8-mer other than the implanted pattern appears more than once.[9] After counting all 8-mer occurrences in figure 4.2 (c) we will observe that, although most 8-mers appear in the sample just once (with a few appearing twice), there is one 8-mer that appears in the sample suspiciously many times—seven or more. This overrepresented 8-mer is the pattern P we are trying to find.

Unlike our simple implanted patterns above, DNA uses a more inventive notion of regulatory motifs by allowing for mutations at some nucleotide positions [fig. 4.2 (d)]. For example, table 4.2 shows eighteen different NF-κB motifs; notice that, although none of them are the consensus binding site sequence **TCGGGGATTTCC**, each one is not substantially different. When the implanted pattern P is allowed to mutate, reconstructing P becomes more

---

9. The probability that any 8-mer appears in the sample is less than $280/4^8 \approx 0.004$

CGGGGCTGGGTCGTCACATTCCCCTTTCGATA
TTTGAGGGTGCCCAATAACCAAAGCGGACAAA
GGGATGCCGTTTGACGACCTAAATCAACGGCC
AAGGCCAGGAGCGCCTTTGCTGGTTCTACCTG
AATTTTCTAAAAAGATTATAATGTCGGTCCTC
CTGCTGTACAACTGAGATCATGCTGCTTCAAC
TACATGATCTTTTGTGGATGAGGGAATGATGC

(a) Seven random sequences.

CGGGGCT<u>ATGCAACT</u>GGGTCGTCACATTCCCCTTTCGATA
TTTGAGGGTGCCCAATAA<u>ATGCAACT</u>CCAAAGCGGACAAA
GG<u>ATGCAACT</u>GATGCCGTTTGACGACCTAAATCAACGGCC
AAGG<u>ATGCAACT</u>CCAGGAGCGCCTTTGCTGGTTCTACCTG
AATTTTCTAAAAAGATTATAATGTCGGTCC<u>ATGCAACT</u>TC
CTGCTGTACAACTGAGATCATGCTGC<u>ATGCAACT</u>TTCAAC
TACATGATCTTTTG<u>ATGCAACT</u>TGGATGAGGGAATGATGC

(b) The same DNA sequences with the implanted
pattern ATGCAACT.

CGGGGCTATGCAACTGGGTCGTCACATTCCCCTTTCGATA
TTTGAGGGTGCCCAATAAATGCAACTCCAAAGCGGACAAA
GGATGCAACTGATGCCGTTTGACGACCTAAATCAACGGCC
AAGGATGCAACTCCAGGAGCGCCTTTGCTGGTTCTACCTG
AATTTTCTAAAAAGATTATAATGTCGGTCCATGCAACTTC
CTGCTGTACAACTGAGATCATGCTGCATGCAACTTTCAAC
TACATGATCTTTTGATGCAACTTGGATGAGGGAATGATGC

(c) Same as (b), but hiding the implant locations. Suddenly this problem looks difficult to solve.

CGGGGCT<u>ATcCAgCT</u>GGGTCGTCACATTCCCCTTTCGATA
TTTGAGGGTGCCCAATAA<u>ggGCAACT</u>CCAAAGCGGACAAA
GG<u>ATGgAtCT</u>GATGCCGTTTGACGACCTAAATCAACGGCC
AAGG<u>AaGCAACc</u>CCAGGAGCGCCTTTGCTGGTTCTACCTG
AATTTTCTAAAAAGATTATAATGTCGGTCC<u>tTGgAACT</u>TC
CTGCTGTACAACTGAGATCATGCTGC<u>ATGCcAtT</u>TTCAAC
TACATGATCTTTTG<u>ATGgcACT</u>TGGATGAGGGAATGATGC

(d) Same as (b), but with the implanted pattern ATG-
CAACT randomly mutated in two positions; no two
implanted instances are the same. If we hide the locations as in (c), the difficult problem becomes nearly
impossible.

Figure 4.2   DNA sequences with implanted motifs.

Table 4.2   A small collection of putative NF-κB binding sites.

```
T C G G G G A T T T C A
A C G G G G A T T T T T
T C G G T A C T T T A C
T T G G G G A C T T T T
C C G G T G A T T C C C
G C G G G G A A T T T C
T C G G G G A T T C C T
T C G G G G A T T C C T
T A G G G G A A C T A C
T C G G G T A T A A A C
T C G G G G G T T T T T
C C G G T G A C T T A C
C C A G G G A C T C C C
A A G G G G A C T T C C
T T G G G G A C T T T T
T T T G G G A G T C C C
T C G G T G A T T T C C
T A G G G G A A G A C C
```

A:	2	3	1	0	0	1	16	3	1	2	4	1
T:	12	3	1	0	4	1	0	9	15	11	5	6
G:	1	0	16	18	14	16	1	1	1	0	0	0
C:	3	12	0	0	0	0	1	5	1	5	9	11

```
T C G G G G A T T T C C
```

complicated, since the 8-mer count does not reveal the pattern. In fact, the string **ATGCAACT** does not even appear in figure 4.2 (d), but the seven mutated versions of it appear at position 8 in the first sequence, position 19 in the second sequence, 3 in the third, 5 in the fourth, 31 in the fifth, 27 in the sixth, and 15 in the seventh.

In order to unambiguously formulate the motif finding problem, we need to define precisely what we mean by "motif." Relying on a single string to represent a motif often fails to represent the variation of the pattern in real biological sequences, as in figure 4.2 (d). A more flexible representation of a motif uses a profile matrix.

Consider a set of t DNA sequences, each of which has n nucleotides. Select one position in each of these t sequences, thus forming an array s =

```
                                    CGGGGCTATcCAgCTGGGTCGTCACATTCCCCTT...
                        TTTGAGGGTGCCCAATAAggGCAACTCCAAAGCGGACAAA
                                GGATGgAtCTGATGCCGTTTGACGACCTA...
                              AAGGAaGCAACcCCAGGAGCGCCTTTGCTGG...
        AATTTTCTAAAAAGATTATAATGTCGGTCCtTGgAACTTC
        CTGCTGTACAACTGAGATCATGCTGCATGCcAtTTTCAAC
                    TACATGATCTTTTGATGgcACTTGGATGAGGGAATGATGC
```

(a) Superposition of the seven highlighted 8-mers from figure 4.2 (d).

		A	T	C	C	A	G	C	T
		G	G	G	C	A	A	C	T
		A	T	G	G	A	T	C	T
Alignment		A	A	G	C	A	A	C	C
		T	T	G	G	A	A	C	T
		A	T	G	C	C	A	T	T
		A	T	G	G	C	A	C	T
Profile	**A**	5	1	0	0	5	5	0	0
	T	1	5	0	0	0	1	1	6
	G	1	1	6	3	0	1	0	0
	C	0	0	1	4	2	0	6	1
Consensus		A	T	G	C	A	A	C	T

(b) The alignment matrix, profile matrix and consensus string formed from the 8-mers starting at positions $\mathbf{s} = (8, 19, 3, 5, 31, 27, 15)$ in figure 4.2 (d).

Figure 4.3 From DNA sample, to alignment matrix, to profile, and, finally, to consensus string. If $\mathbf{s} = (8, 19, 3, 5, 31, 27, 15)$ is an array of starting positions for 8-mers in figure 4.2 (d), then $Score(\mathbf{s}) = 5 + 5 + 6 + 4 + 5 + 5 + 6 + 6 = 42$.

$(s_1, s_2, \ldots, s_t)$, with $1 \leq s_i \leq n - l + 1$. The l-mers starting at these positions can be compiled into a $t \times l$ *alignment matrix* whose (i, j)th element is the nucleotide in the $s_i + j - 1$th element in the ith sequence (fig. 4.3). Based on the alignment matrix, we can compute the $4 \times l$ *profile matrix* whose (i, j)th element holds the number of times nucleotide i appears in column j of the alignment matrix, where i varies from 1 to 4. The profile matrix, or *profile*, illustrates the variability of nucleotide composition at each position for a particular choice of l-mers. For example, the positions 3, 7, and 8 are highly conserved, while position 4 is not. To further summarize the profile matrix, we can form a *consensus string* from the most popular element in each column of the alignment matrix, which is the nucleotide with the largest entry in the profile matrix. Figure 4.3 shows the alignment matrix for $\mathbf{s} = (8, 19, 3, 5, 31, 27, 15)$, the corresponding profile matrix, and the resulting consensus string ATGCAACT.

By varying the starting positions in $\mathbf{s}$, we can construct a large number of different profile matrices from a given sample. We need some way of grading them against each other. Some profiles represent high conservation of a pattern while others represent no conservation at all. An imprecise formulation of the Motif Finding problem is to find the starting positions $\mathbf{s}$ corresponding to the most conserved profile. We now develop a specific measure of conservation, or strength, of a profile.

4.6 The Motif Finding Problem

If $\mathbf{P}(\mathbf{s})$ denotes the profile matrix corresponding to starting positions $\mathbf{s}$, then we will use $M_{\mathbf{P}(\mathbf{s})}(j)$ to denote the largest count in column j of $\mathbf{P}(\mathbf{s})$. For the profile $\mathbf{P}(\mathbf{s})$ in figure 4.3, $M_{\mathbf{P}(\mathbf{s})}(1) = 5$, $M_{\mathbf{P}(\mathbf{s})}(2) = 5$, and $M_{\mathbf{P}(\mathbf{s})}(8) = 6$. Given starting positions $\mathbf{s}$, the *consensus score* is defined to be $Score(\mathbf{s}, DNA) = \sum_{j=1}^{l} M_{\mathbf{P}(\mathbf{s})}(j)$. For the starting positions in figure 4.3, $Score(\mathbf{s}, DNA) = 5 + 5 + 6 + 4 + 5 + 5 + 6 + 6 = 42$. $Score(\mathbf{s}, DNA)$ can be used to measure the strength of a profile corresponding to the starting positions $\mathbf{s}$. A consensus score of $l \cdot t$ corresponds to the best possible alignment, in which each row of a column has the same letter. A consensus score of $\frac{lt}{4}$, however, corresponds to the worst possible alignment, which has an equal mix of all nucleotides in each column. In its simplest form, the Motif Finding problem can be formulated as selecting starting positions $\mathbf{s}$ from the sample that maximize $Score(\mathbf{s}, DNA)$.[10]

Motif Finding Problem:
*Given a set of DNA sequences, find a set of l-mers, one from each
sequence, that maximizes the consensus score.*

> **Input:** A $t \times n$ matrix of DNA, and l, the length of the pattern
> to find.
>
> **Output:** An array of t starting positions $\mathbf{s} = (s_1, s_2, \ldots, s_t)$
> maximizing $Score(\mathbf{s}, DNA)$.

Another view onto this problem is to reframe the Motif Finding problem
as the problem of finding a *median string*. Given two l-mers v and w, we can
compute the *Hamming distance* between them, $d_H(v, w)$, as the number of positions that differ in the two strings. For example, $d_H(\mathsf{ATTGTC}, \mathsf{ACTCTC}) =$
2:

$$
\begin{array}{cccccc}
\mathsf{A} & \mathsf{T} & \mathsf{T} & \mathsf{G} & \mathsf{T} & \mathsf{C} \\
: & \mathsf{X} & : & \mathsf{X} & : & : \\
\mathsf{A} & \mathsf{C} & \mathsf{T} & \mathsf{C} & \mathsf{T} & \mathsf{C}
\end{array}
$$

Now suppose that $\mathbf{s} = (s_1, s_2, \ldots, s_t)$ is an array of starting positions, and
that v is some l-mer. We will abuse our notation a bit and use $d_H(v, \mathbf{s})$ to
denote the *total Hamming distance* between v and the l-mers starting at posi-
tions $\mathbf{s}$: $d_H(v, \mathbf{s}) = \sum_{i=1}^{t} d_H(v, s_i)$, where $d_H(v, s_i)$ is the Hamming distance
between v and the l-mer that starts at s_i in the ith DNA sequence. We will
use $TotalDistance(v, DNA) = \min_{\mathbf{s}}(d_H(v, \mathbf{s}))$ to denote the minimum possible total Hamming distance between a given string v and any set of starting
positions in the DNA. Finding $TotalDistance(v, DNA)$ is a simple problem:
first one has to find the best match for v in the first DNA sequence (i.e., a position minimizing $d_H(v, s_1)$ for $1 \leq s_1 \leq n - l + 1$), then the best match in the

---

10. Another approach is to maximize the *entropy* of the corresponding profile. Let $\mathbf{P}(\mathbf{s}) = (p_{i,j})$,
where $p_{i,j}$ is the count at element (i, j) of the $4 \times l$ profile matrix. Entropy is defined as

$$
\sum_{j=1}^{l} \sum_{i=1}^{4} \frac{p_{i,j}}{t} \log \frac{p_{i,j}}{t}
$$

where t is the number of sequences in the DNA sample. Although entropy is a more statistically
adequate measure of profile strength than the consensus score, for the sake of simplicity we use
the consensus score in the examples below.

A	T	C	C	A	G	C	T
G	G	G	C	A	A	C	T
A	T	G	G	A	T	C	T
A	A	G	C	A	A	C	C
T	T	G	G	A	A	C	T
A	T	G	C	C	A	T	T
A	T	G	G	C	A	C	T

Figure 4.4 Calculating the total Hamming distance for the consensus string ATG-CAACT (the alignment is the same as in figure 4.3). The bold letters show the consensus sequence; the total Hamming distance can be calculating as the number of nonbold letters.

second one, and so on. That is, the minimum is taken over all possible starting positions s. Finally, we define the *median string* for DNA as the string v that minimizes $TotalDistance(v, DNA)$; this minimization is performed over all 4^l strings v of length l.

We can formulate the problem of finding a median string in DNA sequences as follows.

Median String Problem:
Given a set of DNA sequences, find a median string.

 Input: A $t \times n$ matrix DNA, and l, the length of the pattern to find.

 Output: A string v of l nucleotides that minimizes $TotalDistance(v, DNA)$ over all strings of that length.

Notice that this is a double minimization: we are finding a string v that minimizes $TotalDistance(v, DNA)$, which is in turn the smallest distance among all choices of starting points s in the DNA sequences. That is, we are calculating

$$\underset{\substack{\text{all choices of} \\ l\text{-mers } v}}{\min} \quad \underset{\substack{\text{all choices of} \\ \text{starting positions s}}}{\min} \quad d_H(v, \mathbf{s}).$$

Despite the fact that the Median String problem is a minimization problem and the Motif Finding problem is a maximization problem, the two problems are computationally equivalent. Let s be a set of starting positions with consensus score $Score(\mathbf{s}, DNA)$, and let w be the consensus string of the corresponding profile. Then

$$d_H(w, \mathbf{s}) = lt - Score(\mathbf{s}, DNA).$$

For example, in figure 4.4, the Hamming distance between the consensus string w and each of the seven implanted patterns is 2, and $d_H(w, \mathbf{s}) = 2 \cdot 7 = 7 \cdot 8 - 42$.

The consensus string minimizes $d_H(v, \mathbf{s})$ over all choices of v, i.e.,

$$d_H(w, \mathbf{s}) = \min_{\text{all choices of } v} d_H(v, \mathbf{s}) = lt - Score(\mathbf{s}, DNA)$$

Since t and l are constants, the smallest value of d_H can also be obtained by maximizing $Score(\mathbf{s}, DNA)$ over all choices of s:

$$\min_{\text{all choices of } \mathbf{s}} \min_{\text{all choices of } v} d_H(v, \mathbf{s}) = lt - \max_{\text{all choices of } \mathbf{s}} Score(\mathbf{s}, DNA).$$

The problem on the left is the Median String problem while the problem on the right is the Motif Finding problem.

In other words, the consensus string for the solution of the Motif Finding problem is the median string for the input DNA sample. The median string for DNA can be used to generate a profile that solves the Motif Finding problem, by searching in each of the t sequences for the substring with the smallest Hamming distance from the median string.

We introduce this formulation of the Median String problem to give more efficient alternative motif finding algorithms below.

4.7   Search Trees

In both the Median String problem and the Motif Finding problem we have to sift through a large number of alternatives to find the best one but we so far lack the algorithmic tools to do so. For example, in the Motif Finding problem we have to consider all $(n - l + 1)^t$ possible starting positions s:

$$
\begin{array}{rrrrr}
(& 1, & 1, \ \dots, & 1, & 1 \ ) \\
(& 1, & 1, \ \dots, & 1, & 2 \ ) \\
(& 1, & 1, \ \dots, & 1, & 3 \ ) \\
& & \vdots & & \\
(& 1, & 1, \ \dots, & 1, & n-l+1 \ ) \\
(& 1, & 1, \ \dots, & 2, & 1 \ ) \\
(& 1, & 1, \ \dots, & 2, & 2 \ ) \\
(& 1, & 1, \ \dots, & 2, & 3 \ ) \\
& & \vdots & & \\
(& 1, & 1, \ \dots, & 2, & n-l+1 \ ) \\
& & \vdots & & \\
(& n-l+1, & n-l+1, \ \dots, & n-l+1, & 1 \ ) \\
(& n-l+1, & n-l+1, \ \dots, & n-l+1, & 2 \ ) \\
(& n-l+1, & n-l+1, \ \dots, & n-l+1, & 3 \ ) \\
& & \vdots & & \\
(& n-l+1, & n-l+1, \ \dots, & n-l+1, & n-l+1 \ )
\end{array}
$$

For the Median String problem we need to consider all 4^l possible l-mers:

$$
\begin{array}{c}
\text{AA} \cdots \text{AA} \\
\text{AA} \cdots \text{AT} \\
\text{AA} \cdots \text{AG} \\
\text{AA} \cdots \text{AC} \\
\text{AA} \cdots \text{TA} \\
\text{AA} \cdots \text{TT} \\
\text{AA} \cdots \text{TG} \\
\text{AA} \cdots \text{TC} \\
\vdots \\
\text{CC} \cdots \text{GG} \\
\text{CC} \cdots \text{GC} \\
\text{CC} \cdots \text{CA} \\
\text{CC} \cdots \text{CT} \\
\text{CC} \cdots \text{CG} \\
\text{CC} \cdots \text{CC}
\end{array}
$$

Figure 4.5   All 4-mers in the alphabet of $\{1, 2\}$.

We note that this latter progression is equivalent to the following one if we let 1 stand for A, 2 for T, 3 for G, and 4 for C:

$$(1, 1, \ldots, 1, 1)$$
$$(1, 1, \ldots, 1, 2)$$
$$(1, 1, \ldots, 1, 3)$$
$$(1, 1, \ldots, 1, 4)$$
$$(1, 1, \ldots, 2, 1)$$
$$(1, 1, \ldots, 2, 2)$$
$$(1, 1, \ldots, 2, 3)$$
$$(1, 1, \ldots, 2, 4)$$
$$\vdots$$
$$(4, 4, \ldots, 3, 3)$$
$$(4, 4, \ldots, 3, 4)$$
$$(4, 4, \ldots, 4, 1)$$
$$(4, 4, \ldots, 4, 2)$$
$$(4, 4, \ldots, 4, 3)$$
$$(4, 4, \ldots, 4, 4)$$

In general, we want to consider all k^L L-mers in a k-letter alphabet. For the Motif Finding problem, $k = n-l+1$, whereas for the Median String problem, $k = 4$. Figure 4.5 shows all 2^4 4-mers in the two-letter alphabet of 1 and 2. Given an L-mer from a k-letter alphabet, the subroutine NEXTLEAF (below) demonstrates how to jump from an L-mer $\mathbf{a} = (a_1 a_2 \cdots a_L)$ to the next L-mer in the progression. Exactly why this algorithm is called NEXTLEAF will become clear shortly.

NEXTLEAF(**a**, L, k)
```
1   for  i ← L to 1
2       if  a_i < k
3           a_i ← a_i + 1
4           return a
5       a_i ← 1
6   return a
```

NEXTLEAF operates in a way that is very similar to the natural process of counting. In most cases, $(a_1, a_2, \ldots, a_L)$ is followed by $(a_1, a_2, \ldots, a_L + 1)$. However, when $a_L = k$, the next invocation of NEXTLEAF will reset a_L to 1 and add 1 to a_{L-1}—compare this to the transition from 3719 to 3720 in counting. However, when there is a long string of the value k on the right-hand side of **a**, the algorithm needs to reset them all to 1—compare this with the transition from 239999 to 240000. When all entries in **a** are k, the algorithm wraps around and returns $(1, 1, \ldots, 1)$, which is one way we can tell that we are finished examining L-mers. In the case that $L = 10$, NEXTLEAF is exactly like counting decimal numbers, except that we use "digits" from 1 to 10, rather than from 0 to 9.

The following algorithm, ALLLEAVES, simply uses NEXTLEAF to output all the 4-mers in the order shown in figure 4.5.

ALLLEAVES(L, k)
```
1   a ← (1, …, 1)
2   while  forever
3       output a
4       a ← NEXTLEAF(a, L, k)
5       if  a = (1, 1, …, 1)
6           return
```

Even though line 2 of this algorithm seems as though it would loop forever, since NEXTLEAF will eventually loop around to $(1, 1, \ldots, 1)$, the **return** in line 6 will get reached and it will eventually stop.

Computer scientists often represent all L-mers as *leaves in a tree*, as in figure 4.6. L-mer trees will have L levels (excluding the topmost *root* level), and each vertex has k children. L-mers form leaves at the lowest level of the tree, $(L - 1)$-mers form the next level up, and $(L - 2)$-mers a level above that, and so on. For example, the vertices on the third level of the tree represent the eight different 3-mers: $(1, 1, 1)$, $(1, 1, 2)$, $(1, 2, 1)$, $(1, 2, 2)$, $(2, 1, 1)$, $(2, 1, 2)$,

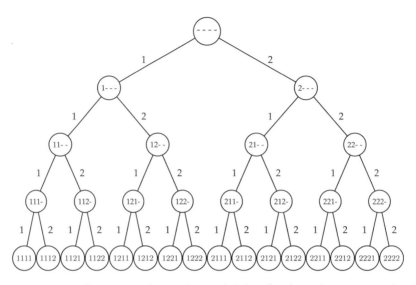

Figure 4.6 All 4-mers in the two-letter alphabet $\{1, 2\}$ can be represented as leaves in a tree.

$(2, 2, 1)$, and $(2, 2, 2)$.

The tree in figure 4.6 with $L = 4$ and $k = 2$ has 31 vertices.[11] Note that in a tree with L levels and k children per vertex, each leaf is equivalent to an array **a** of length L in which each element a_i takes on one of k different values. In turn, this is equivalent to an L-long string from an alphabet of size k, which is what we have been referring to as an L-mer. We will consider all of these representations to be equivalent. Internal vertices, on the other hand, can be represented as a pair of items: a list **a** of length L and an integer i that specifies the vertex's level. The entries $(a_1, a_2, \ldots, a_i)$ uniquely identify a vertex at level i; we will rely on the useful fact that the representation of an internal vertex is the *prefix* that is common to all of the leaves underneath it.

To represent all possible starting positions for the Motif Finding problem, we can construct the tree with $L = t$ levels[12] and $k = n - l + 1$ children per vertex. For the Median String problem, $L = l$ and $k = 4$. The astute reader may realize that the internal vertices of the tree are somewhat meaningless

---

11. In general, a tree with k children per vertex has k^i vertices at level i (every vertex at level $i - 1$ gives birth to k children); the total number of vertices in the tree is then $\sum_{i=0}^{L} k^i$, while the number of leaves is only k^L.

12. As a reminder, t is the number of DNA sequences, n is the length of each one, and l is the length of the profile we would like to find.

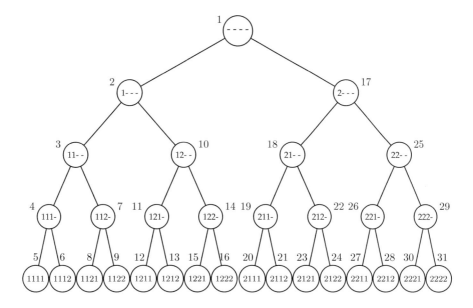

```
PREORDER(v)
1  output v
2  if v has children
3      PREORDER( left child of v )
4      PREORDER( right child of v )
```

Figure 4.7 The order of traversing all vertices in a tree. The recursive algorithm PREORDER demonstrates how the vertices were numbered.

for the purposes of finding motifs, since they do not represent a sensible choice of starting positions in *all* of the t sequences. For this reason, we would like a method of scanning only the leaves of a tree and ignore the internal vertices. In doing this, it will probably appear that we have only complicated matters: we deliberately constructed a tree that contains internal vertices and now we will summarily ignore them. However, using the tree representation will allow us to use the branch-and-bound technique to improve upon brute force algorithms.

Listing the leaves of a tree is straightforward, but listing all vertices (i.e., all leaves and all internal vertices) is somewhat trickier. We begin at level 0 (the root) and then consider each of its k children in order. For each child, we again consider each of its k children and so on. Figure 4.7 shows the order of traversing vertices for a tree with $L = 4$ and $k = 2$ and also gives an elegant recursive algorithm to perform this process. The sequence of vertices that PREORDER($root$) would return on the tree of $L = 4$ and $k = 2$ would be as follows:

$$
\begin{array}{l}
(-,-,-,-) \\
(1,-,-,-) \\
(1,1,-,-) \\
(1,1,1,-) \\
(1,1,1,1) \\
(1,1,1,2) \\
(1,1,2,-) \\
(1,1,2,1) \\
(1,1,2,2) \\
(1,2,-,-) \\
(1,2,1,-) \\
(1,2,1,1) \\
(1,2,1,2) \\
(1,2,2,-) \\
(1,2,2,1) \\
(1,2,2,2) \\
(2,-,-,-) \\
(2,1,-,-) \\
(2,1,1,-) \\
(2,1,1,1) \\
(2,1,1,2) \\
(2,1,2,-) \\
(2,1,2,1) \\
(2,1,2,2) \\
(2,2,-,-) \\
(2,2,1,-) \\
(2,2,1,1) \\
(2,2,1,2) \\
(2,2,2,-) \\
(2,2,2,1) \\
(2,2,2,2)
\end{array}
$$

Traversing the complete tree iteratively is implemented in the NEXTVERTEX algorithm, below. NEXTVERTEX takes vertex $\mathbf{a} = (a_1, \dots, a_L)$ at level i as an input and returns the next vertex in the tree. In reality, at level i NEXTVER-TEX only uses the values $a_1, \ldots, a_i$ and ignores $a_{i+1}, \ldots, a_L$. NEXTVERTEX takes inputs that are similar to NEXTLEAF, with the exception that the "current leaf" is now the "current vertex," so it uses the parameter i for the level on which the vertex lies. Given $\mathbf{a}$, L, i, and k, NEXTVERTEX returns the next vertex in the tree as the pairing of an array and a level. The algorithm will return a level number of 0 when the traversal is complete.

NEXTVERTEX$(\mathbf{a}, i, L, k)$
1   **if** $i < L$
2        $a_{i+1} \leftarrow 1$
3   **return** $(\mathbf{a}, i + 1)$
4   **else**
5   **for** $j \leftarrow L$ **to** 1
6   **if** $a_j < k$
7                $a_j \leftarrow a_j + 1$
8   **return** $(\mathbf{a}, j)$
9   **return** $(\mathbf{a}, 0)$

When $i < L$, NEXTVERTEX$(\mathbf{a}, i, L, k)$ moves down to the next lower level and explores that subtree of $\mathbf{a}$. If $i = L$, NEXTVERTEX either moves along the lowest level as long as $a_L < k$ or jumps back up in the tree.

We alluded above to using this tree representation to help reduce work in brute force search algorithms. The general branch-and-bound approach will allow us to ignore any children (or grandchildren, great-grandchildren, and so on) of a vertex if we can show that they are all uninteresting. If none of the descendents of a vertex could possibly have a better score than the best leaf that has already been explored, then there really is no point descending into the children of that vertex. At each vertex we calculate a bound–the most wildly optimistic score of any leaves in the subtree rooted at that vertex— and then decide whether or not to consider its children. In fact, the strategy is named branch-and-bound for exactly this reason: at each point we calculate a bound and then decide whether or not to branch out further (figure 4.8).

Branching-and-bounding requires that we can skip an entire subtree rooted at an arbitrary vertex. The subroutine NEXTVERTEX is not up to this task, but the algorithm BYPASS, below, allows us to skip the subtree rooted at vertex $(\mathbf{a}, i)$. If we skip a vertex at level i of the tree, we can just increment a_i (unless $a_i = k$, in which case we need to jump up in the tree). The algorithm BYPASS takes the same type of input and produces the same type of output as NEXTLEAF.

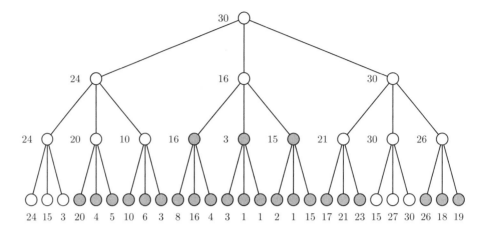

Figure 4.8 A tree that has uninteresting subtrees. The numbers next to a leaf represent the "score" for that *L*-mer. Scores at internal vertices represent the maximum score in the subtree rooted at that vertex. To improve the brute force algorithm, we want to "prune" (ignore) subtrees that do not contain high-scoring leaves. For example, since the score of the very first leaf is 24, it does not make sense to analyze the 4th, 5th, or 6th leaves whose scores are 20, 4, and 5, respectively. Therefore, the subtree containing these vertices can be ignored.

BYPASS($\mathbf{a}, i, L, k$)
1   **for** $j \leftarrow i$ **to** 1
2   **if** $a_j < k$
3             $a_j \leftarrow a_j + 1$
4   **return** $(\mathbf{a}, j)$
5   **return** $(\mathbf{a}, 0)$

We pause to remark that the iterative version of tree navigation that we present here is equivalent to the standard recursive approach that would be found in an introductory algorithms text for computer scientists. Rather than rob you of this discovery, the problems at the end of this chapter explore this relationship in more detail. Simply transforming the list of alternatives that need to be searched into a search tree makes many brute force algorithms blatantly obvious; even better, a sensible branch-and-bound strategy will often become clear.

4.8   Finding Motifs

The brute force approach to solve the Motif Finding problem looks through all possible starting positions.

BRUTEFORCEMOTIFSEARCH(DNA, t, n, l)
1   $bestScore \leftarrow 0$
2   **for each** $(s_1, \ldots, s_t)$ **from** $(1, \ldots, 1)$ **to** $(n - l + 1, \ldots, n - l + 1)$
3   **if** $Score(\mathbf{s}, DNA) > bestScore$
4             $bestScore \leftarrow Score(\mathbf{s}, DNA)$
5   **bestMotif** $\leftarrow (s_1, s_2, \ldots, s_t)$
6   **return bestMotif**

There are $n - l + 1$ choices for the first index (s_1) and for each of those, there are $n - l + 1$ choices for the second index (s_2). For each of those choices, there are $n - l + 1$ choices for the third index, and so on. Therefore, the overall number of positions is $(n - l + 1)^t$, which is exponential in t, the number of sequences. For each $\mathbf{s}$, the algorithm calculates $Score(\mathbf{s}, DNA)$, which requires $O(l)$ operations. Thus, the overall complexity of the algorithm is evaluated as $O(ln^t)$.

The only remaining question is how to write line 2 using standard pseudocode operations. This is particularly easy if we use NEXTLEAF from the previous section. In this case, $L = n - l + 1$ and $k = t$. Rewriting BRUTEFORCEMOTIFSEARCH in this way we arrive at BRUTEFORCEMOTIFSEARCHAGAIN.

BRUTEFORCEMOTIFSEARCHAGAIN(DNA, t, n, l)
1   $\mathbf{s} \leftarrow (1, 1, \ldots, 1)$
2   $bestScore \leftarrow Score(\mathbf{s}, DNA)$
3   **while** forever
4        $\mathbf{s} \leftarrow$ NEXTLEAF($\mathbf{s}, t, n - l + 1$)
5   **if** $Score(\mathbf{s}, DNA) > bestScore$
6             $bestScore \leftarrow Score(\mathbf{s}, DNA)$
7   **bestMotif** $\leftarrow (s_1, s_2, \ldots, s_t)$
8   **if** $s = (1, 1, \ldots, 1)$
9   **return bestMotif**

Finally, to prepare for the branch-and-bound strategy, we will want the equivalent version, SIMPLEMOTIFSEARCH, which uses NEXTVERTEX to explore each leaf.

SIMPLEMOTIFSEARCH(DNA, t, n, l)
1   $\mathbf{s} \leftarrow (1, \ldots, 1)$
2   $bestScore \leftarrow 0$
3   $i \leftarrow 1$
4   **while** $i > 0$
5   **if** $i < t$
6           $(\mathbf{s}, i) \leftarrow$ NEXTVERTEX($\mathbf{s}, i, t, n - l + 1$)
7   **else**
8   **if** $Score(\mathbf{s}, DNA) > bestScore$
9               $bestScore \leftarrow Score(\mathbf{s}, DNA)$
10   **bestMotif** $\leftarrow (s_1, s_2, \ldots, s_t)$
11          $(\mathbf{s}, i) \leftarrow$ NEXTVERTEX($\mathbf{s}, i, t, n - l + 1$)
12 **return bestMotif**

Observe that some sets of starting positions can be ruled out immediately without iterating over them, based simply on the most optimistic estimate of their score. For example, if the first i of t starting positions [i.e., $(s_1, s_2, \ldots, s_i)$] form a "weak" profile, then it may not be necessary to even consider any starting positions in the sequences $i + 1, i + 2, \ldots, t$, since the resulting profile could not possibly be better than the highest-scoring profile already found.

Given a set of starting positions $\mathbf{s} = (s_1, s_2, \ldots, s_t)$, define the *partial consensus score*, $Score(\mathbf{s}, i, DNA)$, to be the consensus score of the $i \times l$ alignment matrix involving only the first i rows of DNA corresponding to starting positions $(s_1, s_2, \ldots, s_i, -, -, \ldots, -)$. In this case, a $-$ indicates that we have not chosen any value for that entry in $\mathbf{s}$. If we have the partial consensus score for $s_1, \ldots, s_i$, even in the best of circumstances the remaining $t - i$ rows can only improve the consensus score by $(t - i) \cdot l$; therefore, the score of any alignment matrix with the first i starting positions $(s_1, \ldots, s_i)$ could be at most $Score(\mathbf{s}, i, DNA) + (t - i) \cdot l$. This implies that if $Score(\mathbf{s}, i, DNA) + (t - i) \cdot l$ is less than the currently best score, $bestScore$, then it does not make sense to explore any of the remaining $t - i$ sequences in the sample—with this choice of $(s_1, \ldots, s_i)$—it would obviously result in wasted effort. Therefore, the bound $Score(\mathbf{s}, i, DNA) + (t - i) \cdot l$ could save us the trouble of looking at $(n - l + 1)^{t-i}$ leaves.

BRANCHANDBOUNDMOTIFSEARCH(DNA, t, n, l)
1    $\mathbf{s} \leftarrow (1, \ldots, 1)$
2    $bestScore \leftarrow 0$
3    $i \leftarrow 1$
4   **while** $i > 0$
5   **if** $i < t$
6            $optimisticScore \leftarrow Score(\mathbf{s}, i, DNA) + (t - i) \cdot l$
7   **if** $optimisticScore < bestScore$
8                $(\mathbf{s}, i) \leftarrow$ BYPASS$(\mathbf{s}, i, t, n - l + 1)$
9   **else**
10                $(\mathbf{s}, i) \leftarrow$ NEXTVERTEX$(\mathbf{s}, i, t, n - l + 1)$
11   **else**
12   **if** $Score(\mathbf{s}, DNA) > bestScore$
13                $bestScore \leftarrow Score(\mathbf{s})$
14   **bestMotif** $\leftarrow (s_1, s_2, \ldots, s_t)$
15            $(\mathbf{s}, i) \leftarrow$ NEXTVERTEX$(\mathbf{s}, i, t, n - l + 1)$
16   **return bestMotif**

Though this branch-and-bound strategy improves our algorithm for some problem instances, we have not improved the worst-case efficiency: you can design a sample with an implanted pattern that requires exponential time to find.

4.9   Finding a Median String

We mentioned above that the Median String problem gives us an alternate approach to finding motifs. If we apply the brute force technique to solve this problem, we arrive at the following algorithm[13]:

---

13. The parameters t, n, and l in this algorithm are needed to compute the value of TOTALDISTANCE($word, DNA$).   A more detailed pseudocode would use TOTALDISTANCE($word, DNA, t, n, l$) but we omit these details for brevity.

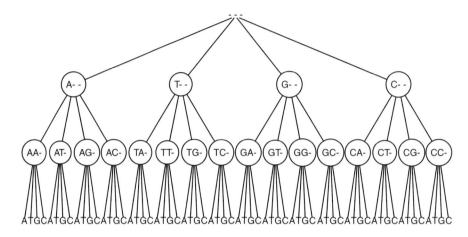

Figure 4.9 A search tree for the Median String problem. Each branching point can give rise to only four children, as opposed to the $n-l+1$ children in the Motif Finding problem.

BRUTEFORCEMEDIANSEARCH(DNA, t, n, l)
1   $bestWord \leftarrow$ AAA $\cdots$ AA
2   $bestDistance \leftarrow \infty$
3   **for** each l-mer $word$ **from** AAA...A **to** TTT...T
4   **if** TOTALDISTANCE$(word, DNA) < bestDistance$
5               $bestDistance \leftarrow$ TOTALDISTANCE$(word, DNA)$
6               $bestWord \leftarrow word$
7   **return** $bestWord$

BRUTEFORCEMEDIANSEARCH considers each of 4^l nucleotide strings of length l and computes TOTALDISTANCE at every step. Given $word$, we can calculate $TotalDistance(word, DNA)$ in a single pass over DNA (i.e., in $O(nt)$ time), rather than by considering all possible starting points in the DNA sample. Therefore, BRUTEFORCEMEDIANSEARCH has running time $O(4^l \cdot n \cdot t)$, which compares favorably with the $O(ln^t)$ of SIMPLEMOTIFSEARCH. A typical motif has a length (l) ranging from eight to fifteen nucleotides, while the typical size of upstream regions that are analyzed have length (n) ranging from 500 to 1000 nucleotides. BRUTEFORCEMEDIANSEARCH is a practical algorithm for finding short motifs while SIMPLEMOTIFSEARCH is not.

We will proceed along similar lines to construct a branch-and-bound strategy for BRUTEFORCEMEDIANSTRING as we did in the transformation of

BRUTEFORCEMOTIFSEARCH into SIMPLEMOTIFSEARCH. We can modify the median string search to explore the entire tree of all l-nucleotide strings (see figure 4.9) rather than only the leaves of that tree as in BRUTEFORCEMEDIANSEARCH. A vertex at level i in this tree represents a nucleotide string of length i, which can be viewed as the i-long prefix of every leaf below that vertex. SIMPLEMEDIANSEARCH assumes that nucleotides A, C, G, T are coded as numerals (1, 2, 3, 4); for example, the assignment in line 1 sets s to the l-mer $(1, 1, \ldots, 1)$, corresponding to the nucleotide string AA ... A.

SIMPLEMEDIANSEARCH(DNA, t, n, l)
```
 1   s ← (1, 1, ..., 1)
 2   bestDistance ← ∞
 3   i ← 1
 4   while i > 0
 5       if i < l
 6             (s, i) ← NEXTVERTEX(s, i, l, 4)
 7       else
 8             word ← nucleotide string corresponding to (s₁, s₂, ... sₗ)
 9             if TOTALDISTANCE(word, DNA) < bestDistance
10                 bestDistance ← TOTALDISTANCE(word, DNA)
11                 bestWord ← word
12             (s, i) ← NEXTVERTEX(s, i, l, 4)
13   return bestWord
```

In accordance with the branch-and-bound strategy, we find a bound for $TotalDistance(word, DNA)$ at each vertex. It should be clear that if the total distance between the i-prefix of $word$ and DNA is larger than the smallest seen so far for one of the leaves (nucleotide strings of length l), then there is no point investigating subtrees of the vertex corresponding to that i-prefix of $word$; all extensions of this prefix into an l-mer will have at least the same total distance and probably more. This is what forms our branch-and-bound strategy. The bound in BRANCHANDBOUNDMEDIANSEARCH relies on the optimistic scenario that there *could* be some extension to the prefix that matches every string in the sample, which would add 0 to the total distance calculation.

BranchAndBoundMedianSearch(DNA, t, n, l)

```
 1   s ← (1, 1, . . . , 1)
 2   bestDistance ← ∞
 3   i ← 1
 4   while i > 0
 5       if i < l
 6           prefix ← nucleotide string corresponding to (s₁, s₂, . . . , sᵢ)
 7           optimisticDistance ← TotalDistance(prefix, DNA)
 8           if optimisticDistance > bestDistance
 9               (s, i) ← Bypass(s, i, l, 4)
10           else
11               (s, i) ← NextVertex(s, i, l, 4)
12       else
13           word ← nucleotide string corresponding to (s₁, s₂, . . . sₗ)
14           if TotalDistance(word, DNA) < bestDistance
15               bestDistance ← TotalDistance(word, DNA)
16               bestWord ← word
17           (s, i) ← NextVertex(s, i, l, 4)
18   return bestWord
```

The naive bound in BranchAndBoundMedianSearch is overly generous, and it is possible to design more aggressive bounds (this is left as a problem at the end of the chapter). As usual with branch-and-bound algorithms, BranchAndBoundMedianSearch provides no improvement in the worst-case running time but often results in a practical speedup.

4.10   Notes

In 1965, Werner Arber (32) discovered restriction enzymes, conjecturing that they cut DNA at positions where specific nucleotide patterns occur. In 1970, Hamilton Smith (95) verified Arber's hypothesis by showing that the *Hind*II restriction enzyme cuts DNA in the middle of palindromes **GTGCAC** or **GTTAAC**. Other restriction enzymes have similar properties, but cut DNA at different patterns. Dan Nathans pioneered the application of restriction enzymes to genetics and constructed the first ever restriction map in 1973 (26). All three were awarded the Nobel Prize in 1978.

The PartialDigest algorithm for the construction of restriction maps was proposed by Steven Skiena and colleagues in 1990 (94). In 1994 Zheng Zhang (114) came up with a "difficult" instance of PDP that requires an exponential time to solve using the PARTIALDIGEST algorithm. In 2002, Maurice Nivat and colleagues described a polynomial algorithm to solve the PDP (30).

Studies of gene regulation were pioneered by François Jacob and Jacques Monod in the 1950s. They identified genes (namely, regulatory genes) whose proteins (transcription factors) have as their sole function the regulation of other genes. Twenty years later it was shown that these transcription factors bind specifically in the upstream areas of the genes they regulate and recognize certain patterns (motifs) in DNA. It was later discovered that, in some cases, transcription factors may bind at a distance and regulate a gene from very far (tens of thousands of nucleotides) away.

Computational approaches to motif finding were pioneered by Michael Waterman (89), Gary Stormo (46) and their colleagues in the mid-1980s. Profiles were introduced by Gary Stormo and colleagues in 1982 (101) and further developed by Michael Gribskov and colleagues in 1987 (43). Although the naive exhaustive motif search described in this chapter is too slow, there exist fast and practical branch-and-bound approaches to motif finding (see Marsan and Sagot, 2000 (72), Eskin and Pevzner, 2002 (35)).

Gary Stormo, born 1950 in South Dakota, is currently a professor in the Department of Genetics at Washington University in St. Louis. Stormo went to Caltech as a physics major, but switched to biology in his junior year. Although that was only at an undergraduate level, the strong introduction to the physical sciences and math helped prepare him for the opportunities that came later. He has a PhD in Molecular Biology from the University of Colorado at Boulder. His principal research interests center around the analysis of gene regulation and he was an early advocate of using computers to infer regulatory motifs and understand gene regulation.

He went to the University of Colorado in Boulder as a graduate student and quickly got excited about understanding gene regulation, working in the lab of Larry Gold. During his graduate career, methods for sequencing DNA were developed so he suddenly had many examples of regulatory sites that he could compare to each other, and could also compare to the mutants that he had collected. Together with Tom Schneider he set out to write a collection of programs for various kinds of analysis on the sequences that were available. At the time neither the algorithms nor the math were very difficult; even quite simple approaches were new and useful. He did venture into some artificial intelligence techniques that took some effort to understand, but the biggest challenge was that they had to do everything themselves. GenBank didn't exist yet so they had to develop their own databases to keep all of the DNA sequences and their annotation, and they even had to type most of them in by hand (with extensive error checking) because in those days most sequences were simply published in journals.

As part of his thesis work he developed profiles (also called position weight matrices) as a better representation of regulatory sites than simple consensus sequences. He had published a few different ways to derive the profiles matrices, depending on the types of data available and the use to be made of them. But he was looking for a method to discover the matrix if one only knew a sample of DNA sequences that had the regulatory sites some-where within them at unknown positions, the problem that is now known as the Motif Finding problem. A few years earlier Michael Waterman had published an algorithm for discovering a consensus motif from a sample of DNA sequences, and Stormo wanted to do the same thing with a profile representation. The problem has two natural aspects to it, how to find the correct alignment of regulatory sites without examining all possible align-ments, and how to evaluate different alignments so as to choose the best. For the evaluation step he used the entropy-based information content measure from Tom Schneider's thesis because it had nice statistical properties and they had shown that, with some simplifying assumptions, it was directly re-lated to the binding energy of the protein to the sites. In retrospect is seems almost trivial, but at the time it took them considerable effort to come up with the approach that is employed in the greedy CONSENSUS program.

Stormo knew that the idea would work, and of course it did, so long as the problem wasn't too difficult—the pattern had to have sufficient information content to stand out from the background. He knew this would be a very useful tool, although at the time nobody anticipated DNA array experiments which make it even more useful because one can get samples of putatively coregulated genes so much more easily. Of course these data have more noise than originally thought, so the algorithms have had to become more robust.

One of Stormo's most enjoyable scientific experiences came soon after he got his PhD and began working on a collaborative project with his adviser, Larry Gold and Pete von Hippel at the University of Oregon. Gold's lab had previously shown that the gene in the T4 phage known as "32" (which partic-ipates in replication, recombination, and repair) also regulated its own syn-thesis at the translational level. von Hippel's group had measured the bind-ing parameters of the protein, while another group had recently sequenced the gene and its regulatory region. By combining the binding parameters of the protein with analysis of the sequence, including a comparison to other gene sequences, they were able to provide a model for the protein's activity in gene regulation. A few years later Stormo got to help fill in some more details of the model through a comparison of the regulatory region from the closely related phages T2 and T6 and showed that there was a conserved pseudoknot structure that acted as a nucleation site for the autogenous bind-ing. Stormo says:

> This was very satisfying because of how all of the different aspects of the problem, from biophysical measurements to genetics to sequence

analysis came together to describe a really interesting example of gene regulation.

Discoveries can come in many different ways, and the most important thing is to be ready for them. Some people will pick a particular problem and work on it very hard, bringing all of the tools available, even inventing new ones, to try and solve it. Another way is to look for connections between different problems, or methods in one field that can be applied to problems in another field. Gary thinks this interdisciplinary approach is particularly useful in bioinformatics, although the hard work focused on specific problems is also important. His research style has always been to follow his interests which can easily wander from an initial focus. He feels that if he had followed a more consistent line of work he could have made more progress in certain areas, but he really enjoys reading widely and working on problems where he can make a contribution, even if they are outside his major research areas.

> I think the regulation of gene expression will continue to be an important problem for a long time. Although significant progress has been made, there are still lots of connections to be made between transcription factors and the genes they regulate. Plus lots of gene regulation happens post-transcriptionally and we are just beginning to look at that in a systematic way. The ultimate goal of really understanding the complete regulatory networks is a major challenge. I also think evolutionary biology will be an increasingly important topic for understanding the diversity of life on the planet.

4.11   Problems

Problem 4.1

Write an algorithm that, given a set X, calculates the multiset ΔX.

Problem 4.2

Consider partial digest

$$L = \{1, 1, 1, 2, 2, 3, 3, 3, 4, 4, 5, 5, 6, 6, 6, 9, 9, 10, 11, 12, 15\}.$$

Solve the Partial Digest problem for L (i.e., find X such that $\Delta X = L$).

Problem 4.3

Write an algorithm that, given an n-element set, generates all m-element subsets of this set. For example, the set $\{1, 2, 3, 4\}$ has six two-element subsets $\{1, 2\}$, $\{1, 3\}$, $\{1, 4\}$, $\{2, 3\}$, $\{2, 4\}$, and $\{3, 4\}$. How long will your algorithm take to run? Can it be done faster?

Problem 4.4

Write an algorithm that, given an n-element multiset, generates all m-element subsets of this set. For example, the set $\{1, 2, 2, 3\}$ has four two-element subsets $\{1, 2\}$, $\{1, 3\}$, $\{2, 3\}$, and $\{2, 2\}$. How long will your algorithm take to run? Can it be done faster?

Problem 4.5

Prove that the sets $U \oplus V = \{u+v : u \in U, v \in V\}$ and $U \ominus V = \{u-v : u \in U, v \in V\}$ are homometric for any two sets U and V.

Problem 4.6

Given a multiset of integers $A = \{a_i\}$, we call the polynomial $A(x) = \sum_i x^{a_i}$ the *generating function* for A. Verify that the generating function for ΔA is $\Delta A(x) = A(x)A(x^{-1})$. Given generating functions for U and V, find generating functions for $A = U \oplus V$ and $B = U \ominus V$. Compare generating functions for ΔA and ΔB. Are they the same?

Problem 4.7

Write pseudocode for the PARTIALDIGEST algorithm that has fewer lines than the one presented in the text.

Problem 4.8

Find a set ΔX with the smallest number of elements that could have arisen from more than one X, not counting shifts and reflections.

Double Digest mapping is a restriction mapping technique that is even simpler (experimentally) than a partial digest but uses two different restriction enzymes. In this approach, biologists digest DNA in such a way that only fragments between *consecutive* sites are formed (fig. 4.10). One way to construct a double digest map is to measure the fragment lengths (but not the order) from a complete digestion of the DNA by each of the two enzymes singly, and then by the two

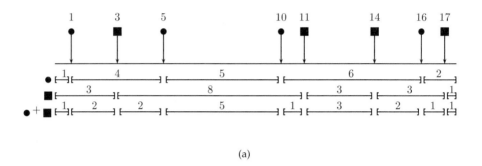

(a)

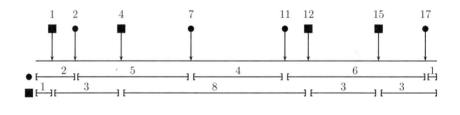

(b)

Figure 4.10   Restriction map of two restriction enzymes. When the digest is performed with each restriction enzyme separately and then with both enzymes combined, you may be able to reconstruct the original restriction map. The single digests $\{1, 2, 4, 5, 6\}$ and $\{1, 3, 3, 3, 8\}$ as well as the double digest $\{1, 1, 1, 1, 2, 2, 2, 3, 5\}$ allow one to uniquely reconstruct the restriction map (a). There are many other restriction maps that yield the same single digests, but produce a different double digest (b).

enzymes applied together. The problem of determining the positions of the cuts from fragment length data is known as the *Double Digest problem*, or *DDP*.

Figure 4.10 shows "DNA" cut by two restriction enzymes, A (shown by a circle) and B (shown by a square). Through gel electrophoresis experiments with these two restriction enzymes, a biologist can obtain information about the sizes of the restriction fragments generated by each individually. However, there are many orderings (maps) corresponding to these fragment lengths. To find out which of the maps is the correct one, biologists cleave DNA by *both* enzymes at the same time; this procedure is known as a *double digest*. The two maps presented in figure 4.10 produce the same single digests A and B but different double digests $A + B$. The double digest that fits experimental data corresponds to the correct map. The Double Digest problem is to find a physical map, given three sets of fragment lengths: A, B, and $A + B$.

Problem 4.9

Devise a brute force algorithm for the DDP and suggest a branch-and-bound approach to improve its performance.

Another technique used to build restriction maps leads to the *Probed Partial Digest problem (PPDP)*. In this method DNA is partially digested with a restriction enzyme, thus generating a collection of DNA fragments between every two cutting sites. After this, a labeled probe that attaches to the DNA between two cutting sites is hybridized to the partially digested DNA, and the sizes of fragments to which the probe hybridizes are measured. In contrast to the PDP, where the input consists of *all* partial fragments, the input for the PPDP consists of all partial fragments that contain a given point (this point corresponds to the position of the labeled probe). The problem is to reconstruct the positions of the sites from the multiset of measured lengths.

In the PPDP, we assume that the labeled probe is hybridized at position 0 and that A is the set of restriction sites with negative coordinates while B is the set of restriction sites with positive coordinates. The probed partial digest experiment provides the multiset $\{b - a \; : \; a \in A, b \in B\}$ and the problem is to find A and B given this set.

Problem 4.10

Design a brute force algorithm for the PPDP and suggest a branch-and-bound approach to improve its performance.

Problem 4.11

The search trees in the text are *complete k-ary trees*: each vertex that is not a leaf has exactly k children. It is also *balanced*: the number of edges in the path from the root to any leaf is the same (this is sometimes referred to as the *height* of the tree). Find a closed-form expression for the total number of vertices in a complete and balanced k-ary tree of height L.

Problem 4.12

Given a long *text* string T and one shorter *pattern* string s, find the first occurrence of s in T (if any). What is the complexity of your algorithm?

Problem 4.13

Given a long *text* string T, one shorter *pattern* string s, and an integer k, find the first occurrence in T of a string (if any) s' such that $d_H(s, s') \leq k$. What is the complexity of your algorithm?

Problem 4.14

Implement an algorithm that counts the number of occurrences of each l-mer in a string of length n. Run it over a bacterial genome and construct the distribution of l-mer frequencies. Compare this distribution to that of a random string of the same length as the bacterial genome.

Problem 4.15

The following algorithm is a cousin of one of the motif finding algorithms we have considered in this chapter. Identify which algorithm is a cousin of ANOTHERMOTIF-SEARCH and find the similarities and differences between these two algorithms.

ANOTHERMOTIFSEARCH(DNA, t, n, l)
1   $\mathbf{s} \leftarrow (1, 1, \ldots, 1)$
2   $\mathbf{bestMotif} \leftarrow$ FINDINSEQ($\mathbf{s}, 1, t, n, l$)
3   **return** $\mathbf{bestMotif}$

FINDINSEQ($\mathbf{s}, currentSeq, t, n, l$)
1   $bestScore \leftarrow 0$
2   **for** $j \leftarrow 1$ **to** $n - l + 1$
3       $s_{currentSeq} \leftarrow j$
4   **if** $currentSeq \neq t$
5           $\mathbf{s} \leftarrow$ FINDINSEQ($\mathbf{s}, currentSeq + 1, t, n, l$)
6   **if** $Score(\mathbf{s}) > bestScore$
7           $bestScore \leftarrow Score(\mathbf{s})$
8           $\mathbf{bestMotif} \leftarrow \mathbf{s}$
9   **return** $\mathbf{bestMotif}$

Problem 4.16

The following algorithm is a cousin of one of the motif finding algorithms we have considered in this chapter. Identify which algorithm is a cousin of YETANOTHERMO-TIFSEARCH and find the similarities and differences between these two algorithms.

YETANOTHERMOTIFSEARCH(DNA, t, n, l)
1   $\mathbf{s} \leftarrow (1, 1, \ldots, 1)$
2   **bestMotif** $\leftarrow$ FIND($\mathbf{s}, 1, t, n, l$)
3   **return bestMotif**

FIND($\mathbf{s}, currentSeq, t, n, l$)
  1   $i \leftarrow currentSeq$
  2   $bestScore \leftarrow 0$
  3   **for** $j \leftarrow 1$ **to** $n - l + 1$
  4       $s_i \leftarrow j$
  5       $bestPossibleScore \leftarrow Score(\mathbf{s}, i) + (t - i) \cdot l$
  6   **if** $bestPossibleScore > bestScore$
  7   **if** $currentSeq \neq t$
  8               $\mathbf{s} \leftarrow$ FIND($\mathbf{s}, currentSeq + 1, t, n, l$)
  9   **if** $Score(\mathbf{s}) > bestScore$
10               $bestScore \leftarrow Score(\mathbf{s})$
11   **bestMotif** $\leftarrow \mathbf{s}$
12   **return bestMotif**

Problem 4.17

Derive a tighter bound for the branch-and-bound strategy for the Median String problem. *Hint:* Split an l-mer w into two parts, u and v. Use $TotalDistance(u, DNA) + TotalDistance(v, DNA)$ to bound $TotalDistance(w, DNA)$.

5 *Greedy Algorithms*

The algorithm USCHANGE in chapter 2 is an example of a greedy strategy: at each step, the cashier would only consider the largest denomination smaller than (or equal to) M. Since the goal was to minimize the number of coins returned to the customer, this seemed like a sensible strategy: you would never use five nickels in place of one quarter. A generalization of USCHANGE, BETTERCHANGE also used what seemed like the best option and did not consider any others, which is what makes an algorithm "greedy." Unfortunately, BETTERCHANGE actually returned incorrect results in some cases because of its short-sighted notion of "good." This is a common characteristic of greedy algorithms: they often return suboptimal results, but take very little time to do so. However, there are a lucky few greedy algorithms that find optimal rather than suboptimal solutions.

5.1 Genome Rearrangements

Waardenburg's syndrome is a genetic disorder resulting in hearing loss and pigmentary abnormalities, such as two differently colored eyes. The disease was named after the Dutch ophthalmologist who first noticed that people with two differently colored eyes frequently had hearing problems as well. In the early 1990s, biologists narrowed the search for the gene implicated in Waardenburg's syndrome to human chromosome 2, but its exact location remained unknown for some time. There was another clue that shed light on the gene associated with Waardenburg's syndrome, that drew attention to chromosome 2: for a long time, breeders scrutinized mice for mutants, and one of these, designated *splotch*, had pigmentary abnormalities like patches of white spots, similar to those in humans with Waardenburg's syndrome. Through breeding, the *splotch* gene was mapped to one of the mouse chro-

Mouse X Chromosome

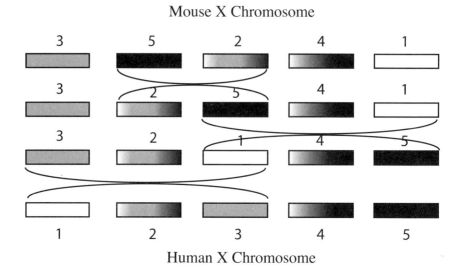

Human X Chromosome

Figure 5.1 Transformation of the mouse gene order into the human gene order on the X chromosome (only the five longest synteny blocks are shown here).

mosomes. As gene mapping proceeded it became clear that there are groups of genes in mice that appear in the same order as they do in humans: these genes are likely to be present in the same order in a common ancestor of humans and mice—the ancient mammalian genome. In some ways, the human genome is just the mouse genome cut into about 300 large genomic fragments, called *synteny blocks*, that have been pasted together in a different order. Both sequences are just two different shufflings of the ancient mammalian genome. For example, chromosome 2 in humans is built from fragments that are similar to mouse DNA residing on chromosomes 1, 2, 3, 5, 6, 7, 10, 11, 12, 14, and 17. It is no surprise, then, that finding a gene in mice often leads to clues about the location of the related gene in humans.

Every genome rearrangement results in a change of gene ordering, and a series of these rearrangements can alter the genomic architecture of a species. Analyzing the rearrangement history of mammalian genomes is a challenging problem, even though a recent analysis of human and mouse genomes implies that fewer than 250 genomic rearrangements have occurred since the divergence of humans and mice approximately 80 million years ago. Every study of genome rearrangements involves solving the combinatorial puzzle of finding a series of rearrangements that transform one genome into another. Figure 5.1 presents a *rearrangement scenario* in which the mouse X chromosome is transformed into the human X chromosome.[1] The elementary rearrangement event in this scenario is the flipping of a genomic segment, called a *reversal*, or an *inversion*. One can consider other types of evolutionary events but in this book we only consider reversals, the most common evolutionary events.

Biologists are interested in the most parsimonious evolutionary scenario, that is, the scenario involving the smallest number of reversals. While there is no guarantee that this scenario represents an actual evolutionary sequence, it gives us a lower bound on the number of rearrangements that have occurred and indicates the similarity between two species.[2]

Even for the small number of synteny blocks shown, it is not so easy to verify that the three evolutionary events in figure 5.1 represent a *shortest* series of reversals transforming the mouse gene order into the human gene order on the X chromosome. The exhaustive search technique that we presented in the previous chapter would hardly work for rearrangement studies since the number of variants that need to be explored becomes enormous for more than ten synteny blocks. Below, we explore two greedy approaches that work to differing degrees of success.

5.2 Sorting by Reversals

In their simplest form, rearrangement events can be modeled by a series of reversals that transform one genome into another. The order of genes (rather, of synteny blocks) in a genome can be represented by a permutation[3]

---

1. Extreme conservation of genes on X chromosomes across mammalian species provides an opportunity to study the evolutionary history of X chromosomes independently of the rest of the genomes, since the gene content of X chromosomes has barely changed throughout mammalian evolution. However, the order of genes on X chromosomes has been disrupted several times. In other words, genes that reside on the X chromosome stay on the X chromosome (but their order may change). All other chromosomes may exchange genes, that is, a gene can move from one chromosome to another.
2. In fact, a sequence of reversals that transforms the X chromosome of mouse into the X chromosome of man does not even represent an evolutionary sequence, since humans are not descended from the present-day mouse. However, biologists believe that the architecture of the X chromosome in the human-mouse ancestor is about the same as the architecture of the human X chromosome.
3. A permutation of a sequence of n numbers is just a reordering of that sequence. We will always use permutations of consecutive integers: for example, $2\,1\,3\,4\,5$ is a permutation of $1\,2\,3\,4\,5$.

$\pi = \pi_1\pi_2\cdots\pi_n$. The order of synteny blocks on the X chromosome in humans is represented in figure 5.1 by $(1,2,3,4,5)$, while the ordering in mice is $(3,5,2,4,1)$.[4]

A reversal $\rho(i,j)$ has the effect of reversing the order of synteny blocks

$$\pi_i\pi_{i+1}\cdots\pi_{j-1}\pi_j$$

In effect, this transforms

$$\pi = \pi_1\cdots\pi_{i-1}\underrightarrow{\pi_i\pi_{i+1}\cdots\pi_{j-1}\pi_j}\pi_{j+1}\cdots\pi_n$$

into

$$\pi \cdot \rho(i,j) = \pi_1\cdots\pi_{i-1}\underleftarrow{\pi_j\pi_{j-1}\cdots\pi_{i+1}\pi_i}\pi_{j+1}\cdots\pi_n$$

For example, if $\pi = 1\,2\,\underline{4\,3\,7\,5}\,6$, then $\pi \cdot \rho(3,6) = 1\,2\,\underline{5\,7\,3\,4}\,6$. With this representation of a genome, and a rigorous definition of an evolutionary event, we are in a position to formulate the computational problem that mimics the biological rearrangement process.

Reversal Distance Problem:

Given two permutations, find a shortest series of reversals that transforms one permutation into another.

   Input: Permutations π and σ.

   Output: A series of reversals $\rho_1, \rho_2, \ldots, \rho_t$ transforming π into σ (i.e., $\pi \cdot \rho_1 \cdot \rho_2 \cdots \rho_t = \sigma$), such that t is minimum.

We call t the *reversal distance between π and σ*, and write $d(\pi, \sigma)$ to denote the reversal distance for a given π and σ. In practice, one usually selects the second genome's order as a gold standard, and arbitrarily sets σ to be the *identity permutation* $1\,2\,\cdots\,n$. The Sorting by Reversals problem is similar to the Reversal Distance problem, except that it requires only one permutation as input.

---

4. In reality, genes and synteny blocks have directionality, reflecting whether they reside on the direct strand or the reverse complement strand of the DNA. In other words, the synteny block order in an organism is really represented by a *signed* permutation. However, in this section we ignore the directionality of the synteny blocks for simplicity.

Sorting by Reversals Problem:
Given a permutation, find a shortest series of reversals that transforms it into the identity permutation.

Input: Permutation π.

Output: A series of reversals $\rho_1, \rho_2, \ldots, \rho_t$ transforming π into the identity permutation such that t is minimum.

In this case, we call t the *reversal distance of* π and denote it as $d(\pi)$. When sorting a permutation $\pi = 1\,2\,3\,6\,4\,5$, it hardly makes sense to move the already-sorted first three elements of π. If we define *prefix*(π) to be the number of already-sorted elements of π, then a sensible strategy for sorting by reversals is to increase *prefix*(π) at every step. This approach sorts π in 2 steps: $1\,2\,3\,\underline{6\,4}\,5 \rightarrow 1\,2\,3\,4\,\underline{6\,5} \rightarrow 1\,2\,3\,4\,5\,6$. Generalizing this leads to an algorithm that sorts a permutation by repeatedly moving its ith element to the ith position.[5]

SIMPLEREVERSALSORT(π)
1   **for** $i \leftarrow 1$ **to** $n-1$
2       $j \leftarrow$ position of element i in π (i.e., $\pi_j = i$)
3   **if** $j \neq i$
4           $\pi \leftarrow \pi \cdot \rho(i,j)$
5   **output** π
6   **if** π is the identity permutation
7   **return**

SIMPLEREVERSALSORT is an example of a greedy algorithm that chooses the "best" reversal at every step. However, the notion of "best" here is rather short-sighted—simply increasing *prefix*(π) does not guarantee the smallest number of reversals. For example, SIMPLEREVERSALSORT takes five steps to sort $6\,1\,2\,3\,4\,5$:

$$\underline{6\,1}\,2\,3\,4\,5 \rightarrow 1\,\underline{6\,2}\,3\,4\,5 \rightarrow 1\,2\,\underline{6\,3}\,4\,5 \rightarrow 1\,2\,3\,\underline{6\,4}\,5 \rightarrow 1\,2\,3\,4\,\underline{6\,5} \rightarrow 1\,2\,3\,4\,5\,6$$

However, the same permutation can be sorted in just two steps:

$$\underline{6\,1\,2\,3\,4\,5} \rightarrow \underline{5\,4\,3\,2\,1}\,6 \rightarrow 1\,2\,3\,4\,5\,6.$$

---

5. Note the superficial similarity of this algorithm to SELECTIONSORT in chapter 2.

Therefore, we can confidently say that SIMPLEREVERSALSORT is not a correct algorithm, in the strict sense of chapter 2. In fact, despite its commonsense appeal, SIMPLEREVERSALSORT is a terrible algorithm since it takes $n-1$ steps to sort the permutation $\pi = n\,1\,2\,\ldots\,(n-1)$ even though $d(\pi) = 2$.

Even before biologists faced genome rearrangement problems, computer scientists studied the related Sorting by Prefix Reversals problem, also known as the Pancake Flipping problem: given an arbitrary permutation π, find $d_{pref}(\pi)$, which is the minimum number of reversals of the form $\rho(1, i)$ sorting π. The Pancake Flipping problem was inspired by the following "real-life" situation described by (the fictitious) Harry Dweighter:

> The chef in our place is sloppy, and when he prepares a stack of pancakes they come out all different sizes. Therefore, when I deliver them to a customer, on the way to a table I rearrange them (so that the smallest winds up on top, and so on, down to the largest at the bottom) by grabbing several from the top and flipping them over, repeating this (varying the number I flip) as many times as necessary. If there are n pancakes, what is the maximum number of flips that I will ever have to use to rearrange them?

An analog of SIMPLEREVERSALSORT will sort every permutation by at most $2(n-1)$ prefix reversals. For example, one can sort 1 2 3 6 4 5 by 4 prefix reversals ($\underline{1\,2\,3}\,6\,4\,5 \rightarrow \underline{6\,3\,2\,1}\,4\,5 \rightarrow \underline{5\,4\,1\,2\,3}\,6 \rightarrow \underline{3\,2\,1}\,4\,5\,6 \rightarrow 1\,2\,3\,4\,5\,6$) but it is not clear whether there exists an even shorter series of prefix reversals to sort this permutation. William Gates, an undergraduate student at Harvard in the mid-1970s, and Christos Papadimitriou, a professor at Harvard in the mid-1970s, now at Berkeley, made the first attempt to solve this problem and proved that any permutation can be sorted by at most $\frac{5}{3}(n+1)$ prefix reversals. However, the Pancake Flipping problem remains unsolved.

5.3 Approximation Algorithms

In chapter 2 we mentioned that, for many problems, efficient polynomial algorithms are still unknown and unlikely ever to be found. For such problems, computer scientists often find a compromise in *approximation algorithms* that produce an approximate solution rather than an optimal one.[6] The *approximation ratio of algorithm $\mathcal{A}$ on input π* is defined as $\frac{\mathcal{A}(\pi)}{OPT(\pi)}$, where $\mathcal{A}(\pi)$ is the solution produced by the algorithm $\mathcal{A}$ and $OPT(\pi)$ is the correct (optimal) solution of the problem.[7] The *approximation ratio, or performance guarantee of algorithm $\mathcal{A}$* is defined as its maximum approximation ratio over *all* inputs of size n, that is, as

$$\max_{|\pi|=n} \frac{\mathcal{A}(\pi)}{OPT(\pi)}.$$

We assume that $\mathcal{A}$ is a *minimization* algorithm, i.e., an algorithm that attempts to minimize its objective function. For maximization algorithms, the approximation ratio is

$$\min_{|\pi|=n} \frac{\mathcal{A}(\pi)}{OPT(\pi)}.$$

In essence, an approximation algorithm gives a worst-case scenario of just how far off an algorithm's output can be from some hypothetical perfect algorithm. The approximation ratio of SIMPLEREVERSALSORT is at least $\frac{n-1}{2}$, so a biologist has no guarantee that this algorithm comes anywhere close to the correct solution. For example, if n is 1001, this algorithm could return a series of reversals that is as large as 500 times the optimal. Our goal is

---

6. Approximation algorithms are only relevant to problems that have a numerical objective function like minimizing the number of coins returned to the customer. A problem that does not have such an objective function (like the Partial Digest problem) does not lend itself to approximation algorithms.

7. Technically, an approximation algorithm is not correct, in the sense of chapter 2, since there exists some input that returns a suboptimal (incorrect) output. The approximation ratio gives one an idea of just how incorrect the algorithm can be.

$$0 \; \underset{\leftarrow}{2 \; 1} \; \underset{\rightarrow}{3 \; 4 \; 5} \; \underset{\leftarrow}{8 \; 7 \; 6} \; 9$$

Figure 5.2   Breakpoints, adjacencies, and strips for permutation $2\,1\,3\,4\,5\,8\,7\,6$ (extended by 0 and 9 on the ends). Strips with more than one element are divided into decreasing strips ($\leftarrow$) and increasing strips ($\rightarrow$). The boundary between two nonconsecutive elements (in this case, 02, 13, 58, and 69) is a breakpoint; breakpoints demarcate the boundaries of strips.

to design approximation algorithms with better performance guarantees, for example, an algorithm with an approximation ratio of 2, or even better, 1.01. Of course, an algorithm with an approximation ratio of 1 (by definition, a correct and optimal algorithm) would be the acme of perfection, but such algorithms can be hard to find. As of the writing of this book, the best known algorithm for sorting by reversals has a performance guarantee of 1.375.

5.4   Breakpoints: A Different Face of Greed

We have described a greedy algorithm that attempts to maximize $prefix(\pi)$ in every step, but any chess player knows that greed often leads to wrong decisions. For example, the ability to take a queen in a single step is usually a good sign of a trap. Good chess players use a more sophisticated notion of greed that evaluates a position based on many subtle factors rather than simply on the face value of a piece they can take.

The problem with SIMPLEREVERSALSORT is that $prefix(\pi)$ is a naive measure of our progress toward the identity permutation, and does not accurately reflect how difficult it is to sort a permutation. Below we define *breakpoints* that can be viewed as "bottlenecks" for sorting by reversals. Using the number of breakpoints, rather than $prefix(\pi)$, as the basis of greed leads to a better algorithm for sorting by reversals, in the sense that it produces a solution that is closer to the optimal one.

It will be convenient for us to extend the permutation $\pi_1 \cdots \pi_n$ by $\pi_0 = 0$ and $\pi_{n+1} = n + 1$ on the ends. To be clear, we do not move π_0 or π_{n+1} during the process of sorting. We call a pair of neighboring elements π_i and π_{i+1}, for $0 \leq i \leq n$, an *adjacency* if π_i and π_{i+1} are consecutive numbers; we call

the pair a *breakpoint* if not. The permutation in figure 5.2 has five adjacencies (2 1, 3 4, 4 5, 8 7, and 7 6) and four breakpoints (0 2, 1 3, 5 8, and 6 9). A
permutation on n elements may have as many as $n + 1$ breakpoints (e.g., the
permutation 0 6 1 3 5 7 2 4 8 on seven elements has eight breakpoints) and as
few as 0 (the identity permutation 0 1 2 3 4 5 6 7 8).[8] Every breakpoint corresponds to a pair of elements π_i and π_{i+1} that are neighbors in π but not in
the identity permutation. In fact, the identity permutation is the only permutation with no breakpoints at all. Therefore, the nonconsecutive elements
π_i and π_{i+1} forming a breakpoint must be separated in the process of transforming π to the identity, and we can view sorting by reversals as the process
of eliminating breakpoints. The observation that every reversal can eliminate
at most two breakpoints (one on the left end and another on the right end of
the reversal) immediately implies that $d(\pi) \geq \frac{b(\pi)}{2}$, where $b(\pi)$ is the number
of breakpoints in π. The algorithm BREAKPOINTREVERSALSORT eliminates
as many breakpoints as possible in every step in order to reach the identity
permutation.

BREAKPOINTREVERSALSORT(π)
1   **while** $b(\pi) > 0$
2   Among all reversals, choose reversal ρ minimizing $b(\pi \cdot \rho)$
3           $\pi \leftarrow \pi \cdot \rho$
4   **output** π
5   **return**

One problem with this algorithm is that it is not clear why BREAKPOINTRE-
VERSALSORT is a better approximation algorithm than SIMPLEREVERSAL-
SORT. Moreover, it is not even obvious yet that BREAKPOINTREVERSALSORT
terminates! How can we be sure that removing some breakpoints does not
introduce others, leading to an endless cycle?

We define a *strip* in a permutation π as an interval between two consecutive
breakpoints, that is, as any maximal segment without breakpoints (see figure 5.2). For example, the permutation 0 2 1 3 4 5 8 7 6 9 consists of five strips:
0, 2 1, 3 4 5, 8 7 6, and 9. Strips can be further divided into *increasing* strips
(3 4 5) and *decreasing* strips (2 1) and (8 7 6). Single-element strips can be considered to be either increasing or decreasing, but it will be convenient to

---

8. We remind the reader that we extend permutations by 0 and $n + 1$ on their ends, thus introducing potential breakpoints in the beginning and in the end.

define them as decreasing (except for elements 0 and $n+1$ which will always be classified as increasing strips).

We present the following theorems, first to show that endless cycles of breakpoint removal cannot happen, and then to show that the approximation ratio of the algorithm is 4. While the notion of "theorem" and "proof" might seem overly formal for what is, at heart, a biological problem, it is important to consider that we have modeled the biological process in mathematical terms. We are proving analytically that the algorithm meets certain expectations. This notion of proof without experimentation is very different from what a biologist would view as proof, but it is just as important when working in bioinformatics.

Theorem 5.1 *If a permutation π contains a decreasing strip, then there is a reversal ρ that decreases the number of breakpoints in π, that is, $b(\pi \cdot \rho) < b(\pi)$.*

Proof: Among all decreasing strips in π, choose the strip containing the smallest element k ($k=3$ for permutation $0\,1\,2\,7\,6\,5\,8\,4\,3\,9$). Element $k-1$ in π cannot belong to a decreasing strip, since otherwise we would choose a strip ending at $k-1$ rather than a strip ending at k. Therefore, $k-1$ belongs to an increasing strip; moreover, it is easy to see that $k-1$ terminates this strip (for permutation $0\,1\,2\,7\,6\,5\,8\,4\,3\,9$, $k-1=2$ and 2 is at the right end of the increasing strip $0\,1\,2$). Therefore elements k and $k-1$ correspond to two breakpoints, one at the end of the decreasing strip ending with k and the other at the end of the increasing strip ending in $k-1$. Reversing the segment between k and $k-1$ brings them together, as in $0\,1\,2\,7\,6\,5\,8\,4\,3\,9 \rightarrow 0\,1\,2\,3\,4\,8\,5\,6\,7\,9$, thus reducing the number of breakpoints in π. □

For example, BREAKPOINTREVERSALSORT may perform the following four steps when run on the input $(0\,8\,2\,7\,6\,5\,1\,4\,3\,9)$ in order to reduce the number of breakpoints:

$$
\begin{array}{ll}
(0\ 8\ 2\ 7\ 6\ 5\ 1\ 4\ 3\ 9) & b(\pi)=6 \\
(0\ 2\ 8\ 7\ 6\ 5\ 1\ 4\ 3\ 9) & b(\pi)=5 \\
(0\ 2\ 3\ 4\ 1\ 5\ 6\ 7\ 8\ 9) & b(\pi)=3 \\
(0\ 4\ 3\ 2\ 1\ 5\ 6\ 7\ 8\ 9) & b(\pi)=2 \\
(0\ 1\ 2\ 3\ 4\ 5\ 6\ 7\ 8\ 9) & b(\pi)=0
\end{array}
$$

In this case, BREAKPOINTREVERSALSORT steadily reduces the number of breakpoints in every step of the algorithm. In other cases, (e.g., the permutation $(0\,1\,5\,6\,7\,2\,3\,4\,8\,9)$ without decreasing strips), no reversal reduces the number of breakpoints. In order to overcome this, we can simply find any increasing strip (excluding π_0 and π_{n+1}, of course) and flip it. This creates a decreasing strip and we can proceed.

IMPROVEDBREAKPOINTREVERSALSORT(π)
1   **while** $b(\pi) > 0$
2   **if** π has a decreasing strip
3   Among all reversals, choose reversal ρ minimizing $b(\pi \cdot \rho)$
4   **else**
5   Choose a reversal ρ that flips an increasing strip in π
6       $\pi \leftarrow \pi \cdot \rho$
7   **output** π
8   **return**

The theorem below demonstrates that such "no progress" situations do not happen too often in the course of IMPROVEDBREAKPOINTREVERSALSORT. In fact, the theorem quanitfies exactly how often those situations could possibly occur and provides an approximation ratio guarantee.

Theorem 5.2 IMPROVEDBREAKPOINTREVERSALSORT *is an approximation algorithm with a performance guarantee of at most 4.*

Proof: Theorem 5.1 implies that as long as π has a decreasing strip, IMPROVEDBREAKPOINTREVERSALSORT reduces the number of breakpoints in π. On the other hand, it is easy to see that if all strips in π are increasing, then there might not be a reversal that reduces the number of breakpoints. In this case IMPROVEDBREAKPOINTREVERSALSORT finds a reversal ρ that reverses an increasing strip(s) in π. By reversing an increasing strip, ρ creates a decreasing strip in π implying that IMPROVEDBREAKPOINTREVERSALSORT will be able to reduce the number of strips at the next step. Therefore, for every "no progress" step, IMPROVEDBREAKPOINTREVERSALSORT will make progress at the next step which means that IMPROVEDBREAKPOINTREVERSALSORT eliminates at least one breakpoint in every two steps. In the worst-case scenario, the number of steps in IMPROVEDBREAKPOINTREVERSALSORT is at most $2b(\pi)$ and its approximation ratio is at most $\frac{2b(\pi)}{d(\pi)}$. Since $d(\pi) \geq \frac{b(\pi)}{2}$, IMPROVEDBREAKPOINTREVERSALSORT has a performance guarantee[9] bounded above by $\frac{2b(\pi)}{d(\pi)} \leq \frac{2b(\pi)}{\frac{b(\pi)}{2}} = 4$. $\qquad\square$

---

9. To be clear, we are not claiming that IMPROVEDBREAKPOINTREVERSALSORT will take four times as long, or use four times as much memory as an (unknown) optimal algorithm. We

5.5   A Greedy Approach to Motif Finding

In chapter 4 we saw a brute force algorithm to solve the Motif Finding problem. With a disappointing running time of $O(l \cdot n^t)$, the practical limitation of that algorithm is that we simply cannot run it on biological samples. We choose instead to rely on a faster greedy technique, even though it is not correct (in the sense of chapter 2) and does not result in an algorithm with a good performance guarantee. Despite the fact that this algorithm is an approximation algorithm with an unknown approximation ratio, a popular tool based on this approach developed by Gary Stormo and Gerald Hertz in 1989, CONSENSUS, often produces results that are as good as or better than more complicated algorithms.

GREEDYMOTIFSEARCH scans each DNA sequence only once. Once we have scanned a particular sequence i, we decide which of its l-mer has the best contribution to the partial alignment score $Score(\mathbf{s}, i, DNA)$ for the first i sequences and immediately claim that this l-mer is part of the alignment. The pseudocode is shown below.

GREEDYMOTIFSEARCH(DNA, t, n, l)
1   **bestMotif** $\leftarrow (1, 1, \ldots, 1)$
2   $\mathbf{s} \leftarrow (1, 1, \ldots, 1)$
3   **for** $s_1 \leftarrow 1$ **to** $n - l + 1$
4   **for** $s_2 \leftarrow 1$ **to** $n - l + 1$
5   **if** $Score(\mathbf{s}, 2, DNA) > Score(\textbf{bestMotif}, 2, DNA)$
6                     $BestMotif_1 \leftarrow s_1$
7                     $BestMotif_2 \leftarrow s_2$
8   $s_1 \leftarrow BestMotif_1$
9   $s_2 \leftarrow BestMotif_2$
10 **for** $i \leftarrow 3$ **to** t
11   **for** $s_i \leftarrow 1$ **to** $n - l + 1$
12   **if** $Score(\mathbf{s}, i, DNA) > Score(\textbf{bestMotif}, i, DNA)$
13                    $bestMotif_i \leftarrow s_i$
14        $s_i \leftarrow bestMotif_i$
15 **return bestMotif**

---

are saying that IMPROVEDBREAKPOINTREVERSALSORT will return an answer that contains no more than four times as many steps as an optimal answer. Unfortunately, we cannot determine exactly how far from optimal we are for each particular input, so we have to rely on this upper bound for the approximation ratio.

GREEDYMOTIFSEARCH first finds the two closest l-mers—in the sense of Hamming distance—in sequences 1 and 2 and forms a $2 \times l$ *seed matrix*. This stage requires $l(n - l + 1)^2$ operations. At each of the remaining $t - 2$ iterations GREEDYMOTIFSEARCH extends the seed matrix into a matrix with one more row by scanning the ith sequence (for $3 \leq i \leq t$) each of the remaining $t - 2$ sequences and selecting the one l-mer that has the maximum $Score(\mathbf{s}, i)$. This amounts to roughly $l \cdot (n - l + 1)$ operations in each iteration. Thus, the running time of this algorithm is $O(ln^2 + lnt)$, which is vastly better than the $O(ln^t)$ of SIMPLEMOTIFSEARCH or even the $O(4^l nt)$ of BRUTEFORCEMEDIANSTRING. When t is small compared to n, GREEDYMOTIFSEARCH really behaves as $O(ln^2)$, and the bulk of the time is actually spent locating the l-mers from the first two sequences that are the most similar.

As you can imagine, because the sequences are scanned sequentially, it is possible to construct input instances where GREEDYMOTIFSEARCH will miss the optimal motif. One important difference between the popular **CONSENSUS** motif finding software tool and the algorithm presented here is that **CONSENSUS** can scan the sequences in a random order, thereby making it more difficult to construct inputs that elicit worst-case behavior. Another important difference is that **CONSENSUS** saves a large number (usually at least 1000) of seed matrices at each iteration rather than only the one that GREEDYMOTIFSEARCH saves, making **CONSENSUS** less likely to miss the optimal solution. However, no embellishment of this greedy approach will be guaranteed to find an optimal motif.

5.6 Notes

The analysis of genome rearrangements in molecular biology was pioneered by Theodosius Dobzhansky and Alfred Sturtevant who, in 1936, published a milestone paper (102) presenting a rearrangement scenario for the species of fruit fly. In 1984 Nadeau and Taylor (78) estimated that surprisingly few genomic rearrangements (about 200) had taken place since the divergence of the human and mouse genomes. This estimate, made in the pregenomic era and based on a very limited data set, comes close to the recent postgenomic estimates based on the comparison of the entire human and mouse DNA sequences (85). The computational studies of the Reversal Distance problem were pioneered by David Sankoff in the early 1990s (93). The greedy algorithm based on breakpoint elimination is from a paper (56) by John Kececioglu and David Sankoff. The best currently known algorithm for sorting by reversals has an approximation ratio of 1.375 and was introduced by Piotr Berman, Sridhar Hannenhalli and Marek Karpinski (13). The first algorithmic analysis of the Pancake Flipping problem was the work of William Gates and Christos Papadimitriou in 1979 (40).

The greedy CONSENSUS algorithm was introduced by Gerald Hertz and Gary Stormo, and further improved upon in 1999 in a later paper by the same authors (47).

David Sankoff currently holds the Canada Research Chair in Mathematical Genomics at the University of Ottawa. He studied at McGill University, doing a PhD in Probability Theory with Donald Dawson, and writing a thesis on stochastic models for historical linguistics. He joined the new Centre de recherches mathématiques (CRM) of the University of Montreal in 1969 and was also a professor in the Mathematics and Statistics Department from 1984–2002. He is one of the founding fathers of bioinformatics whose fundamental contributions to the area go back to the early 1970s.

Sankoff was trained in mathematics and physics; his undergraduate summers in the early 1960s, however, were spent in a microbiology lab at the University of Toronto helping out with experiments in the field of virology and whiling away evenings and weekends in the library reading biological journals. It was exciting, and did not require too much background to keep up with the molecular biology literature: the Watson-Crick model was not even ten years old, the deciphering of the genetic code was still incomplete, and mRNA was just being discovered. With this experience, Sankoff had no problems communicating some years later with Robert J. Cedergren, a biochemist with a visionary interest in applying computers to problems in molecular biology.

In 1971, Cedergren asked Sankoff to find a way to align RNA sequences. Sankoff knew little of algorithm design and nothing of discrete dynamic programming, but as an undergraduate he had effectively used the latter in working out an economics problem matching buyers and sellers. The same approach worked with alignment. Bob and David became hooked on the topic, exploring statistical tests for alignment and other problems, fortunately before they realized that Needleman and Wunsch had already published a dynamic programming technique for biological sequence comparison.

A new question that emerged early in the Sankoff and Cedergren work was that of multiple alignment and its pertinence to molecular evolution.

Sankoff was already familiar with phylogeny problems from his work on language families and participation in the early numerical taxonomy meetings (before the schism between the parsimony-promoting cladists, led by Steve Farris, and the more statistically oriented systematists). Combining phylogenetics with sequence comparison led to tree-based dynamic programming for multiple alignment. Phylogenetic problems have cropped up often in Sankoff's research projects over the following decades.

Sankoff and Cedergren also studied RNA folding, applying several passes of dynamic programming to build energy-optimal RNA structures. They did not find the loop-matching reported by Daniel Kleitman's group (later integrated into a general, widely-used algorithm by Michael Zuker), though they eventually made a number of contributions in the 1980s, in particular to the problem of multiple loops and to simultaneous alignment and folding. Sankoff says:

> My collaboration with Cedergen also ran into its share of dead ends. Applying multidimensional scaling to ribosome structure did not lead very far, efforts to trace the origin of the genetic code through the phylogenetic analyses of tRNA sequences eventually petered out, and an attempt at dynamic programming for consensus folding of proteins was a flop.

The early and mid-1970s were nevertheless a highly productive time for Sankoff; he was also working on probabilistic analysis of grammatical variation in natural languages, on game theory models for electoral processes, and various applied mathematics projects in archaeology, geography, and physics. He got Peter Sellers interested in sequence comparison; Sellers later attracted attention by converting the longest common subsequence (LCS) formulation to the edit distance version. Sankoff collaborated with prominent mathematician Vaclav Chvatal on the expected length of the LCS of two random sequences, for which they derived upper and lower bounds. Several generations of probabilists have contributed to narrowing these bounds. Sankoff says:

> Evolutionary biologists Walter Fitch and Steve Farris spent sabbaticals with me at the CRM, as did computer scientist Bill Day, generously adding my name to a series of papers establishing the hardness of various phylogeny problems, most importantly the parsimony problem.

In 1987, Sankoff became a Fellow of the new Evolutionary Biology Program of the Canadian Institute for Advanced Research (CIAR). At the very first meeting of the CIAR program he was inspired by a talk by Monique Turmel on the comparison of chloroplast genomes from two species of algae. This led Sankoff to the comparative genomics–genome rearrangement track that has been his main research line ever since. Originally he took a probabilistic approach, but within a year or two he was trying to develop algorithms and programs for reversal distance. A phylogeny based on the reversal distances among sixteen mitochondrial genomes proved that a strong phylogenetic signal can be conserved in the gene order of even a miniscule genome across many hundreds of millions of years. Sankoff says:

> The network of fellows and scholars of the CIAR program, including Bob Cedergren, Ford Doolittle, Franz Lang, Mike Gray, Brian Golding, Mike Zuker, Claude Lemieux, and others across Canada; and a stellar group of international advisors (such as Russ Doolittle, Michael Smith, Marcus Feldman, Wally Gilbert) and associates (Mike Waterman, Joe Felsenstein, Mike Steel and many others) became my virtual "home department," a source of intellectual support, knowledge, and experience across multiple disciplines and a sounding board for the latest ideas.

> My comparative genomics research received two key boosts in the 1990s. One was the sustained collaboration of a series of outstanding students and postdocs: Guillaume Leduc, Vincent Ferretti, John Kececioglu, Mathieu Blanchette, Nadia El-Mabrouk and David Bryant. The second was my meeting Joe Nadeau; I already knew his seminal paper with Taylor on estimating the number of conserved linkage segments and realized that our interests coincided perfectly while our backgrounds were complementary.

When Nadeau showed up in Montreal for a short-lived appointment in the Human Genetics Department at McGill, it took no more than an hour for him and Sankoff to get started on a major collaborative project. They reformulated the Nadeau-Taylor approach in terms of gene content data, freeing it from physical or genetic distance measurements. The resulting simpler model allowed them to thoroughly explore the mathematical properties of the Nadeau-Taylor model and to experiment with the consequences of deviating from it.

The synergy between the algorithmic and probabilistic aspects of comparative genomics has become basic to how Sankoff understands evolution. The algorithmic is an ambitious attempt at deep inference, based on heavy assumptions and the sophisticated but inflexible mathematics they enable. The probabilistic is more descriptive and less explicitly revelatory of historical process, but the models based on statistics are easily generalized, their hypotheses weakened or strengthened, and their robustness ascertained. In Sankoff's view, it is the playing out of this dialectic that makes the field of whole-genome comparison the most interesting topic of research today and for the near future.

> My approach to research is not highly planned. Not that I don't have a vision about the general direction in which to go, but I have no specific set of tools that I apply as a matter of course, only an intuition about what type of method or model, what database or display, might be helpful. When I am lucky I can proceed from one small epiphany to another, working out some of the details each time, until some clear story emerges. Whether this involves stochastic processes, combinatorial optimization, or differential equations is secondary; it is the biology of the problem that drives its mathematical formulation. I am rarely motivated to research well-studied problems; instead I find myself confronting new problems in relatively unstudied areas; alignment was not a burning preoccupation with biologists or computer scientists when I started working on it, neither was genome rearrangement fifteen years later. I am quite pleased, though sometimes bemused, by the veritable tidal wave of computational biologists and bioinformaticians who have inundated the field where there were only a few isolated researchers thirty or even twenty years ago.

5.7 Problems

Problem 5.1

Suppose you have a maximization algorithm, $\mathcal{A}$, that has an approximation ratio of 4. When run on some input π, $\mathcal{A}(\pi) = 12$. What can you say about the true (correct) answer $OPT = OPT(\pi)$?

- $OPT \geq 3$
- $OPT \leq 3$
- $OPT \geq 12$
- $OPT \leq 12$
- $OPT \geq 48$
- $OPT \leq 48$

Problem 5.2

What is the approximation ratio of the BETTERCHANGE algorithm?

Problem 5.3

Design an approximation algorithm for the Pancake Flipping problem. What is its approximation ratio?

Problem 5.4

Perform the BREAKPOINTREVERSALSORT algorithm with $\pi = 3\ 4\ 6\ 5\ 8\ 1\ 7\ 2$ and show all intermediate permutations (break ties arbitrarily). Since BREAKPOINTREVERSAL-SORT is an approximation algorithm, there may be a sequence of reversals that is shorter than the one found by BREAKPOINTREVERSALSORT. Could you find such a sequence of reversals? Do you know if it is the shortest possible sequence of reversals?

Problem 5.5

Find a permutation with no decreasing strips for which there exists a reversal that reduces the number of breakpoints.

Problem 5.6

Can you find a permutation for which BREAKPOINTREVERSALSORT produces four times as many reversals than the optimal solution of the Reversal Sorting problem?

A DNA molecule is not always shaped like a line segment. Some simple organisms have a circular DNA molecule as a genome, where the molecule has no beginning and no end. These circular genomes can be visualized as a sequence of integers written along the perimeter of a circle. Two circular sequences would be considered equivalent if you could rotate one of the circles and get the same sequence written on the other.

Problem 5.7

Devise an approximation algorithm to sort a circular genome by reversals (i.e., transform it to the identity circular permutation). Evaluate the algorithm's performance guarantee.

Problem 5.8

Devise a better algorithm (i.e., one with a better approximation ratio) for the Sorting by Reversals problem.

The *swap sorting* of permutation π is a transformation of π into the identity permutation by exchanges of adjacent elements. For example, $3142 \rightarrow 1342 \rightarrow 1324 \rightarrow 1234$ is a three-step swap sorting of permutation 3124.

Problem 5.9

Design an algorithm for swap sorting that uses the minimum number of swaps to sort a permutation.

Problem 5.10

Design an algorithm for swap sorting that uses the minimum number of swaps to sort a *circular* permutation.

Problem 5.11

How many permutations on n elements have a single breakpoint? How many permutations have exactly two breakpoints? How many permutations have exactly three breakpoints?

Given permutations π and σ, a breakpoint between π and σ is defined as a pair of adjacent elements π_i and π_{i+1} in π that are separated in σ. For example, if $\pi = 143256$ and $\sigma = 123465$, then $\pi_1 = 1$ and $\pi_2 = 4$ in π form a breakpoint between π and σ since 1 and 4 are separated in σ. The number of breakpoints between π=01432567 and σ=01234657 is three (14, 25 and 67), while the number of breakpoints between σ and π is also three (12, 46 and 57).

Problem 5.12

Prove that the number of breakpoints between π and σ equals the number of breakpoints between σ and π.

Problem 5.13

Given permutations $\pi^1 = 124356$, $\pi^2 = 143256$ and $\pi^3 = 123465$, compute the number of breakpoints between: (1) π^1 and π^2, (2) π^1 and π^3, and (3) π^2 and π^3.

Problem 5.14

Given the three permutations π^1, π^2, and π^3 from the previous problem, find an *ancestral permutation* σ which minimizes the total breakpoint distance $\sum_{i=1}^{3} br(\pi^i, \sigma)$ between all three genomes and σ ($br(\pi^i, \sigma)$ is the number of breakpoints between π^i and σ).

Problem 5.15

Given three permutations π^1, π^2, and π^3 from the previous problem, find an ancestral permutation σ which minimizes the total reversal distance $\sum_{i=1}^{3} d(\pi^i, \sigma)$ between all three genomes and σ.

Analysis of genome rearrangements in multiple genomes corresponds to the following *Multiple Breakpoint Distance problem:* : given a set of permutations $\pi^1, \ldots, \pi^k$, find an ancestral permutation σ such that $\sum_{i=1,k} br(\pi^i, \sigma)$ is minimal, where $br(\pi^i, \sigma)$ is the number of breakpoints between π^i and σ.

Problem 5.16

Design a greedy algorithm for the Multiple Breakpoint Distance problem and evaluate its approximation ratio.

Problem 5.17

Alice and Bob have been assigned the task of implementing the BREAKPOINTREVERSALSORT approximation algorithm.

- Bob wants to get home early so he decides to *naively* implement the algorithm, without putting any thought into performance improvements. What is the running time of his program?

- Alice makes some changes to the algorithm and claims her algorithm achieves the same approximation ratio as Bob's (4) and runs in time $O(n^2)$. Give the pseudocode for Alice's algorithm.

- Not to be outdone, Bob gets a copy of Alice's algorithm, and makes an improvement of his own. He claims that in the case where every strip is increasing, he can guarantee that there will be a decreasing strip in each of the next *two* steps (rather than one as in BREAKPOINTREVERSALSORT). Bob believes that this will give his new algorithm a better approximation ratio than the previous algorithms. What is Bob's improvement, and what approximation ratio does it achieve?

Problem 5.18

Design an input for the GREEDYMOTIFSEARCH algorithm that causes the algorithm to output an incorrect result. That is, create a sample that has a strong pattern that is missed because of the greedy nature of the algorithm. If *optimalScore* is the score of the strongest motif in the sample and *greedyScore* is the score returned by GREEDYMOTIFSEARCH, how large can *optimalScore/greedyScore* be?

6 Dynamic Programming Algorithms

We introduced dynamic programming in chapter 2 with the Rocks problem. While the Rocks problem does not appear to be related to bioinformatics, the algorithm that we described is a computational twin of a popular alignment algorithm for sequence comparison. Dynamic programming provides a framework for understanding DNA sequence comparison algorithms, many of which have been used by biologists to make important inferences about gene function and evolutionary history. We will also apply dynamic programming to gene finding and other bioinformatics problems.

6.1 The Power of DNA Sequence Comparison

After a new gene is found, biologists usually have no idea about its function. A common approach to inferring a newly sequenced gene's function is to find similarities with genes of known function. A striking example of such a biological discovery made through a similarity search happened in 1984 when scientists used a simple computational technique to compare the newly discovered cancer-causing ν-sis oncogene with all (at the time) known genes. To their astonishment, the cancer-causing gene matched a normal gene involved in growth and development called platelet-derived growth factor (PDGF).[1] After discovering this similarity, scientists became suspicious that cancer might be caused by a normal growth gene being switched on at the wrong time—in essence, a good gene doing the right thing at the wrong time.

---

1. *Oncogenes* are genes in viruses that cause a cancer-like transformation of infected cells. Oncogene ν-sis in the *simian sarcoma virus* causes uncontrolled cell growth and leads to cancer in monkeys. The seemingly unrelated *growth factor* PDGF is a protein that stimulates cell growth.

Another example of a successful similarity search was the discovery of the cystic fibrosis gene. Cystic fibrosis is a fatal disease associated with abnormal secretions, and is diagnosed in children at a rate of 1 in 3900. A defective gene causes the body to produce abnormally thick mucus that clogs the lungs and leads to lifethreatening lung infections. More than 10 million Americans are unknowing and symptomless carriers of the defective cystic fibrosis gene; each time two carriers have a child, there is a 25% chance that the child will have cystic fibrosis.

In 1989 the search for the cystic fibrosis gene was narrowed to a region of 1 million nucleotides on the chromosome 7, but the exact location of the gene remained unknown. When the area around the cystic fibrosis gene was sequenced, biologists compared the region against a database of all known genes, and discovered similarities between some segment within this region and a gene that had already been discovered, and was known to code for *adenosine triphosphate (ATP) binding proteins.*[2] These proteins span the cell membrane multiple times as part of the ion transport channel; this seemed a plausible function for a cystic fibrosis gene, given the fact that the disease involves sweat secretions with abnormally high sodium content. As a result, the similarity analysis shed light on a damaged mechanism in faulty cystic fibrosis genes.

Establishing a link between cancer-causing genes and normal growth genes and elucidating the nature of cystic fibrosis were only the first success stories in sequence comparison. Many applications of sequence comparison algorithms quickly followed, and today bioinformatics approaches are among the dominant techniques for the discovery of gene function.

This chapter describes algorithms that allow biologists to reveal the similarity between different DNA sequences. However, we will first show how dynamic programming can yield a faster algorithm to solve the Change problem.

6.2   The Change Problem Revisited

We introduced the Change problem in chapter 2 as the problem of changing an amount of money M into the smallest number of coins from denominations $\mathbf{c} = (c_1, c_2, \ldots, c_d)$. We showed that the naive greedy solution used by cashiers everywhere is not actually a correct solution to this problem, and ended with a correct—though slow—brute force algorithm. We will con-

---

2. ATP binding proteins provide energy for many reactions in the cell.

sider a slightly modified version of the Change problem, in which we do not concern ourselves with the actual combination of coins that make up the optimal change solution. Instead, we only calculate the smallest number of coins needed (it is easy to modify this algorithm to also return the coin combination that achieves that number).

Suppose you need to make change for 77 cents and the only coin denominations available are 1, 3, and 7 cents. The best combination for 77 cents will be one of the following:

- the best combination for $77 - 1 = 76$ cents, plus a 1-cent coin;

- the best combination for $77 - 3 = 74$ cents, plus a 3-cent coin;

- the best combination for $77 - 7 = 70$ cents, plus a 7-cent coin.

For 77 cents, the best combination would be the smallest of the above three choices. The same logic applies to 76 cents (best of 75, 73, or 69 cents), and so on (fig. 6.1). If $bestNumCoins_M$ is the smallest number of coins needed to change M cents, then the following recurrence relation holds:

$$bestNumCoins_M = \min \begin{cases} bestNumCoins_{M-1} + 1 \\ bestNumCoins_{M-3} + 1 \\ bestNumCoins_{M-7} + 1 \end{cases}$$

In the more general case of d denominations $\mathbf{c} = (c_1, \ldots, c_d)$:

$$bestNumCoins_M = \min \begin{cases} bestNumCoins_{M-c_1} + 1 \\ bestNumCoins_{M-c_2} + 1 \\ \vdots \\ bestNumCoins_{M-c_d} + 1 \end{cases}$$

This recurrence motivates the following algorithm:

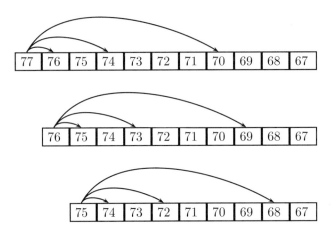

Figure 6.1   The relationships between optimal solutions in the Change problem. The smallest number of coins for 77 cents depends on the smallest number of coins for 76, 74, and 70 cents; the smallest number of coins for 76 cents depends on the smallest number of coins for 75, 73, and 69 cents, and so on.

RECURSIVECHANGE($M, \mathbf{c}, d$)
1   **if** $M = 0$
2   **return** 0
3   $bestNumCoins \leftarrow \infty$
4   **for** $i \leftarrow 1$ **to** d
5   **if** $M \geq c_i$
6             $numCoins \leftarrow$ RECURSIVECHANGE($M - c_i, \mathbf{c}, d$)
7   **if** $numCoins + 1 < bestNumCoins$
8                  $bestNumCoins \leftarrow numCoins + 1$
9   **return** $bestNumCoins$

The sequence of calls that RECURSIVECHANGE makes has a feature in common with the sequence of calls made by RECURSIVEFIBONACCI, namely, that RECURSIVECHANGE recalculates the optimal coin combination for a given amount of money repeatedly. For example, the optimal coin combination for 70 cents is recomputed repeatedly nine times over and over as $(77 - 7)$, $(77 - 3 - 3 - 1)$, $(77 - 3 - 1 - 3)$, $(77 - 1 - 3 - 3)$, $(77 - 3 - 1 - 1 - 1 - 1)$, $(77 - 1 - 3 - 1 - 1 - 1)$, $(77 - 1 - 1 - 3 - 1 - 1)$, $(77 - 1 - 1 - 1 - 3 - 1)$, $(77 - 1 - 1 - 1 - 1 - 3)$, and $(77 - 1 - 1 - 1 - 1 - 1 - 1 - 1)$. The optimal coin combination for 20 cents will be recomputed billions of times rendering RECURSIVECHANGE impractical.

To improve RECURSIVECHANGE, we can use the same strategy as we did for the Fibonacci problem—all we really need to do is use the fact that the solution for M relies on solutions for $M - c_1$, $M - c_2$, and so on, and then reverse the order in which we solve the problem. This allows us to leverage previously computed solutions to form solutions to larger problems and avoid all this recomputation.

Instead of trying to find the minimum number of coins to change M cents, we attempt the superficially harder task of doing this for *each* amount of money, m, from 0 to M. This appears to require more work, but in fact, it simplifies matters. The following algorithm with running time $O(Md)$ calculates $bestNumCoins_m$ for increasing values of m. This works because the best number of coins for some value m depends only on values less than m.

DPCHANGE$(M, \mathbf{c}, d)$
1 $bestNumCoins_0 \leftarrow 0$
2 **for** $m \leftarrow 1$ **to** M
3      $bestNumCoins_m \leftarrow \infty$
4   **for** $i \leftarrow 1$ **to** d
5   **if** $m \geq c_i$
6   **if** $bestNumCoins_{m-c_i} + 1 < bestNumCoins_m$
7                  $bestNumCoins_m \leftarrow bestNumCoins_{m-c_i} + 1$
8 **return** $bestNumCoins_M$

The key difference between RECURSIVECHANGE and DPCHANGE is that the first makes d recursive calls to compute the best change for M (and each of these calls requires a lot of work!), while the second analyzes the d already precomputed values to almost instantly compute the new one. As surprising as it may sound, simply reversing the order of computations in figure 6.1 makes a dramatic difference in efficiency (fig. 6.2).

We stress again the difference between the complexity of a problem and the complexity of an algorithm. In particular, we initially showed an $O(M^d)$ algorithm to solve the Change problem, and there did not appear to be any easy way to remedy this situation. Yet the DPCHANGE algorithm provides a simple $O(Md)$ solution. Conversely, a minor modification of the Change problem renders the problem very difficult. Suppose you had a limited number of each denomination and needed to change M cents using no more than the provided supply of each coin. Since you have fewer possible choices in

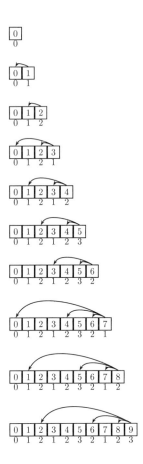

Figure 6.2 The solution for 9 cents ($bestNumCoins_9$) depends on 8 cents, 6 cents and 2 cent, but the smallest number of coins can be obtained by computing $bestNumCoins_m$ for $0 \leq m \leq 9$.

this new problem, it would seem to require even less time than the original Change problem, and that a minor modification to DPCHANGE would work. However, this is not the case and this problem turns out to be very difficult.

6.3   The Manhattan Tourist Problem

We will further illustrate dynamic programming with a surprisingly useful toy problem, called the Manhattan Tourist problem, and then build on this intuition to describe DNA sequence alignment.

Imagine a sightseeing tour in the borough of Manhattan in New York City, where a group of tourists are determined to walk from the corner of 59th Street and 8th Avenue to the Chrysler Building at 42nd Street and Lexington Avenue. There are many attractions along the way, but assume for the moment that the tourists want to see as many attractions as possible. The tourists are allowed to move either to the south or to the east, but even so, they can choose from many different paths (exactly how many is left as a problem at the end of the chapter). The upper path in figure 6.3 will take the tourists to the Museum of Modern Art, but they will have to miss Times Square; the bottom path will allow the tourists to see Times Square, but they will have to miss the Museum of Modern Art.

The map above can also be represented as a gridlike structure (figure 6.4) with the numbers next to each line (called *weights*) showing the number of attractions on every block. The tourists must decide among the many possible paths between the northwesternmost point (called the *source vertex*) and the southeasternmost point (called the *sink vertex*). The weight of a path from the source to the sink is simply the sum of weights of its edges, or the overall number of attractions. We will refer to this kind of construct as a *graph*, the intersections of streets we will call *vertices*, and the streets themselves will be *edges* and have a weight associated with them. We assume that horizontal edges in the graph are oriented to the east like $\rightarrow$ while vertical edges are oriented to the south like $\downarrow$. A *path* is a continuous sequence of edges, and the *length of a path* is the sum of the edge weights in the path.[3] A more detailed discussion of graphs can be found in chapter 8.

Although the upper path in figure 6.3 is better than the bottom one, in the sense that the tourists will see more attractions, it is not immediately clear if there is an even better path in the grid. The Manhattan Tourist problem is to find the path with the maximum number of attractions,[4] that is, a *longest path*

---

3. We emphasize that the length of paths in the graph represent the overall number of attractions on this path and has nothing to do with the real length of the path (in miles), that is, the distance the tourists travel.
4. There are many interesting museums and architectural landmarks in Manhattan. However, it is impossible to please everyone, so one can change the relative importance of the types of attractions by modulating the weights on the edges in the graph. This flexibility in assigning weights will become important when we discuss *scoring matrices* for sequence comparison.

(a path of maximum overall weight) in the grid.

Manhattan Tourist Problem:
Find a longest path in a weighted grid.

Input: A weighted grid G with two distinguished vertices:
a *source* and a *sink*.

Output: A longest path in G from *source* to *sink*.

Note that, since the tourists only move south and east, any grid positions
west or north of the source are unusable. Similarly, any grid positions south
or east of the sink are unusable, so we can simply say that the source vertex
is at $(0, 0)$ and that the sink vertex at (n, m) defines the southeasternmost
corner of the grid. In figure 6.4 $n = m = 4$, but n does not always have
to equal m. We will use the grid shown in figure 6.4, rather than the one
corresponding to the map of Manhattan in figure 6.3 so that you can see a
nontrivial example of this problem.

The brute force approach to the Manhattan Tourist problem is to search
among all paths in the grid for the longest path, but this is not an option
for even a moderately large grid. Inspired by the previous chapter you may
be tempted to use a greedy strategy. For example, a sensible greedy strategy would be to choose between two possible directions (south or east) by
comparing how many attractions tourists would see if they moved one block
south instead of moving one block east. This greedy strategy may provide re-
warding sightseeing experience in the beginning but, a few blocks later, may
bring you to an area of Manhattan you really do not want to be in. In fact,
no known greedy strategy for the Manhattan Tourist problem provides an
optimal solution to the problem. Had we followed the (obvious) greedy algorithm, we would have chosen the following path, corresponding to twenty
three attractions.[5]

---

5. We will show that the optimal number is, in fact, thirty-four.

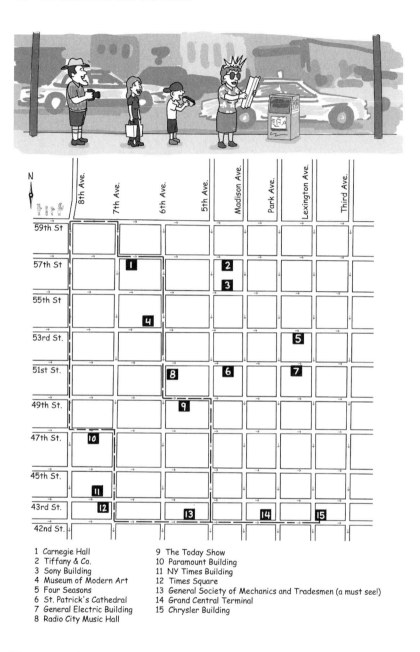

1 Carnegie Hall
2 Tiffany & Co.
3 Sony Building
4 Museum of Modern Art
5 Four Seasons
6 St. Patrick's Cathedral
7 General Electric Building
8 Radio City Music Hall

9   The Today Show
10 Paramount Building
11 NY Times Building
12 Times Square
13 General Society of Mechanics and Tradesmen (a must see!)
14 Grand Central Terminal
15 Chrysler Building

Figure 6.3   A city somewhat like Manhattan, laid out on a grid with one-way streets. You may travel only to the east or to the south, and you are currently at the north-westernmost point (source) and need to travel to the southeasternmost point (sink). Your goal is to visit as many attractions as possible.

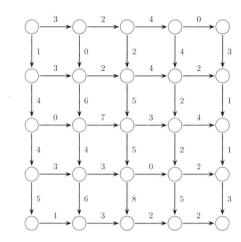

Figure 6.4   Manhattan represented as a graph with weighted edges.

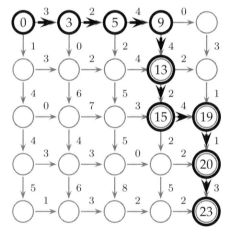

Instead of solving the Manhattan Tourist problem directly, that is, finding the longest path from *source* $(0, 0)$ to *sink* (n, m), we solve a more general problem: find the longest path from *source* to an arbitrary vertex (i, j) with $0 \leq i \leq n$, $0 \leq j \leq m$. We will denote the length of such a best path as $s_{i,j}$, noticing that $s_{n,m}$ is the weight of the path that represents the solution to the

Manhattan Tourist problem. If we only care about the longest path between $(0,0)$ and (n,m)—the Manhattan Tourist problem—then we have to answer one question, namely, what is the best way to get from *source* to *sink*. If we solve the general problem, then we have to answer $n \times m$ questions: what is the best way to get from *source* to anywhere. At first glance it looks like we have just created $n \times m$ different problems (computing (i,j) with $0 \le i \le n$ and $0 \le j \le m$) instead of a single one (computing $s_{n,m}$), but the fact that solving the more general problem is as easy as solving the Manhattan Tourist problem is the basis of dynamic programming. Note that DPCHANGE also generalized the problems that it solves by finding the optimal number of coins for *all* values less than or equal to M.

Finding $s_{0,j}$ (for $0 \le j \le m$) is not hard, since in this case the tourists do not have any flexibility in their choice of path. By moving strictly to the east, the weight of the path $s_{0,j}$ is the sum of weights of the first j city blocks. Similarly, $s_{i,0}$ is also easy to compute for $0 \le i \le n$, since the tourists move only to the south.

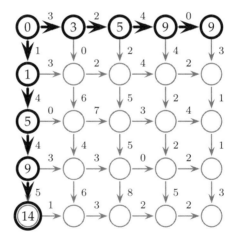

Now that we have figured out how to compute $s_{0,1}$ and $s_{1,0}$, we can compute $s_{1,1}$. The tourists can arrive at $(1,1)$ in only two ways: either by traveling south from $(0,1)$ or east from $(1,0)$. The weight of each of these paths is

- $s_{0,1}$ + weight of the edge (block) between (0,1) and (1,1);

- $s_{1,0}$ + weight of the edge (block) between (1,0) and (1,1).

Since the goal is to find the longest path to, in this case, $(1, 1)$, we choose the larger of the above two quantities: $3 + 0$ and $1 + 3$. Note that since there are no other ways to get to grid position $(1, 1)$, we have found the longest path from $(0, 0)$ to $(1, 1)$.

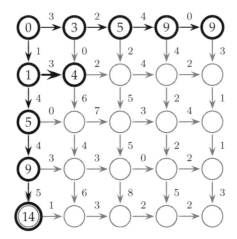

We have just found $s_{1,1}$. Similar logic applies to $s_{2,1}$, and then to $s_{3,1}$, and so on; once we have calculated $s_{i,0}$ for all i, we can calculate $s_{i,1}$ for all i.

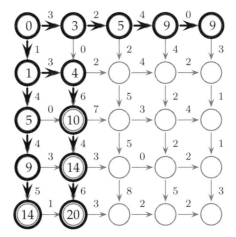

Once we have calculated $s_{i,1}$ for all i, we can use the same idea to calculate $s_{i,2}$ for all i, and so on. For example, we can calculate $s_{1,2}$ as follows.

$$s_{1,2} = \max \begin{cases} s_{1,1} + \text{weight of the edge between (1,1) and (1,2)} \\ s_{0,2} + \text{weight of the edge between (0,2) and (1,2)} \end{cases}$$

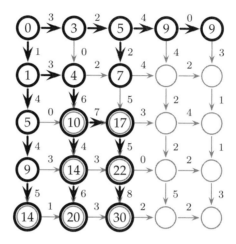

In general, having the entire column $s_{*,j}$ allows us to compute the next whole column $s_{*,j+1}$. The observation that the only way to get to the intersection at (i, j) is either by moving south from intersection $(i - 1, j)$ or by moving east from the intersection $(i, j - 1)$ leads to the following recurrence:

$$s_{i,j} = \max \begin{cases} s_{i-1,j} + \text{weight of the edge between } (i - 1, j) \text{ and } (i, j) \\ s_{i,j-1} + \text{weight of the edge between } (i, j - 1) \text{ and } (i, j) \end{cases}$$

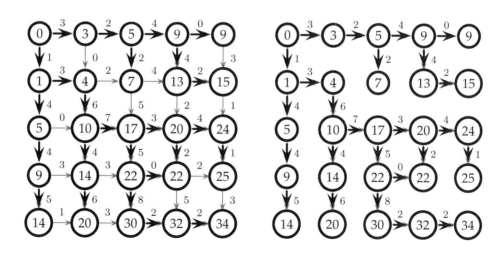

This recurrence allows us to compute every score $s_{i,j}$ in a single sweep of the grid. The algorithm MANHATTANTOURIST implements this procedure. Here, $\overset{\downarrow}{w}$ is a two-dimensional array representing the weights of the grid's edges that run north to south, and $\vec{w}$ is a two-dimensional array representing the weights of the grid's edges that run west to east. That is, $\overset{\downarrow}{w}_{i,j}$ is the weight of the edge between $(i, j-1)$ and (i, j); and $\vec{w}_{i,j}$ is the weight of the edge between $(i, j-1)$ and (i, j).

MANHATTANTOURIST($\overset{\downarrow}{\mathbf{w}}, \overset{\rightarrow}{\mathbf{w}}, n, m$)
1   $s_{0,0} \leftarrow 0$
2   **for** $i \leftarrow 1$ **to** n
3       $s_{i,0} \leftarrow s_{i-1,0} + \overset{\downarrow}{w}_{i,0}$
4   **for** $j \leftarrow 1$ **to** m
5       $s_{0,j} \leftarrow s_{0,j-1} + \vec{w}_{0,j}$
6   **for** $i \leftarrow 1$ **to** n
7   **for** $j \leftarrow 1$ **to** m
8           $s_{i,j} \leftarrow \max \begin{cases} s_{i-1,j} + \overset{\downarrow}{w}_{i,j} \\ s_{i,j-1} + \vec{w}_{i,j} \end{cases}$
9   **return** $s_{n,m}$

Lines 1 through 5 set up the *initial conditions* on the matrix s, and line 8 corresponds to the *recurrence* that allows us to fill in later table entries based on earlier ones. Most of the dynamic programming algorithms we will develop in the context of DNA sequence comparison will look just like MANHATTANTOURIST with only minor changes. We will generally just arrive at a recurrence like line 8 and call it an algorithm, with the understanding that the actual implementation will be similar to MANHATTANTOURIST.[6]

Many problems in bioinformatics can be solved efficiently by the application of the dynamic programming technique, once they are cast as traveling in a Manhattan-like grid. For example, development of new sequence comparison algorithms often amounts to building an appropriate "Manhattan" that adequately models the specifics of a particular biological problem, and by defining the block weights that reflect the costs of mutations from one DNA sequence to another.

---

6. MANHATTANTOURIST computes the length of the longest path in the grid, but does not give the path itself. In section 6.5 we will describe a minor modification to the algorithm that returns not only the optimal length, but also the optimal path.

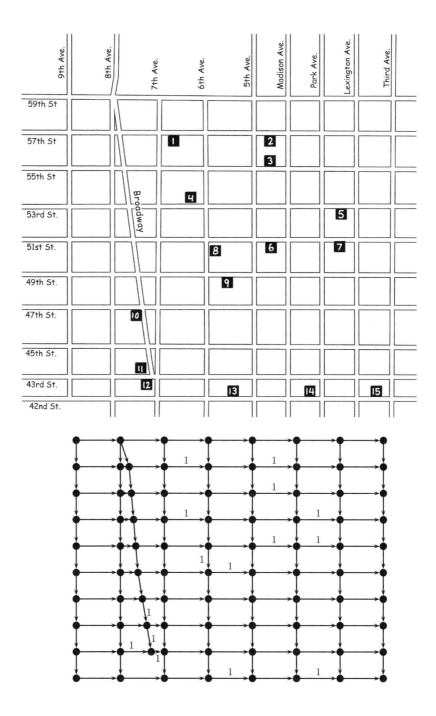

Figure 6.5   A city somewhat more like Manhattan than figure 6.4 with the complicating issue of a street that runs diagonally across the grid. Broadway cuts across several blocks. In the case of the Manhattan Tourist problem, it changes the optimal path (the optimal path in this new city has six attractions instead of five).

Unfortunately, Manhattan is not a perfectly regular grid. Broadway cuts across the borough (figure 6.5). We would like to solve a generalization of the Manhattan Tourist problem for the case in which the street map is not a regular rectangular grid. In this case, one can model any city map as a graph with vertices corresponding to the intersections of streets, and edges corresponding to the intervals of streets between the intersections. For the sake of simplicity we assume that the city blocks correspond to *directed* edges, so that the tourist can move only in the direction of the edge and that the resulting graph has no *directed cycles*.[7] Such graphs are called *directed acyclic graphs*, or *DAGs*. We assume that every edge has an associated weight (e.g., the number of attractions) and represent a graph G as a pair of two sets, V for vertices and E for edges: $G = (V, E)$. We number vertices from 1 to $|V|$ with a single integer, rather than a row-column pair as in the Manhattan problem. This does not change the generic dynamic programming algorithm other than in notation, but it allows us to represent imperfect grids. An edge from E can be specified in terms of its origin vertex u and its destination vertex v as (u, v). The following problem is simply a generalization of the Manhattan Tourist problem that is able to deal with arbitrary DAGs rather than with perfect grids.

Longest Path in a DAG Problem:
Find a longest path between two vertices in a weighted DAG.

Input: A weighted DAG G with *source* and *sink* vertices.

Output: A longest path in G from *source* to *sink*.

Not surprisingly, the Longest Path in a DAG problem can also be solved by dynamic programming. At every vertex, there may be multiple edges that "flow in" and multiple edges that "flow out." In the city analogy, any intersection may have multiple one-way streets leading in, and some other number of one-way streets exiting. We will call the number of edges entering a vertex (i.e., the number of inbound streets) the *indegree* of the vertex (i.e., intersection), and the number of edges leaving a vertex (i.e., the number of outbound streets) the *outdegree* of the vertex.

In the nicely regular case of the Manhattan problem, most vertices had

---

7. A directed cycle is a path from a vertex back to itself that respects the directions of edges. If the resulting graph contained a cycle, a tourist could start walking along this cycle revisiting the same attractions many times. In this case there is no "best" solution since a tourist may increase the number of visited attractions indefinitely.

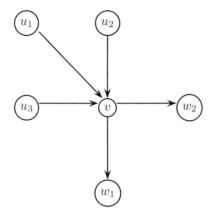

Figure 6.6   A graph with six vertices. The vertex v has indegree 3 and outdegree 2. The vertices u_1, u_2 and u_3 are all predecessors of v, and w_1 and w_2 are successors of v.

indegree 2 and outdegree 2, except for the vertices along the boundaries of the grid. In the more general DAG problem, a vertex can have an arbitrary indegree and outdegree. We will call u a *predecessor* to vertex v if $(u, v) \in E$— in other words, a predecessor of a vertex is any vertex that can be reached by traveling backwards along an inbound edge. Clearly, if v has indegree k, it has k predecessors.

Suppose a vertex v has indegree 3, and the set of predecessors of v is $\{u_1, u_2, u_3\}$ (figure 6.6). The longest path to v can be computed as follows:

$$
s_v = \max \begin{cases} s_{u_1} + \text{ weight of edge from } u_1 \text{ to } v \\ s_{u_2} + \text{ weight of edge from } u_2 \text{ to } v \\ s_{u_3} + \text{ weight of edge from } u_3 \text{ to } v \end{cases}
$$

In general, one can imagine a rather hectic city plan, but the recurrence relation remains simple, with the score s_v of the vertex v defined as follows.

$$
s_v = \max_{u \in Predecessors(v)} (s_u + \text{ weight of edge from } u \text{ to } v)
$$

Here, $Predecessors(v)$ is the set of all vertices u such that u is a predecessor of v. Since every edge participates in only a single recurrence, the running

Figure 6.7   The "Dressing in the Morning problem" represented by a DAG. Some of us have more trouble than others.

time of the algorithm is defined by the number of edges in the graph.[8] The one hitch to this plan for solving the Longest Path problem in a DAG is that one must decide on the order in which to visit the vertices while computing s. This ordering is important, since by the time vertex v is analyzed, the values s_u for *all* its predecessors must have been computed. Three popular strategies for exploring the perfect grid are displayed in figure 6.9, column by column, row by row, and diagonal by diagonal. These exploration strategies correspond to different *topological orderings* of the DAG corresponding to the perfect grid. An ordering of vertices $v_1, \ldots, v_n$ of a DAG is called *topological* if every edge (v_i, v_j) of the DAG connects a vertex with a smaller index to a vertex with a larger index, that is, $i < j$. Figure 6.7 represents a DAG that corresponds to a problem that we each face every morning. Every DAG has a topological ordering (fig. 6.8); a problem at the end of this chapter asks you to prove this fact.

---

8. A graph with vertex set V can have at most $|V|^2$ edges, but graphs arising in sequence comparison are usually sparse, with many fewer edges.

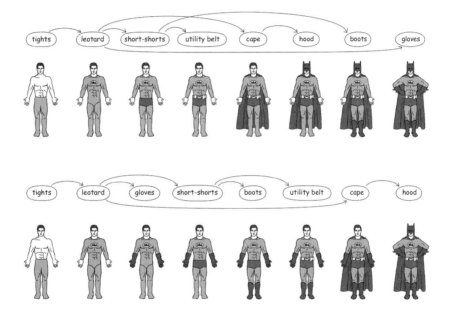

Figure 6.8   Two different ways of getting dressed in the morning corresponding to two different topological orderings of the graph in figure 6.7.

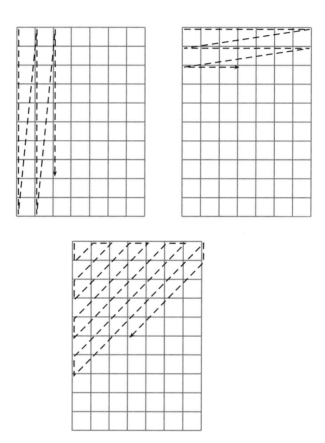

Figure 6.9 Three different strategies for filling in a dynamic programming array. The first fills in the array column by column: earlier columns are filled in before later ones. The second fills in the array row by row. The third method fills array entries along the diagonals and is useful in parallel computation.

6.4   Edit Distance and Alignments

So far, we have been vague about what we mean by "sequence similarity" or "distance" between DNA sequences. Hamming distance (introduced in chapter 4), while important in computer science, is not typically used to compare DNA or protein sequences. The Hamming distance calculation rigidly assumes that the ith symbol of one sequence is already *aligned* against the ith symbol of the other. However, it is often the case that the ith symbol in one sequence corresponds to a symbol at a different—and unknown—position in the other. For example, mutation in DNA is an evolutionary process: DNA replication errors cause substitutions, insertions, and deletions of nucleotides, leading to "edited" DNA texts. Since DNA sequences are subject to insertions and deletions, biologists rarely have the luxury of knowing in advance whether the ith symbol in one DNA sequence corresponds to the ith symbol in the other.

As figure 6.10 (a) shows, while strings **ATATATAT** and **TATATATA** are very different from the perspective of Hamming distance, they become very similar if one simply moves the second string over one place to align the $(i+1)$-st letter in **ATATATAT** against the ith letter in **TATATATA** for $1 \leq i \leq 7$. Strings **ATATATAT** and **TATAAT** present another example with more subtle similarities. Figure 6.10 (b) reveals these similarities by aligning position 2 in **ATATATAT** against position 1 in **TATAAT**. Other pairs of aligned positions are 3 against 2, 4 against 3, 5 against 4, 7 against 5, and 8 against 6 (positions 1 and 6 in **ATATATAT** remain unaligned).

In 1966, Vladimir Levenshtein introduced the notion of the *edit distance* between two strings as the minimum number of editing operations needed to transform one string into another, where the edit operations are insertion of a symbol, deletion of a symbol, and substitution of one symbol for another. For example, **TGCATAT** can be transformed into **ATCCGAT** with five editing operations, shown in figure 6.11. This implies that the edit distance between **TGCATAT** and **ATCCGAT** is at most 5. Actually, the edit distance between them is 4 because you can transform one to the other with one move fewer, as in figure 6.12.

Unlike Hamming distance, edit distance allows one to compare strings of different lengths. Oddly, Levenshtein introduced the definition of edit distance but never described an algorithm for actually finding the edit distance between two strings. This algorithm has been discovered and rediscovered many times in applications ranging from automated speech recognition to, obviously, molecular biology. Although the details of the algorithms are

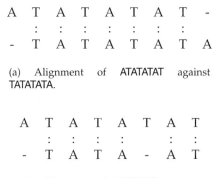

(a) Alignment of **ATATATAT** against **TATATATA**.

(b) Alignment of **ATATATAT** against **TATAAT**.

Figure 6.10 Alignment of **ATATATAT** against **TATATATA** and of **ATATATAT** against **TATAAT**.

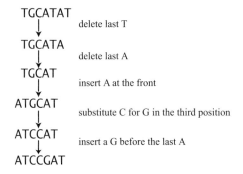

Figure 6.11 Five edit operations can take **TGCATAT** into **ATCCGAT**.

slightly different across the various applications, they are all based on dynamic programming.

The *alignment* of the strings **v** (of n characters) and **w** (of m characters, with m not necessarily the same as n) is a two-row matrix such that the first row contains the characters of **v** in order while the second row contains the characters of **w** in order, where spaces may be interspersed throughout the strings in different places. As a result, the characters in each string appear in order, though not necessarily adjacently. We also assume that no column

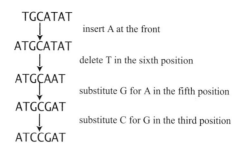

Figure 6.12   Four edit operations can also take TGCATAT into ATCCGAT.

of the alignment matrix contains spaces in both rows, so that the alignment may have at most $n + m$ columns.

A	T	-	G	T	T	A	T	-
A	T	C	G	T	-	A	-	C

Columns that contain the same letter in both rows are called *matches*, while columns containing different letters are called *mismatches*. The columns of the alignment containing one space are called *indels*, with the columns containing a space in the top row called *insertions* and the columns with a space in the bottom row *deletions*. The alignment shown in figure 6.13 (top) has five matches, zero mismatches, and four indels. The number of matches plus the number of mismatches plus the number of indels is equal to the length of the alignment matrix and must be smaller than $n + m$.

Each of the two rows in the alignment matrix is represented as a string interspersed by space symbols "−"; for example AT−GTTAT− is a representation of the row corresponding to **v** = ATGTTAT, while ATCGT−A−C is a representation of the row corresponding to **w** = ATCGTAC. Another way to represent the row AT−GTTAT− is 1 2 2 3 4 5 6 7 7, which shows the number of symbols of **v** present up to a given position. Similarly, ATCGT−A−C is represented as 1 2 3 4 5 5 6 6 7. When both rows of an alignment are represented in this way (fig. 6.13, top), the resulting matrix is

$$\binom{0}{0}\binom{1}{1}\binom{2}{2}\binom{2}{3}\binom{3}{4}\binom{4}{5}\binom{5}{5}\binom{6}{6}\binom{7}{6}\binom{7}{7}$$

Each column in this matrix is a coordinate in a two-dimensional $n \times m$ grid;

the entire alignment is simply a path

$$(0,0) \to (1,1) \to (2,2) \to (2,3) \to (3,4) \to (4,5) \to (5,5) \to (6,6) \to (7,6) \to (7,7)$$

from $(0,0)$ to (n,m) in that grid (again, see figure 6.13). This grid is similar to the Manhattan grid that we introduced earlier, where each entry in the grid looks like a city block. The main difference is that here we can move along the diagonal. We can construct a graph, this time called the *edit graph*, by introducing a vertex for every intersection of streets in the grid, shown in figure 6.13. The edit graph will aid us in calculating the edit distance.

Every alignment corresponds to a path in the edit graph, and every path in the edit graph corresponds to an alignment where every edge in the path corresponds to one column in the alignment (fig. 6.13). Diagonal edges in the path that end at vertex (i,j) in the graph correspond to the column $\begin{pmatrix} v_i \\ w_j \end{pmatrix}$, horizontal edges correspond to $\begin{pmatrix} - \\ w_j \end{pmatrix}$, and vertical edges correspond to $\begin{pmatrix} v_i \\ - \end{pmatrix}$. The alignment above can be drawn as follows.

$$
\begin{matrix}
\text{A} & \text{T} & - & \text{G} & \text{T} & \text{T} & \text{A} & \text{T} & - \\
\begin{pmatrix} 0 \\ 0 \end{pmatrix} \begin{pmatrix} 1 \\ 1 \end{pmatrix} \begin{pmatrix} 2 \\ 2 \end{pmatrix} \begin{pmatrix} 2 \\ 3 \end{pmatrix} \begin{pmatrix} 3 \\ 4 \end{pmatrix} \begin{pmatrix} 4 \\ 5 \end{pmatrix} \begin{pmatrix} 5 \\ 5 \end{pmatrix} \begin{pmatrix} 6 \\ 6 \end{pmatrix} \begin{pmatrix} 7 \\ 6 \end{pmatrix} \begin{pmatrix} 7 \\ 7 \end{pmatrix} \\
\text{A} & \text{T} & \text{G} & \text{C} & \text{T} & - & \text{A} & - & \text{C}
\end{matrix}
$$

Analyzing the merit of an alignment is equivalent to analyzing the merit of the corresponding path in the edit graph. Given any two strings, there are a large number of different alignment matrices and corresponding paths in the edit graph. Some of these have a surplus of mismatches and indels and a small number of matches, while others have many matches and few indels and mismatches. To determine the relative merits of one alignment over another, we rely on the notion of a scoring function, which takes as input an alignment matrix (or, equivalently, a path in the edit graph) and produces a score that determines the "goodness" of the alignment. There are a variety of scoring functions that we could use, but we want one that gives higher scores to alignments with more matches. The simplest functions score a column as a positive number if both letters are the same, and as a negative number if the two letters are different. The score for the whole alignment is the sum of the individual column scores. This scoring scheme amounts to

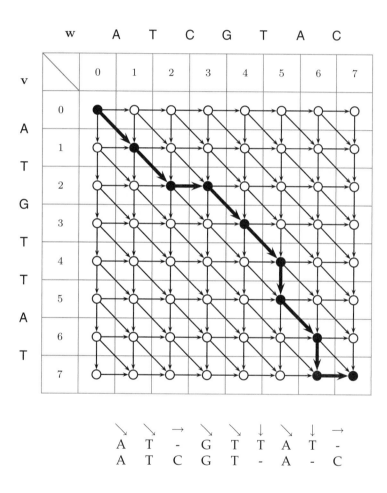

Figure 6.13 An alignment grid for **v** = ATGTTAT and **w** = ATCGTAC. Every alignment corresponds to a path in the alignment grid from $(0, 0)$ to (n, m), and every path from $(0, 0)$ to (n, m) in the alignment grid corresponds to an alignment.

assigning weights to the edges in the edit graph.

By choosing different scoring functions, we can solve different string comparison problems. If we choose the very simple scoring function of "+1 for a match, 0 otherwise," then the problem becomes that of finding the longest common subsequence between two strings, which is discussed below. Before describing how to calculate Levenshtein's edit distance, we develop the Longest Common Subsequence problem as a warm-up.

6.5   Longest Common Subsequences

The simplest form of a sequence similarity analysis is the Longest Common Subsequence (LCS) problem, where we eliminate the operation of substitution and allow only insertions and deletions. A *subsequence* of a string $\mathbf{v}$ is simply an (ordered) sequence of characters (not necessarily consecutive) from $\mathbf{v}$. For example, if $\mathbf{v} = $ ATTGCTA, then AGCA and ATTA are subsequences of $\mathbf{v}$ whereas TGTT and TCG are not.[9] A *common* subsequence of two strings is a subsequence of both of them. Formally, we define the *common subsequence* of strings $\mathbf{v} = v_1 \dots v_n$ and $\mathbf{w} = w_1 \dots w_m$ as a sequence of positions in $\mathbf{v}$,

$$1 \leq i_1 < i_2 < \cdots < i_k \leq n$$

and a sequence of positions in $\mathbf{w}$,

$$1 \leq j_1 < j_2 < \cdots < j_k \leq m$$

such that the symbols at the corresponding positions in $\mathbf{v}$ and $\mathbf{w}$ coincide:

$$v_{i_t} = w_{j_t} \text{ for } 1 \leq t \leq k.$$

For example, TCTA is a common to both ATCTGAT and TGCATA.

Although there are typically many common subsequences between two strings $\mathbf{v}$ and $\mathbf{w}$, some of which are longer than others, it is not immediately obvious how to find the longest one. If we let $s(\mathbf{v}, \mathbf{w})$ be the length of the longest common subsequence of $\mathbf{v}$ and $\mathbf{w}$, then the edit distance between $\mathbf{v}$ and $\mathbf{w}$—under the assumption that only insertions and deletions are allowed—is $d(\mathbf{v}, \mathbf{w}) = n + m - 2s(\mathbf{v}, \mathbf{w})$, and corresponds to the mini-

---

9. The difference between a sub*sequence* and a sub*string* is that a substring consists only of consecutive characters from $\mathbf{v}$, while a subsequence may pick and choose characters from $\mathbf{v}$ as long as their ordering is preserved.

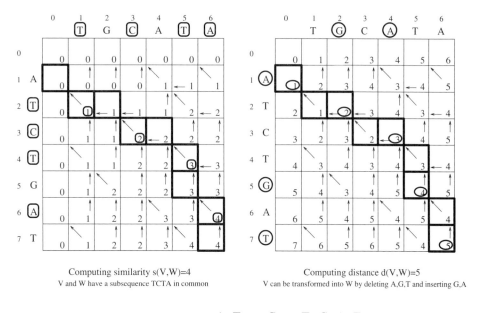

Computing similarity s(V,W)=4
V and W have a subsequence TCTA in common

Computing distance d(V,W)=5
V can be transformed into W by deleting A,G,T and inserting G,A

Alignment:
```
A T - C - T G A T
- T G C A T - A -
```

Figure 6.14   Dynamic programming algorithm for computing the longest common subsequence.

mum number of insertions and deletions needed to transform **v** into **w**. Figure 6.14 (bottom) presents an LCS of length 4 for the strings **v** = ATCTGAT and **w** = TGCATA and a shortest sequence of two insertions and three deletions transforming **v** into **w** (shown by "-" in the figure). The LCS problem follows.

Longest Common Subsequence Problem:
Find the longest subsequence common to two strings.

   Input: Two strings, **v** and **w**.

   Output: The longest common subsequence of **v** and **w**.

What do the LCS problem and the Manhattan Tourist problem have in common? Every common subsequence corresponds to an alignment with no

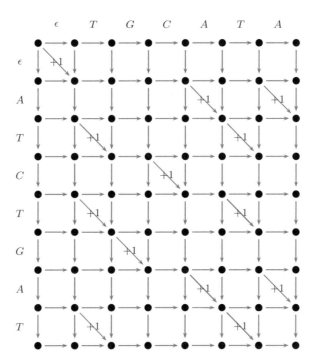

Figure 6.15   An LCS edit graph.

mismatches. This can be obtained simply by removing all diagonal edges from the edit graph whose characters do not match, thus transforming it into a graph like that shown in figure 6.15. We further illustrate the relationship between the Manhattan Tourist problem and the LCS Problem by showing that these two problems lead to very similar recurrences.

Define $s_{i,j}$ to be the length of an LCS between $v_1 \ldots v_i$, the i-prefix of $\mathbf{v}$ and $w_1 \ldots w_j$, the j-prefix of $\mathbf{w}$. Clearly, $s_{i,0} = s_{0,j} = 0$ for all $1 \leq i \leq n$ and

$1 \leq j \leq m$. One can see that $s_{i,j}$ satisfies the following recurrence:

$$s_{i,j} = \max \begin{cases} s_{i-1,j} \\ s_{i,j-1} \\ s_{i-1,j-1} + 1, & \text{if } v_i = w_j \end{cases}$$

The first term corresponds to the case when v_i is not present in the LCS of the i-prefix of $\mathbf{v}$ and j-prefix of $\mathbf{w}$ (this is a deletion of v_i); the second term corresponds to the case when w_j is not present in this LCS (this is an insertion of w_j); and the third term corresponds to the case when both v_i and w_j are present in the LCS (v_i *matches* w_j). Note that one can "rewrite" these recurrences by adding some zeros here and there as

$$s_{i,j} = \max \begin{cases} s_{i-1,j} + 0 \\ s_{i,j-1} + 0 \\ s_{i-1,j-1} + 1, & \text{if } v_i = w_j \end{cases}$$

This recurrence for the LCS computation is like the recurrence given at the end of the section 6.3, if we were to build a particularly gnarly version of Manhattan and gave horizontal and vertical edges weights of 0, and set the weights of diagonal (matching) edges equal to +1 as in figure 6.15.

In the following, we use **s** to represent our dynamic programming table, the data structure that we use to fill in the dynamic programming recurrence. The length of an LCS between **v** and **w** can be read from the element (n, m) of the dynamic programming table, but to reconstruct the LCS from the dynamic programming table, one must keep some additional information about which of the three quantities, $s_{i-1,j}$, $s_{i,j-1}$, or $s_{i-1,j-1} + 1$, corresponds to the maximum in the recurrence for $s_{i,j}$. The following algorithm achieves this goal by introducing *backtracking pointers* that take one of the three values $\leftarrow$, $\uparrow$, or $\nwarrow$. These specify which of the above three cases holds, and are stored in a two-dimensional array **b** (see figure 6.14).

LCS($\mathbf{v}, \mathbf{w}$)
1   **for** $i \leftarrow 0$ **to** n
2        $s_{i,0} \leftarrow 0$
3   **for** $j \leftarrow 1$ **to** m
4        $s_{0,j} \leftarrow 0$
5   **for** $i \leftarrow 1$ **to** n
6   **for** $j \leftarrow 1$ **to** m

7            $s_{i,j} \leftarrow \max \begin{cases} s_{i-1,j} \\ s_{i,j-1} \\ s_{i-1,j-1} + 1, & \text{if } v_i = w_j \end{cases}$

8            $b_{i,j} \leftarrow \begin{cases} \text{``} \uparrow \text{''} & \text{if } s_{i,j} = s_{i-1,j} \\ \text{``} \leftarrow \text{''} & \text{if } s_{i,j} = s_{i,j-1} \\ \text{``} \nwarrow \text{''}, & \text{if } s_{i,j} = s_{i-1,j-1} + 1 \end{cases}$

9   **return** ($s_{n,m}, \mathbf{b}$)

The following recursive program prints out the longest common subsequence using the information stored in **b**. The initial invocation that prints the solution to the problem is PRINTLCS($\mathbf{b}, \mathbf{v}, n, m$).

PRINTLCS($\mathbf{b}, \mathbf{v}, i, j$)
 1   **if** $i = 0$ **or** $j = 0$
 2   **return**
 3   **if** $b_{i,j} = \text{``} \nwarrow \text{''}$
 4   PRINTLCS($\mathbf{b}, \mathbf{v}, i - 1, j - 1$)
 5   **print** v_i
 6   **else**
 7   **if** $b_{i,j} = \text{``} \uparrow \text{''}$
 8   PRINTLCS($\mathbf{b}, \mathbf{v}, i - 1, j$)
 9   **else**
10   PRINTLCS($\mathbf{b}, \mathbf{v}, i, j - 1$)

The dynamic programming table in figure 6.14 (left) presents the computation of the similarity score $s(\mathbf{v}, \mathbf{w})$ between $\mathbf{v}$ and $\mathbf{w}$, while the table on the right presents the computation of the edit distance between $\mathbf{v}$ and $\mathbf{w}$ under the assumption that insertions and deletions are the only allowed operations. The edit distance $d(\mathbf{v}, \mathbf{w})$ is computed according to the initial conditions $d_{i,0} = i$, $d_{0,j} = j$ for all $1 \leq i \leq n$ and $1 \leq j \leq m$ and the following recurrence:

$$d_{i,j} = \min \begin{cases} d_{i-1,j} + 1 \\ d_{i,j-1} + 1 \\ d_{i-1,j-1}, & \text{if } v_i = w_j \end{cases}$$

6.6 Global Sequence Alignment

The LCS problem corresponds to a rather restrictive scoring that awards 1 for matches and does not penalize indels. To generalize scoring, we extend the k-letter alphabet $\mathcal{A}$ to include the gap character "$-$", and consider an arbitrary $(k+1) \times (k+1)$ *scoring matrix* δ, where k is typically 4 or 20 depending on the type of sequences (DNA or protein) one is analyzing. The score of the column $\binom{x}{y}$ in the alignment is $\delta(x, y)$ and the alignment score is defined as the sum of the scores of the columns. In this way we can take into account scoring of mismatches and indels in the alignment. Rather than choosing a particular scoring matrix and then resolving a restated alignment problem, we will pose a general Global Alignment problem that takes the scoring matrix as input.

Global Alignment Problem:
Find the best alignment between two strings under a given scoring matrix.

Input: Strings **v**, **w** and a scoring matrix δ.

Output: An alignment of **v** and **w** whose score (as defined by the matrix δ) is maximal among all possible alignments of **v** and **w**.

The corresponding recurrence for the score $s_{i,j}$ of an optimal alignment between the i-prefix of **v** and j-prefix of **w** is as follows:

$$s_{i,j} = \max \begin{cases} s_{i-1,j} + \delta(v_i, -) \\ s_{i,j-1} + \delta(-, w_j) \\ s_{i-1,j-1} + \delta(v_i, w_j) \end{cases}$$

When mismatches are penalized by some constant $-\mu$, indels are penalized by some other constant $-\sigma$, and matches are rewarded with $+1$, the resulting score is

$$\#matches - \mu \cdot \#mismatches - \sigma \cdot \#indels$$

The corresponding recurrence can be rewritten as

$$
s_{i,j} = \max \begin{cases}
s_{i-1,j} - \sigma \\
s_{i,j-1} - \sigma \\
s_{i-1,j-1} - \mu, \text{ if } v_i \neq w_j \\
s_{i-1,j-1} + 1, \text{ if } v_i = w_j
\end{cases}
$$

We can again store similar "backtracking pointer" information while calculating the dynamic programming table, and from this reconstruct the alignment. We remark that the LCS problem is the Global Alignment problem with the parameters $\mu = 0$, $\sigma = 0$ (or, equivalently, $\mu = \infty$, $\sigma = 0$).

6.7   Scoring Alignments

While the scoring matrices for DNA sequence comparison are usually defined only by the parameters μ (mismatch penalty) and σ (indel penalty), scoring matrices for sequences in the amino acid alphabet of proteins are quite involved. The common matrices for protein sequence comparison, *point accepted mutations (PAM)* and *block substitution (BLOSUM)*, reflect the frequency with which amino acid x replaces amino acid y in evolutionarily related sequences.

Random mutations of the nucleotide sequence within a gene may change the amino acid sequence of the corresponding protein. Some of these mutations do not drastically alter the protein's structure, but others do and impair the protein's ability to function. While the former mutations usually do not affect the fitness of the organism, the latter often do. Therefore some amino acid substitutions are commonly found throughout the process of molecular evolution and others are rare: Asn, Asp, Glu, and Ser are the most "mutable" amino acids while Cys and Trp are the least mutable. For example, the probability that Ser mutates into Phe is roughly three times greater than the probability that Trp mutates into Phe. Knowledge of the types of changes that are most and least common in molecular evolution allows biologists to construct the amino acid scoring matrices and to produce biologically adequate sequence alignments. As a result, in contrast to nucleotide sequence comparison, the optimal alignments of amino acid sequences may have very few matches (if any) but still represent biologically adequate alignments. The entry of amino acid scoring matrix $\delta(i, j)$ usually reflects how often the amino acid i substitutes the amino acid j in the alignments of related protein sequences. If one is provided with a large set of alignments of related sequences, then computing $\delta(i,j)$ simply amounts to counting how many times the amino acid i is aligned with amino acid j. A "minor" complication is that to build this set of biologically adequate alignments one needs to know the scoring matrix! Fortunately, in many cases the alignment of very similar sequences is so obvious that it can be constructed even without a scoring matrix, thus resolving this predicament. For example, if proteins are 90% identical, even a naive scoring matrix (e.g., a matrix that gives premium $+1$ for matches and penalties -1 for mismatches and indels) would do the job. After these "obvious" alignments are constructed they can be used to compute a scoring matrix δ that can be used iteratively to construct less obvious alignments.

This simplified description hides subtle details that are important in the construction of scoring matrices. The probability of Ser mutating into Phe in proteins that diverged 15 million years ago (e.g., related proteins in mouse and rat) is smaller than the probability of the Ser $\rightarrow$ Phe mutation in proteins that diverged 80 million years ago (e.g., related proteins in mouse and human). This observation implies that the best scoring matrices to compare two proteins depends on how similar these organisms are.

Biologists get around this problem by first analyzing extremely similar proteins, for example, proteins that have, on average, only one mutation per 100 amino acids. Many proteins in human and chimpanzee fulfill this requirement. Such sequences are defined as being *one PAM unit diverged* and to a first approximation one can think of a PAM unit as the amount of time in which an "average" protein mutates 1% of its amino acids. The *PAM 1* scoring matrix is defined from many alignments of extremely similar proteins as follows.

Given a set of base alignments, define $f(i,j)$ as the total number of times amino acids i and j are aligned against each other, divided by the total number of aligned positions. We also define $g(i,j)$ as $\frac{f(i,j)}{f(i)}$, where $f(i)$ is the frequency of amino acid i in all proteins from the data set. $g(i,j)$ defines the probability that an amino acid i mutates into amino acid j within 1 PAM unit. The (i,j) entry of the *PAM 1* matrix is defined as $\delta(i,j) = \log \frac{f(i,j)}{f(i)\cdot f(j)} = \log \frac{g(i,j)}{f(j)}$ ($f(i) \cdot f(j)$ stands for the frequency of aligning amino acid i against amino acid j that one expects simply by chance). The *PAM n* matrix can be defined as the result of applying the PAM 1 matrix n times. If $\mathbf{g}$ is the 20×20 matrix of frequencies $g(i,j)$, then $\mathbf{g}^n$ (multiplying this matrix by itself n times) gives the probability that amino acid i mutates into amino acid j during n PAM units. The (i,j) entry of the PAM n matrix is defined as

$\log \frac{g_{i,j}^n}{f(j)}$.

For large n, the resulting PAM matrices often allow one to find related proteins even when there are practically no matches in the alignment. In this case, the underlying nucleotide sequences are so diverged that their comparison usually fails to find any statistically significant similarities. For example, the similarity between the cancer-causing ν-sis oncogene and the growth factor PDGF would probably have remained undetected had Russell Doolittle and colleagues not transformed the nucleotide sequences into amino acid sequences prior to performing the comparison.

6.8   Local Sequence Alignment

The Global Alignment problem seeks similarities between two entire strings. This is useful when the similarity between the strings extends over their entire length, for example, in protein sequences from the same protein family. These protein sequences are often very conserved and have almost the same length in organisms ranging from fruit flies to humans. However, in many biological applications, the score of an alignment between two substrings of **v** and **w** might actually be larger than the score of an alignment between the entireties of **v** and **w**.

For example, *homeobox* genes, which regulate embryonic development, are present in a large variety of species. Although homeobox genes are very different in different species, one region in each gene—called the *homeodomain*—is highly conserved. The question arises how to find this conserved area and ignore the areas that show little similarity. In 1981 Temple Smith and Michael Waterman proposed a clever modification of the global sequence alignment dynamic programming algorithm that solves the Local Alignment problem.

Figure 6.16 presents the comparison of two hypothetical genes **v** and **w** of the same length with a conserved domain present at the beginning of **v** and at the end of **w**. For simplicity, we will assume that the conserved domains in these two genes are identical and cover one third of the entire length, n, of these genes. In this case, the path from *source* to *sink* capturing the similarity between the homeodomains will include approximately $\frac{2}{3}n$ horizontal edges, $\frac{1}{3}n$ diagonal match edges (corresponding to homeodomains), and $\frac{2}{3}n$ vertical edges. Therefore, the score of this path is

$$-\frac{2}{3}n\sigma + \frac{1}{3}n - \frac{2}{3}n\sigma = n\left(\frac{1}{3} - \frac{4}{3}\sigma\right)$$

However, this path contains so many indels that it is unlikely to be the highest scoring alignment. In fact, biologically irrelevant diagonal paths from the source to the sink will likely have a higher score than the biologically relevant alignment, since mismatches are usually penalized less than indels. The expected score of such a diagonal path is $n(\frac{1}{4} - \frac{3}{4}\mu)$ since every diagonal edge corresponds to a match with probability $\frac{1}{4}$ and mismatch with probability $\frac{3}{4}$. Since $(\frac{1}{3} - \frac{4}{3}\sigma) < (\frac{1}{4} - \frac{3}{4}\mu)$ for many settings of indel and mismatch penalties, the global alignment algorithm will miss the correct solution of the real biological problem, and is likely to output a biologically irrelevant near-diagonal path. Indeed, figure 6.16 bears exactly this observation.

When biologically significant similarities are present in certain parts of DNA fragments and are not present in others, biologists attempt to maximize the alignment score $s(v_i \dots v_{i'}, w_j \dots w_{j'})$, over all substrings $v_i \dots v_{i'}$ of **v** and $w_j \dots w_{j'}$ of **w**. This is called the Local Alignment problem since the alignment does not necessarily extend over the entire string length as it does in the Global Alignment problem.

Local Alignment Problem:
Find the best local alignment between two strings.

Input: Strings **v** and **w** and a scoring matrix δ.

Output: Substrings of **v** and **w** whose global alignment, as defined by δ, is maximal among all global alignments of all substrings of **v** and **w**.

The solution to this seemingly harder problem lies in the realization that the Global Alignment problem corresponds to finding the longest local path between vertices $(0,0)$ and (n,m) in the edit graph, while the Local Alignment problem corresponds to finding the longest path among paths between *arbitrary vertices* (i,j) and (i',j') in the edit graph. A straightforward and inefficient approach to this problem is to find the longest path between every pair of vertices (i,j) and (i',j'), and then to select the longest of these computed paths.[10] Instead of finding the longest path from every vertex (i,j) to every other vertex (i',j'), the Local Alignment problem can be reduced to finding the longest paths from the *source* (0,0) to every other vertex by

---

10. This will result in a very slow algorithm with $O(n^4)$ running time: there are roughly n^2 pairs of vertices (i,j) and computing local alignments starting at each of them typically takes $O(n^2)$ time.

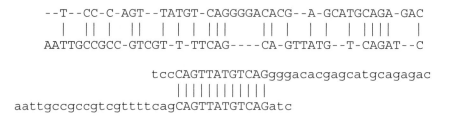

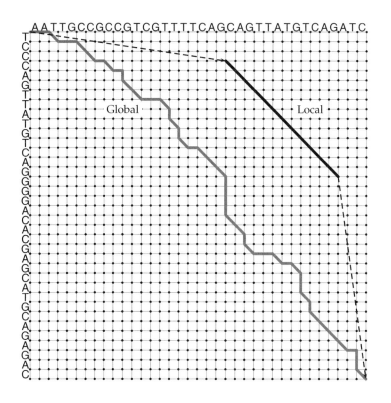

Figure 6.16   (a) Global and (b) local alignments of two hypothetical genes that each have a conserved domain. The local alignment has a much worse score according to the global scoring scheme, but it correctly locates the conserved domain.

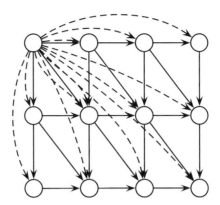

Figure 6.17 The Smith-Waterman local alignment algorithm introduces edges of weight 0 (here shown with dashed lines) from the source vertex $(0,0)$ to every other vertex in the edit graph.

adding edges of weight 0 in the edit graph. These edges make the source vertex (0,0) a predecessor of every vertex in the graph and provide a "free ride" from the source to any other vertex (i,j). A small difference in the following recurrence reflects this transformation of the edit graph (shown in figure 6.17):

$$s_{i,j} = \max \begin{cases} 0 \\ s_{i-1,j} + \delta(v_i, -) \\ s_{i,j-1} + \delta(-, w_j) \\ s_{i-1,j-1} + \delta(v_i, w_j) \end{cases}$$

The largest value of $s_{i,j}$ over the whole edit graph represents the score of the best local alignment of **v** and **w**; recall that in the Global Alignment problem, we simply looked at $s_{n,m}$. The difference between local and global alignment is illustrated in figure 6.16 (top).

Optimal local alignment reports only the longest path in the edit graph. At the same time, several local alignments may have biological significance and methods have been developed to find the k best nonoverlapping local alignments. These methods are particularly important for comparison of multidomain proteins that share similar blocks that have been shuffled in one protein compared to another. In this case, a single local alignment representing all significant similarities may not exist.

6.9   Alignment with Gap Penalties

Mutations are usually caused by errors in DNA replication. Nature frequently deletes or inserts entire substrings as a unit, as opposed to deleting or inserting individual nucleotides. A *gap* in an alignment is defined as a contiguous sequence of spaces in one of the rows. Since insertions and deletions of substrings are common evolutionary events, penalizing a gap of length x as $-\sigma x$ is cruel and unusual punishment. Many practical alignment algorithms use a softer approach to gap penalties and penalize a gap of x spaces by a function that grows slower than the sum of penalties for x indels.

To this end, we define *affine gap penalties* to be a linearly weighted score for large gaps. We can set the score for a gap of length x to be $-(\rho + \sigma x)$, where $\rho > 0$ is the penalty for the introduction of the gap and $\sigma > 0$ is the penalty for each symbol in the gap (ρ is typically large while σ is typically small). Though this may seem to be complicating our alignment approach, it turns out that the edit graph representation of the problem is robust enough to accommodate it.

Affine gap penalties can be accommodated by adding "long" vertical and horizontal edges in the edit graph (e.g., an edge from (i, j) to $(i + x, j)$ of length $-(\rho + \sigma x)$ and an edge from (i, j) to $(i, j + x)$ of the same length) from each vertex to every other vertex that is either east or south of it. We can then apply the same algorithm as before to compute the longest path in this graph. Since the number of edges in the edit graph for affine gap penalties increases, at first glance it looks as though the running time for the alignment algorithm also increases from $O(n^2)$ to $O(n^3)$, where n is the longer of the two string lengths.[11] However, the following three recurrences keep the running time down:

$$\overset{\downarrow}{s}_{i,j} = \max \begin{cases} \overset{\downarrow}{s}_{i-1,j} - \sigma \\ s_{i-1,j} - (\rho + \sigma) \end{cases}$$

$$\vec{s}_{i,j} = \max \begin{cases} \vec{s}_{i,j-1} - \sigma \\ s_{i,j-1} - (\rho + \sigma) \end{cases}$$

---

11. The complexity of the corresponding Longest Path in a DAG problem is defined by the number of edges in the graph. Adding long horizontal and vertical edges imposed by affine gap penalties increases the number of edges by a factor of n.

$$s_{i,j} = \max \begin{cases} s_{i-1,j-1} + \delta(v_i, w_j) \\ \overset{\downarrow}{s}_{i,j} \\ \vec{s}_{i,j} \end{cases}$$

The variable $\overset{\downarrow}{s}_{i,j}$ computes the score for alignment between the i-prefix of **v** and the j-prefix of **w** ending with a deletion (i.e., a gap in **w**), while the variable $\vec{s}_{i,j}$ computes the score for alignment ending with an insertion (i.e., a gap in **v**). The first term in the recurrences for $\overset{\downarrow}{s}_{i,j}$ and $\vec{s}_{i,j}$ corresponds to extending the gap, while the second term corresponds to initiating the gap. Essentially, $\overset{\downarrow}{s}_{i,j}$ and $\vec{s}_{i,j}$ are the scores of optimal paths that arrive at vertex (i, j) via vertical and horizontal edges correspondingly.

Figure 6.18 further explains how alignment with affine gap penalties can be reduced to the Manhattan Tourist problem in the appropriate city grid. In this case the city is built on three levels: the bottom level built solely with vertical ↓ edges with weight $-\sigma$; the middle level built with diagonal edges of weight $\delta(v_i, w_j)$; and the upper level, which is built from horizontal edges → with weight $-\sigma$. The lower level corresponds to gaps in sequence **w**, the middle level corresponds to matches and mismatches, and the upper level corresponds to gaps in sequence **v**. Also, in this graph there are two edges from each vertex $(i, j)_{middle}$ at the middle level that connect this vertex with vertex $(i + 1, j)_{lower}$ at the lower level and with vertex $(i, j + 1)_{upper}$ at the upper level. These edges model a start of the gap and have weight $-(\rho + \sigma)$. Finally, one has to introduce zero-weight edges connecting vertices $(i, j)_{lower}$ and $(i, j)_{upper}$ with vertex $(i, j)_{middle}$ at the middle level (these edges model the end of the gap). In effect, we have created a rather complicated graph, but the same algorithm works with it.

We have now introduced a number of pairwise sequence comparison problems and shown that they can all be solved by what is essentially the same dynamic programming algorithm applied to a suitably built Manhattan-style city. We will now consider other applications of dynamic programming in bioinformatics.

6.10   Multiple Alignment

The goal of protein sequence comparison is to discover structural or functional similarities among proteins. Biologically similar proteins may not exhibit a strong sequence similarity, but we would still like to recognize resem-

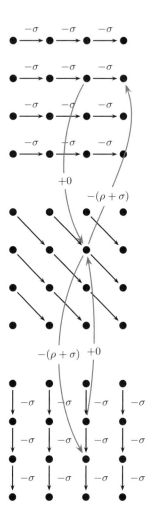

Figure 6.18   A three-level edit graph for alignment with affine gap penalties. Every vertex (i, j) in the middle level has one outgoing edge to the upper level, one outgoing edge to the lower level, and one incoming edge each from the upper and lower levels.

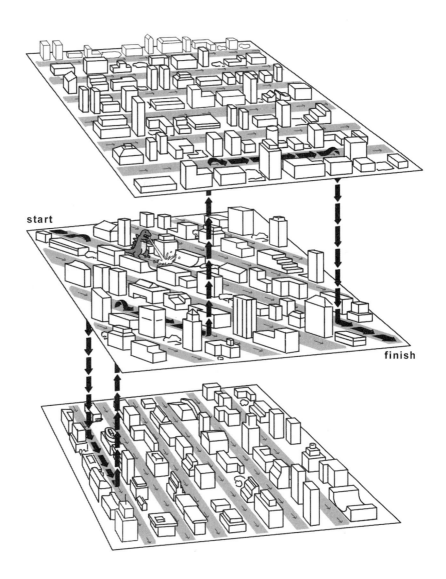

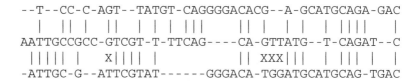

Figure 6.19   Multiple alignment of three sequences.

blance even when the sequences share only weak similarities.[12] If sequence similarity is weak, pairwise alignment can fail to identify biologically related sequences because weak pairwise similarities may fail statistical tests for significance. However, simultaneous comparison of many sequences often allows one to find similarities that are invisible in pairwise sequence comparison.

Let $\mathbf{v}_1, \ldots, \mathbf{v}_k$ be k strings of length $n_1, \ldots, n_k$ over an alphabet $\mathcal{A}$. Let $\mathcal{A}'$ denote the extended alphabet $\mathcal{A} \bigcup \{-\}$, where '$-$' denotes the space character (reserved for insertions and deletions). A *multiple alignment* of strings $\mathbf{v}_1, \ldots, \mathbf{v}_k$ is specified by a $k \times n$ matrix A, where $n \geq \max_{1 \leq i \leq k} n_i$. Each element of the matrix is a member of $\mathcal{A}'$, and each row i contains the characters of $\mathbf{v_i}$ in order, interspersed with $n - n_i$ spaces (figure 6.19). We also assume that every column of the multiple alignment matrix contains at least one symbol from $\mathcal{A}$, that is, no column in a multiple alignment contains only spaces. The multiple alignment matrix we have constructed is a generalization of the pairwise alignment matrix to $k > 2$ sequences. The score of a multiple alignment is defined to be the sum of scores of the columns, with the optimal alignment being the one that maximizes the score. Just as it was in section 4.5, the consensus of an alignment is a string of the most common characters in each column of the multiple alignment. At this point, we will use a very general scoring function that is defined by a k-dimensional matrix δ of size $|\mathcal{A}'| \times \ldots \times |\mathcal{A}'|$ that describes the scores of all possible combinations of k symbols from $\mathcal{A}'$.[13]

A straightforward dynamic programming algorithm in the k-dimensional edit graph formed from k strings solves the Multiple Alignment problem.

---

12. Sequences that code for proteins that perform the same function are likely to be somehow related but it may be difficult to decide whether this similarity is significant or happens just by chance.
13. This is a k-dimensional scoring matrix rather than the two-dimensional $|\mathcal{A}'| \times |\mathcal{A}'|$ matrix for pairwise alignment (which is a multiple alignment with $k = 2$).

For example, suppose that we have three sequences **u**, **v**, and **w**, and that we want to find the "best" alignment of all three. Every multiple alignment of three sequences corresponds to a path in the three-dimensional Manhattan-like edit graph. In this case, one can apply the same logic as we did for two dimensions to arrive at a dynamic programming recurrence, this time with more terms to consider. To get to vertex (i, j, k) in a three-dimensional edit graph, you could come from any of the following predecessors (note that $\delta(x, y, z)$ denotes the score of a column with letters x, y, and z, as in figure 6.20):

1. $(i - 1, j, k)$ for score $\delta(u_i, -, -)$

2. $(i, j - 1, k)$ for score $\delta(-, v_j, -)$

3. $(i, j, k - 1)$ for score $\delta(-, -, w_k)$

4. $(i - 1, j - 1, k)$ for score $\delta(u_i, v_j, -)$

5. $(i - 1, j, k - 1)$ for score $\delta(u_i, -, w_k)$

6. $(i, j - 1, k - 1)$ for score $\delta(-, v_j, w_k)$

7. $(i - 1, j - 1, k - 1)$ for score $\delta(u_i, v_j, w_k)$

We create a three-dimensional dynamic programming array s and it is easy to see that the recurrence for $s_{i,j,k}$ in the three-dimensional case is similar to the recurrence in the two-dimensional case (fig. 6.21). Namely,

$$
s_{i,j,k} = \max \begin{cases}
s_{i-1,j,k} & +\delta(v_i, -, -) \\
s_{i,j-1,k} & +\delta(-, w_j, -) \\
s_{i,j,k-1} & +\delta(-, -, u_k) \\
s_{i-1,j-1,k} & +\delta(v_i, w_j, -) \\
s_{i-1,j,k-1} & +\delta(v_i, -, u_k) \\
s_{i,j-1,k-1} & +\delta(-, w_j, u_k) \\
s_{i-1,j-1,k-1} & +\delta(v_i, w_j, u_k)
\end{cases}
$$

Unfortunately, in the case of k sequences, the running time of this approach is $O((2n)^k)$, so some improvements of the exact algorithm, and many heuristics for suboptimal multiple alignments, have been proposed. A good heuristic would be to compute all $\binom{k}{2}$ optimal pairwise alignments between every pair of strings and then combine them together in such a way that pairwise alignments induced by the multiple alignment are close to the optimal

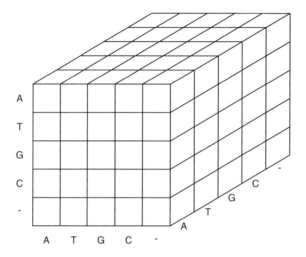

Figure 6.20   The scoring matrix, δ, used in a three-sequence alignment.

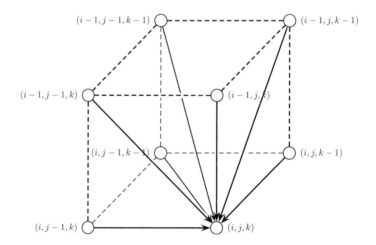

Figure 6.21   A cell in the alignment graph between three sequences.

ones. Unfortunately, it is not always possible to combine optimal pairwise alignments into a multiple alignment since some pairwise alignments may be incompatible. For example, figure 6.22 (a) shows three sequences whose optimal pairwise alignment can be combined into a multiple alignment, whereas (b) shows three sequences that cannot be combined. As a result, some multiple alignment algorithms attempt to combine some compatible subset of optimal pairwise alignments into a multiple alignment.

Another approach to do this uses one particularly strong pairwise alignment as a building block for the multiple k-way alignment, and iteratively adds one string to the growing multiple alignment. This greedy *progressive multiple alignment* heuristic selects the pair of strings with greatest similarity and merges them together into a new string following the principle "once a gap, always a gap."[14] As a result, the multiple alignment of k sequences is reduced to the multiple alignment of $k-1$ sequences. The motivation for the choice of the closest strings at the early steps of the algorithm is that close strings often provide the most reliable information about a real alignment. Many popular iterative multiple alignment algorithms, including the tool **CLUSTAL**, use similar strategies.

Although progressive multiple alignment algorithms work well for very close sequences, there are no performance guarantees for this approach. The problem with progressive multiple alignment algorithms like **CLUSTAL** is that they may be misled by some spuriously strong pairwise alignment, in effect, a bad seed. If the very first two sequences picked for building multiple alignment are aligned in a way that is incompatible with the optimal multiple alignment, the error in this initial pairwise alignment will propagate all the way through to the whole multiple alignment. Many multiple alignment algorithms have been proposed, and even with systematic deficiencies such as the above they remain quite useful in computational biology.

We have described multiple alignment for k sequences as a generalization of the Pairwise Alignment problem, which assumed the existence of a k-dimensional scoring matrix δ. Since such k-dimensional scoring matrices are not very practical, we briefly describe two other scoring approaches that are more biologically relevant. The choice of the scoring function can drastically affect the quality of the resulting alignment, and no single scoring approach is perfect in all circumstances.

The columns of a multiple alignment of k sequences describe a path of

---

14. Essentially, this principle states that once a gap has been introduced into the alignment it will never close, even if that would lead to a better overall score.

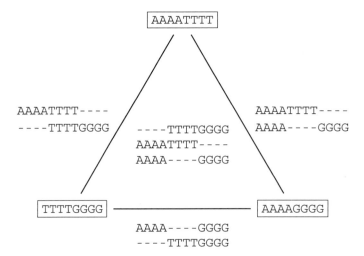

(a) Compatible pairwise alignments

(b) Incompatible pairwise alignments

Figure 6.22 Given three sequences, it might be possible to combine their pairwise alignment into a multiple alignment (a), but it might not be (b).

edges in a k-dimensional version of the Manhattan gridlike edit graph. The weights of these edges are determined by the scoring function δ. Intuitively, we want to assign higher scores to the columns with a low variation in letters, such that high scores correspond to highly conserved sequences. For example, in the *Multiple Longest Common Subsequence* problem, the score of a column is set to 1 if all the characters in the column are the same, and 0 if even one character disagrees.

In the more statistically motivated *entropy* approach, the score of a multiple alignment is defined as the sum of the entropies of the columns, which are defined to be[15]

$$\sum_{x \in \mathcal{A}'} p_x \log p_x$$

where p_x is the frequency of letter $x \in \mathcal{A}'$ in a given column. In this case, the more conserved the column, the larger the entropy score. For example, a column that has each of the 4 nucleotides present $\frac{k}{4}$ times will have an entropy score of $4\frac{1}{4} \log \frac{1}{4} = -2$, while a completely conserved column (as in the multiple LCS problem) would have entropy 0. Finding the longest path in the k-dimensional edit graph corresponds to finding the multiple alignment with the largest entropy score.

While entropy captures some statistical notion of a good alignment, it can be hard to design efficient algorithms that optimize this scoring function. Another popular scoring approach is the *Sum-of-Pairs score (SP-score)*. Any multiple alignment A of k sequences $\mathbf{v_1}, \ldots, \mathbf{v_k}$ forces a pairwise alignment between any two sequences $\mathbf{v_i}$ and $\mathbf{v_j}$ of score $s_A(\mathbf{v_i}, \mathbf{v_j})$.[16] The SP-score for a multiple alignment A is given by $\sum_{1 \leq i < j \leq k} s_A(\mathbf{v_i}, \mathbf{v_j})$. In this definition, the score of an alignment A is built from the scores of all pairs of strings in the alignment.

6.11   Gene Prediction

In 1961 Sydney Brenner and Francis Crick demonstrated that every triplet of nucleotides (codon) in a gene codes for one amino acid in the corresponding protein. They were able to introduce deletions in DNA and observed that deletion of a single nucleotide or two consecutive nucleotides in a gene dramatically alters its protein product. Paradoxically, deleting three consecutive

---

15. The correct way to define entropy is to take the negative of this expression, but the definition above allows us to deal with a maximization rather than a minimization problem.
16. We remark that the resulting "forced" alignment is not necessarily optimal.

nucleotides results in minor changes in the protein. For example, the phrase
THE SLY FOX AND THE SHY DOG (written in triplets) turns into gibber-
ish after deleting one letter (THE SYF OXA NDT HES HYD OG) or two letters (THE SFO XAN DTH ESH YDO G), but makes some sense after deleting three nucleotides THE SOX AND THE SHY DOG. Inspired by this experiment Charles Yanofsky proved that a gene and its protein product are
collinear, that is, the first codon in the gene codes for the first amino acid in
the protein, the second codon codes for the second amino acid (rather than,
say, the seventeenth), and so on. Yanofsky's ingenious experiment was so
influential that nobody even questioned whether codons are represented by
continuous stretches in DNA, and for the subsequent fifteen years biologists
believed that a protein was encoded by a long string of contiguous triplets.
However, the discovery of split human genes in 1977 proved that genes are
often represented by a *collection* of substrings, and raised the computational
problem of predicting the locations of genes in a genome given only the genomic DNA sequence.

The human genome is larger and more complex than bacterial genomes.
This is not particularly surprising since one would expect to find more genes
in humans than in bacteria. However, the genome size of many eukaryotes
does not appear to be related to an organism's genetic complexity; for example, the salamander genome is ten times larger than the human genome. This
apparent paradox was resolved by the discovery that many organisms contain not only genes but also large amounts of so-called *junk DNA* that does
not code for proteins at all. In particular, most human genes are broken into
pieces called *exons* that are separated by this junk DNA. The difference in the
sizes of the salamander and human genomes thus presumably reflects larger
amounts of junk DNA and repeats in the salamander genome.

Split genes are analogous to a magazine article that begins on page 1, continues on page 13, then takes up again on pages 43, 51, 74, 80, and 91, with
pages of advertising appearing in between. We do not understand why these
jumps occur. and a significant portion of the human genome is this junk "advertising" that separates exons.

More confusing is that the jumps between different parts of split genes
are inconsistent from species to species. A gene in an insect edition of the
genome will be organized differently than the related gene in a worm genome.
The number of parts (exons) may be different: the information that appears
in one part in the human edition may be broken up into two in the mouse
version, or vice versa. While the genes themselves are related, they may be
quite different in terms of the parts' structure.

Split genes were first discovered in 1977 in the laboratories of Phillip Sharp and Richard Roberts during studies of the adenovirus. The discovery was such a surprise that the paper by Roberts's group had an unusually catchy title for the journal *Cell*: "An Amazing Sequence Arrangement at the 5' End of Adenovirus 2 Messenger RNA." Sharp's group focused their experiments on an mRNA[17] that encodes a viral protein known as *hexon*. To map the hexon mRNA in the viral genome, mRNA was hybridized to adenovirus DNA and the hybrid molecules were analyzed by electron microscopy. Strikingly, the mRNA-DNA hybrids formed in this experiment displayed three loop structures, rather than the continuous duplex segment suggested by the classic continuous gene model (figure 6.23). Further hybridization experiments revealed that the hexon mRNA is built from four separate fragments of the adenovirus genome. These four continuous segments (called *exons*) in the adenovirus genome are separated by three "junk" fragments called *introns*.

Gene prediction is the problem of locating genes in a genomic sequence. Human genes constitute only 3% of the human genome, and no existing in silico gene recognition algorithm provides completely reliable gene recognition. The intron-exon model of a gene seems to prevail in eukaryotic organisms; prokaryotic organisms (like bacteria) do not have broken genes. As a result, gene prediction algorithms for prokaryotes tend to be somewhat simpler than those for eukaryotes.[18]

There are roughly two categories of approaches that researchers have used for predicting gene location. The statistical approach to gene prediction is to look for features that appear frequently in genes and infrequently elsewhere. Many researchers have attempted to recognize the locations of *splicing signals* at exon-intron junctions.[19] For example, the dinucleotides AG and GT on the left- and right-hand sides of an exon are highly conserved (figure 6.24). In addition, there are other less conserved positions on both sides of the exons. The simplest way to represent such binding sites is by a profile describing the propensities of different nucleotides to occur at different positions. Unfortu-

---

17. At that time, messenger RNA (mRNA) was viewed as a copy of a gene translated into the RNA alphabet. It is used to transfer information from the nuclear genome to the ribosomes to direct protein synthesis.

18. This is not to say that bacterial gene prediction is a trivial task but rather to indicate that eukaryotic gene finding is very difficult.

19. If genes are separated into exons interspersed with introns, then the RNA that is transcribed from DNA (i.e., the complementary copy of a gene) should be longer than the mRNA that is used as a template for protein synthesis. Therefore, some biological process needs to remove the introns in the pre-mRNA and concatenate the exons into a single mRNA string. This process is known as *splicing*, and the resulting mRNA is used as a template for protein synthesis in cytoplasm.

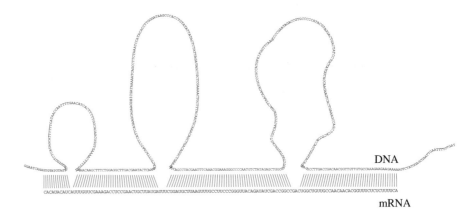

DNA

mRNA

Figure 6.23 An electron microscopy experiment led to the discovery of split genes. When mRNA (below) is hybridized against the DNA that generated it, three distinct loops can be seen (above). Because the loops are present in the DNA and are not present in mRNA, certain parts (introns) must be removed during the process of mRNA formation called splicing.

Figure 6.24 Exons typically are flanked by the dinucleotides AG and GT.

nately, using profiles to detect splice sites has met with limited success since these profiles are quite weak and tend to match frequently in the genome at nonsplice sites. Attempts to improve the accuracy of gene prediction led to the second category of approaches for gene finding: those based on similarity.

The similarity-based approach to gene prediction relies on the observation that a newly sequenced gene has a good chance of being related to one that is already known. For example, 99% of mouse genes have human analogs. However, one cannot simply look for a similar sequence in one organism's genome based on the genes known in another, for the reasons outlined above: both the exon sequence and the exon structure of the related gene in different species are different. The commonality between the related genes in both organisms is that they produce similar proteins. Accordingly, instead of employing a statistical analysis of exons, similarity-based methods attempt to solve a combinatorial puzzle: find a set of substrings (putative exons) in a genomic sequence (say, mouse) whose concatenation fits a known human protein. In this scenario, we suppose we know a human protein, and we want to discover the exon structure of the related gene in the mouse genome. The more sequence data we collect, the more accurate and reliable similarity-based methods become. Consequently, the trend in gene prediction has recently shifted from statistically motivated approaches to similarity-based algorithms.

6.12   Statistical Approaches to Gene Prediction

As mentioned above, statistical approaches to finding genes rely on detecting subtle statistical variations between coding (exons) and non-coding regions. The simplest way to detect potential coding regions is to look at *open reading frames*, or *ORFs*. One can represent a genome of length n as a sequence of $\frac{n}{3}$ codons.[20] The three "stop" codons, (**TAA**, **TAG**, and **TGA**) break this sequence into segments, one between every two consecutive stop codons. The subsegments of these that start from a start codon, **ATG**, are ORFs. ORFs within a single genomic sequence may overlap since there are six possible "reading frames": three on one strand starting at positions 1, 2, and 3, and three on the reverse strand, as shown in figure 6.25.

One would expect to find frequent stop codons in noncoding DNA, since the average number of codons between two consecutive stop codons in "random" DNA should be $\frac{64}{3} \approx 21$.[21] This is much smaller than the number of codons in an average protein, which is roughly 300. Therefore, ORFs longer than some threshold length indicate potential genes. However, gene prediction algorithms based on selecting significantly long ORFs may fail to detect short genes or genes with short exons.

---

20. In fact, there are three such representations for each DNA strand: one starting at position 1, another at 2 (ignoring the first base), and the third one at 3 (ignoring the first two bases).
21. There are $4^3 = 64$ codons, and three of them are Stop codons.

Figure 6.25   The six reading frames for the sequence ATGCTTAGTCTG. The string may be read forward or backward, and there are three frame shifts in each direction.

Many statistical gene prediction algorithms rely on statistical features in protein-coding regions, such as biases in *codon usage*. We can enter the frequency of occurrence of each codon within a given sequence into a 64-element *codon usage array*, as in table 6.1. The codon usage arrays for coding regions are different than the codon usage arrays for non-coding regions, enabling one to use them for gene prediction. For example, in human genes codons CGC and AGG code for the same amino acid (Arg) but have very different frequencies: CGC is 12 times more likely to be used in genes than AGG (table 6.1). Therefore, an ORF that "prefers" CGC over AGG while coding for Arg is a likely candidate gene. One can use a likelihood ratio approach[22] to compute the conditional probabilities of the DNA sequence in a window, under the hypothesis that the window contains a coding sequence, and under the hypothesis that the window contains a noncoding sequence. If we slide this window along the genomic DNA sequence (and calculate the likelihood

---

22. The *likelihood ratio* technique allows one to test the applicability of two distinct hypotheses; when the likelihood ratio is large, the first hypothesis is more likely to be true than the second one.

Table 6.1 The genetic code and codon usage in *Homo sapiens*. The codon for methionine, or **AUG**, also acts as a start codon; all proteins begin with Met. The numbers next to each codon reflects the frequency of that codon's occurrence while coding for an amino acid. For example, among all lysine (**Lys**) residues in all the proteins in a genome, the codon **AAG** generates 25% of them while the codon **AAG** generates 75%. These frequencies differ across species.

	U			C			A			G		
U	UUU Phe	57		UCU Ser	16		UAU Tyr	58		UGU Cys	45
	UUC Phe	43		UCC Ser	15		UAC Tyr	42		UGC Cys	55
	UUA Leu	13		UCA Ser	13		UAA Stp	62		UGA Stp	30
	UUG Leu	13		UCG Ser	15		UAG Stp	8		UGG Trp	100
C	CUU Leu	11		CCU Pro	17		CAU His	57		CGU Arg	37
	CUC Leu	10		CCC Pro	17		CAC His	43		CGC Arg	38
	CUA Leu	4		CCA Pro	20		CAA Gln	45		CGA Arg	7
	CUG Leu	49		CCG Pro	51		CAG Gln	66		CGG Arg	10
A	AUU Ile	50		ACU Thr	18		AAU Asn	46		AGU Ser	15
	AUC Ile	41		ACC Thr	42		AAC Asn	54		AGC Ser	26
	AUA Ile	9		ACA Thr	15		AAA Lys	75		AGA Arg	5
	AUG Met	100		ACG Thr	26		AAG Lys	25		AGG Arg	3
G	GUU Val	27		GCU Ala	17		GAU Asp	63		GGU Gly	34
	GUC Val	21		GCC Ala	27		GAC Asp	37		GGC Gly	39
	GUA Val	16		GCA Ala	22		GAA Glu	68		GGA Gly	12
	GUG Val	36		GCG Ala	34		GAG Glu	32		GGG Gly	15

ratio at each point), genes are often revealed as peaks in the likelihood ratio plots.

An even better coding sensor is the *in-frame hexamer count*[23] proposed by Mark Borodovsky and colleagues. Gene prediction in bacterial genomes also takes advantage of several conserved sequence motifs often found in the regions around the start of transcription. Unfortunately, such sequence motifs are more elusive in eukaryotes.

While the described approaches are successful in prokaryotes, their application to eukaryotes is complicated by the exon-intron structure. The average length of exons in vertebrates is 130 nucleotides, and exons of this length are too short to produce reliable peaks in the likelihood ratio plot while analyzing ORFs because they do not differ enough from random fluctuations to be detectable. Moreover, codon usage and other statistical parameters proba-

---

23. The in-frame hexamer count reflects frequencies of pairs of consecutive codons.

bly have nothing in common with the way the splicing machinery actually recognizes exons. Many researchers have used a more biologically oriented approach and have attempted to recognize the locations of splicing signals at exon-intron junctions. There exists a (weakly) conserved sequence of eight nucleotides at the boundary of an exon and an intron (*donor* splice site) and a sequence of four nucleotides at the boundary of an intron and exon (*acceptor* splice site). Since profiles for splice sites are weak, these approaches have had limited success and have been supplanted by hidden Markov model (HMM) approaches[24] that capture statistical dependencies between sites. A popular example of this latter approach is GENSCAN, which was developed in 1997 by Chris Burge and Samuel Karlin. GENSCAN combines coding region and splicing signal predictions into a single framework. For example, a splice site prediction is more believable if signs of a coding region appear on one side of the site but not on the other. Many such statistics are used in the HMM framework of GENSCAN that merges splicing site statistics, coding region statistics, and motifs near the start of the gene, among others. However, the accuracy of GENSCAN decreases for genes with many short exons or with unusual codon usage.

6.13   Similarity-Based Approaches to Gene Prediction

A similarity-based approach to gene prediction uses previously sequenced genes and their protein products as a template for the recognition of unknown genes in newly sequenced DNA fragments. Instead of employing statistical properties of exons, this method attempts to solve the following combinatorial puzzle: given a known target protein and a genomic sequence, find a set of substrings (candidate exons) of the genomic sequence whose concatenation (splicing) best fits the target.

A naive brute force approach to the spliced alignment problem is to find all local similarities between the genomic sequence and the target protein sequence. Each substring from the genomic sequence that exhibits sufficient similarity to the target protein could be considered a *putative exon*.[25] The putative exons so chosen may lack the canonical exon-flanking dinucleotides AG and GT but we can extend or shorten them slightly to make sure that they are flanked by AG and GT. The resulting set may contain overlapping

---

24. Hidden Markov models are described in chapter 11.
25. Putative here means that the sequence *might* be an exon, even though we have no proof of this.

substrings, and the problem is to choose the best subset of nonoverlapping substrings as a putative exon structure.[26]

We will model a putative exon with a *weighted interval* in the genomic sequence, which is described by three parameters (l, r, w), as in figure 6.26. Here, l is the left-hand position, r is the right-hand position, and w is the weight of the putative exon. The weight w may reflect the local alignment score for the genomic interval against the target protein sequence, or the strength of flanking acceptor and donor sites, or any combination of these and other measures; it reflects the likelihood that this interval is an exon. A *chain* is any set of nonoverlapping weighted intervals. The total weight of a chain is the sum of the weights of the intervals in the chain. A *maximum chain* is a chain with maximum total weight among all possible chains. Below we assume that the weights of all intervals are positive ($w > 0$).

Exon Chaining Problem:

Given a set of putative exons, find a maximum set of nonoverlapping putative exons.

   Input: A set of weighted intervals (putative exons).

   Output: A maximum chain of intervals from this set.

The Exon Chaining problem for n intervals can be solved by dynamic programming in a graph G on $2n$ vertices, n of which represent starting (left) positions of intervals and n of which represent ending (right) positions of intervals, as in figure 6.26. We assume that the set of left and right interval ends is sorted into increasing order and that all positions are distinct, forming an ordered array of vertices $(v_1, \ldots v_{2n})$ in graph G.[27] There are $3n - 1$ edges in this graph: there is an edge between each l_i and r_i of weight w_i for i from 1 to n, and $2n - 1$ additional edges of weight 0 which simply connect adjacent vertices (v_i, v_{i+1}) forming a path in the graph from v_1 to v_{2n}. In the algorithm below, s_i represents the length of the longest path in the graph ending at vertex v_i. Thus, s_{2n} is the solution to the Exon Chaining problem.

---

26. We choose nonoverlapping substrings because exons in real genes do not overlap.
27. In particular, we are assuming that no interval starts exactly where another ends.

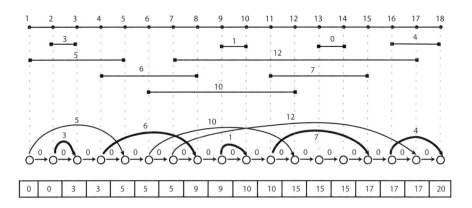

Figure 6.26   A short "genomic" sequence, a set of nine weighted intervals, and the graph used for the dynamic programming solution to the Exon Chaining problem. Five weighted intervals, $(2, 3, 3)$, $(4, 8, 6)$, $(9, 10, 1)$, $(11, 15, 7)$, and $(16, 18, 4)$, shown by bold edges, form an optimal solution to the Exon Chaining problem. The array at the bottom shows the values $s_1, s_2, \ldots, s_{2n}$ generated by the EXONCHANING algorithm.

EXONCHAINING(G, n)
1 **for** $i \leftarrow 1$ **to** $2n$
2      $s_i \leftarrow 0$
3 **for** $i \leftarrow 1$ **to** $2n$
4   **if** vertex v_i in G corresponds to the right end of an interval I
5          $j \leftarrow$ index of vertex for left end of the interval I
6          $w \leftarrow$ weight of the interval I
7          $s_i \leftarrow \max \{s_j + w, s_{i-1}\}$
8   **else**
9          $s_i \leftarrow s_{i-1}$
10 **return** s_{2n}

One shortcoming of this approach is that the endpoints of putative exons are not very well defined, and this assembly method does not allow for any flexibility at these points. More importantly, the optimal chain of intervals may not correspond to any valid alignment. For example, the first interval in the optimal chain may be similar to a suffix of the protein, while the second interval in the optimal chain may be similar to a prefix. In this case, the putative exons corresponding to the valid chain of these two intervals cannot

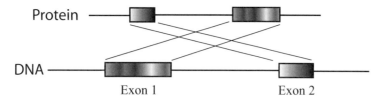

Protein

DNA

Exon 1   Exon 2

Figure 6.27 An infeasible chain that might have a maximal score. The first exon corresponds to a region at the end of the target protein, while the second exon corresponds to a region at the beginning of the target protein. These exons cannot be combined into a valid global DNA-protein alignment.

be combined into a valid alignment (figure 6.27).

6.14 Spliced Alignment

In 1996, Mikhail Gelfand and colleagues proposed the *spliced alignment* approach to find genes in eukaryotes: use a related protein within one genome to reconstruct the exon-intron structure of a gene in another genome. The spliced alignment begins by selecting either all putative exons between potential acceptor and donor sites (e.g., between **AG** and **GT** dinucleotides), or by finding all substrings similar to the target protein, as in the Exon Chaining problem. By filtering this set in a way that attempts not to lose true exons, one is left with a set of candidate exons that may contains many false exons, but definitely contains all the true ones. While it is difficult to distinguish the good (true exons) from the bad (false exons) by a statistical procedure alone, we can use the alignment with the target protein to aid us in our search. In theory, only the true exons will form a coherent representation of a protein.

   Given the set of candidate exons and a target protein sequence, we explore all possible chains (assemblies) of the candidate exon set to find the assembly with the highest similarity score to the target protein. The number of different assemblies may be huge, but the spliced alignment algorithm is able to find the best assembly among all of them in polynomial time. For simplicity we will assume that the protein sequence is expressed in the same alphabet as the genome. Of course, this is not the case in nature, and a problem at the end of this chapter asks you to modify the recurrence relations accordingly.

   Let $G = g_1 \ldots g_n$ be the genomic sequence, $T = t_1 \ldots t_m$ be the target sequence, and $\mathcal{B}$ be the set of candidate exons (blocks). As above, a chain Γ is any sequence of nonoverlapping blocks, and the string formed by a chain is just the concatenation of all the blocks in the chain. We will use Γ^* to denote the string formed by the chain Γ. The chain that we are searching for is the one whose concatenation forms the string with the highest similarity to the target sequence.[28]

Spliced Alignment Problem:

Find a chain of candidate exons in a genomic sequence that best fits a target sequence.

   Input: Genomic sequence G, target sequence T, and a set of candidate exons (blocks) $\mathcal{B}$.

   Output: A chain of candidate exons Γ such that the global alignment score $s(\Gamma^*, T)$ is maximum among all chains of candidate exons from $\mathcal{B}$.

As an example, consider the "genomic" sequence "It was brilliant thrilling morning and the slimy, hellish, lithe doves gyrated and gambled nimbly in the waves" with the set of blocks shown in figure 6.28 (top) by overlapping rectangles. If our target is the famous Lewis Carroll line "'twas brillig, and the slithy toves did gyre and gimble in the wabe" then figure 6.28 illustrates the spliced alignment problem of choosing the best "exons" (or blocks, in this case) that can be assembled into the target.

The spliced alignment problem can be cast as finding a path in a directed acyclic graph [fig. 6.28 (middle)]. Vertices in this graph (shown as rectangles) correspond to blocks (candidate exons), and directed edges connect nonoverlapping blocks. A vertex corresponding to a block B is labeled by a string represented by this block. Therefore, every path in the spliced alignment graph spells out the string obtained by concatenation of labels of its vertices. The weight of a path in this graph is defined as the score of the optimal alignment between the concatenated blocks of this path and the target sequence. Note that we have defined the weight of an entire path in the graph, but we have not defined weights for individual edges. This makes the Spliced Alignment problem different from the standard Longest Path problem. Nevertheless, we can leverage dynamic programming to solve the problem.

---

28. We emphasize the difference between the scoring functions for the Exon Chaining problem and the Spliced Alignment problem. In contrast to the Spliced Alignment problem, the set of nonoverlapping substrings representing the solution of the Exon Chaining problem does not necessarily correspond to a valid alignment between the genomic sequence and the target protein sequence.

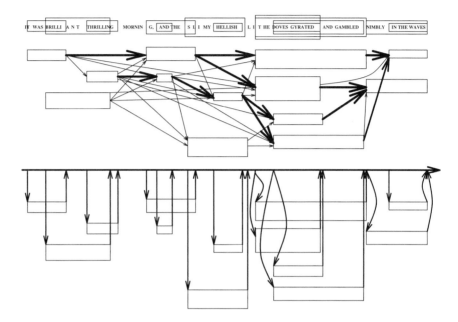

Figure 6.28   The Spliced Alignment problem: four different block assemblies with the best fit to Lewis Carroll's line (top), the corresponding spliced alignment graph (middle), and the transformation of the spliced alignment graph that helps reduce the running time (bottom).

To describe the dynamic programming recurrence for the Spliced Alignment problem, we first need to define the similarity score between the i-prefix of the *spliced alignment graph* in figure 6.28 and the j-prefix of the target sequence T. The difficulty is that there are typically many different i-prefixes of the graph, since there are multiple blocks containing position i.[29]

Let $B = g_{left} \ldots g_i \ldots g_{right}$ be a candidate exon containing position i in genomic sequence G. Define the *i-prefix* of B as $B(i) = g_{left} \ldots g_i$ and $end(B) =$

___
29. For example, there are two i-prefixes ending with E within HELLISH, and four i-prefixes ending in R within GYRATED in figure 6.28.

right (the words *left* and *right* are used here as indices). If the chain $\Gamma = (B_1, B_2, \ldots, B)$ ends at block B, define $\Gamma^*(i)$ to be the concatenation of all candidate exons in the chain up to (and excluding) B, plus all the characters in B up to i. That is, $\Gamma^*(i) = B_1 \circ B_2 \circ \cdots \circ B(i)$.[30] Finally, let

$$S(i, j, B) = \max_{\text{all chains } \Gamma \text{ ending in } B} s(\Gamma^*(i), T(j)).$$

That is, given i, j, and a candidate exon B that covers position i, $S(i, j, B)$ is the score of the optimal spliced alignment between the i-prefix of G and the j-prefix of T under the assumption that this alignment ends in block B.

The following recurrence allows us to efficiently compute $S(i, j, B)$. For the sake of simplicity we consider sequence alignment with linear gap penalties for insertion or deletion equal to $-\sigma$, and use the scoring matrix δ for matches and mismatches.

The dynamic programming recurrence for the Spliced Alignment problem is broken into two cases depending on whether i is the starting vertex of block B or not. In the latter case, the recurrence is similar to the canonical sequence alignment:

$$S(i, j, B) = \max \begin{cases} S(i-1, j, B) - \sigma \\ S(i, j-1, B) - \sigma \\ S(i-1, j-1, B) + \delta(g_i, t_j) \end{cases}$$

On the other hand, if i is the starting position of block B, then

$$S(i, j, B) = \max \begin{cases} S(i, j-1, B) - \sigma \\ \max_{\text{all blocks } B' \text{ preceding } B} S(end(B'), j-1, B') + \delta(g_i, t_j), \\ \max_{\text{all blocks } B' \text{ preceding } B} S(end(B'), j, B') - \sigma, \end{cases}$$

After computing this three-dimensional table $S(i, j, B)$, the score of the optimal spliced alignment is

$$\max_B S(end(B), m, B),$$

where the maximum is taken over all possible blocks. One can further reduce the number of edges in the spliced alignment graph by making a transfor-

---

30. The notation $x \circ y$ denotes concatenation of strings x and y.

mation of the graph in figure 6.28 (middle) into a graph shown in figure 6.28 (bottom). The details of the corresponding recurrences are left as a problem at the end of this chapter.

The above description hides some important details of block generation. The simplest approach to the construction of the candidate exon set is to generate all fragments between potential acceptor sites represented by AG and potential donor sites represented by GT, removing possible exons with stop codons in all three frames. However, this approach does not work well since it generates many short blocks. Experiments with the spliced alignment algorithm have shown that incorrect predictions are frequently associated with the *mosaic effect* caused by very short potential exons. The difficulty is that these short exons can be easily combined to fit any target protein, simply because it is easier to construct a given sentence from a thousand random short strings than from the same number of random long strings. For example, with high probability, the phrase "filtration of candidate exons" can be made up from a sample of a thousand random two-letter strings ("fi," "lt," "ra," etc. are likely to be present in this sample). The probability that the same phrase can be made up from a sample of the same number of random five-letter strings is close to zero (even finding the string "filtr" in this sample is unlikely). This observation explains the mosaic effect: if the number of short blocks is high, chains of these blocks can replace actual exons in spliced alignments, thus leading to predictions with an unusually large number of short exons. To avoid the mosaic effect, the candidate exons should be subjected to some filtering procedure.

6.15   Notes

Although the first dynamic programming algorithm for DNA sequence comparison was published as early as 1970 by Saul Needleman and Christian Wunsch (79), Russell Doolittle and colleagues used heuristic algorithms to establish the similarity between cancer-causing genes and the PDGF gene in 1983 (28). When Needleman and Wunsch published their paper in 1970, they did not know that a very similar algorithm had been published two years earlier in a pioneering paper on automatic speech recognition (105) (though the details of the algorithms are slightly different, they are both variations of dynamic programming). Earlier still, Vladimir Levenshtein introduced the notion of edit distance in 1966 (64), albeit without an algorithm for computing it. The local alignment algorithm introduced by Temple Smith and

Michael Waterman in 1981 (96) quickly became the most popular alignment tool in computational biology. Later Michael Waterman and Mark Eggert developed an algorithm for finding the k best nonoverlapping local alignments (109). The algorithm for alignment with affine gap penalties was the work of Osamu Gotoh (42).

The progressive multiple alignment approach, initially explored by Da-Fei Feng and Russell Doolittle [Feng and Doolittle, 1987 (36)], resulted in many practical algorithms, with CLUSTAL (48) one of the most popular.

The cellular process of splicing was discovered in the laboratories of Phillip Sharp (12) and Richard Roberts (21). Applications of both Markov models and in-frame hexamer count statistics for gene prediction were proposed by Borodovsky and McInnich (14). Chris Burge and Samuel Karlin developed an HMM approach to gene prediction that resulted in the popular GEN-SCAN algorithm in 1997 (20). In 1995 Snyder and Stormo (97) developed a similarity-based gene prediction algorithm that amounts to the solution of a problem that is similar to the Exon Chaining problem. The spliced alignment algorithm was developed by Mikhail Gelfand and colleagues in 1996 (41).

Michael Waterman (born 1942 in Oregon) currently holds an Endowed Associates Chair at the University of Southern California. His BS in Mathematics is from Oregon State University, and his PhD in Statistics and Probability is from Michigan State University. He was named a Guggenheim Fellow in 1995 and was elected to the National Academy of Sciences in 2001. In 2002 he received a Gairdner Foundation Award. He is one of the founding fathers of bioinformatics whose fundamental contributions to the area go back to the 1970s when he worked at Los Alamos National Laboratories. Waterman says:

I went to college to escape what I considered to be a dull and dreary existence of raising livestock on pasture land in western Oregon where my family has lived since 1911. My goal was to find an occupation with a steady income where I could look forward to going to work; this eliminated ranching and logging (which was how I spent my college summers). Research and teaching didn't seem possible or even desirable, but I went on for a PhD because such a job did not appear.

In graduate school at Michigan State I found a wonderful advisor in John Kinney from whom I learned ergodic and information theory. John aimed me at a branch of number theory for a thesis. We were doing statistical properties of the iteration of deterministic functions long before that became a fad. I began using computers to explore iteration, something which puzzled certain of my professors who felt I was wasting time I could be spending proving theorems. After graduation and taking a nonresearch job at a small school in Idaho, my work in iteration led to my first summer visit to Los Alamos National labs. Later I met Temple Smith there in 1973 and was drawn into problems from biology. Later I wrote in my book of New Mexico essays *Skiing the Sun* (107):

I was an innocent mathematician until the summer of 1974. It was then than I met Temple Ferris Smith and for two months was cooped up with him in an office at Los Alamos National Laboratories. That experience transformed my research, my life, and perhaps my sanity. Soon after we met, he pulled out a little blackboard and started lecturing me about biology: what it was, what was important, what was going on. Somewhere in there by implication was what we should work on, but the truth be told he didn't know what that was either. I was totally confused: amino acids, nucleosides, beta sheets. What were these things? Where was the mathematics?

I knew no modern biology, but studying alignment and evolution was quite attractive to me. The most fun was formulating problems, and in my opinion that remains the most important aspect of our subject. Temple and I spent days and weeks trying to puzzle out what we should be working on. Charles DeLisi, a biophysicist who went on to play a key role in jump-starting the Human Genome Project, was in T-10 (theoretical biology) at the lab. When he saw the progress we had made on alignment problems, he came to me and said there was another problem which should interest me. This was the RNA folding problem which was almost untouched. Tinoco had published the idea of making a base-pair matrix for a sequence and that was it. By the fall of 1974 I had seen the neat connection between alignment and folding, and the following summer I wrote a long manuscript that defined the objects of study, established some of their properties, explicitly stated the basic problem of folding (which included free energies for all structural components), and finally gave algorithms for its solution. I had previously wondered what such a discovery might feel like, and it was wonderfully satisfying. However it felt entirely like exploration and not a grand triumph of creation as I had expected. In fact I had always wanted to be an explorer and regretted the end of the American frontier; wandering about this new RNA landscape was a great joy, just as I had thought when I was a child trying to transport myself by daydreams out of my family's fields into some new and unsettled country.

6.16   Problems

In 1879, Lewis Carroll proposed the following puzzle to the readers of *Vanity Fair*: transform one English word into another by going through a series of intermediate English words, where each word in the sequence differs from the next by only one substitution. To transform *head* into *tail* one can use four intermediates: *head* → *heal* → *teal* → *tell* → *tall* → *tail*. We say that two words **v** and **w** are equivalent if **v** can be transformed into **w** by substituting individual letters in such a way that all intermediate words are English words present in an English dictionary.

Problem 6.1

Find an algorithm to solve the following *Equivalent Words problem*.

Equivalent Words Problem:
Given two words and a dictionary, find out whether the words are equivalent.

> **Input:** The dictionary, $\mathcal{D}$ (a set of words), and two words **v** and **w** from the dictionary.
>
> **Output:** A transformation of **v** into **w** by substitutions such that all intermediate words belong to $\mathcal{D}$. If no transformation is possible, output "**v** and **w** are not equivalent."

Given a dictionary $\mathcal{D}$, the *Lewis Carroll distance*, $d_{LC}(\mathbf{v}, \mathbf{w})$, between words **v** and **w** is defined as the smallest number of substitutions needed to transform **v** into **w** in such a way that all intermediate words in the transformation are in the dictionary $\mathcal{D}$. We define $d_{LC}(\mathbf{v}, \mathbf{w}) = \infty$ if **v** and **w** are not equivalent.

Problem 6.2

Find an algorithm to solve the following *Lewis Carroll problem*.

Lewis Carroll Problem:
Given two words and a dictionary, find the Lewis Carroll distance between these words.

> **Input:** The dictionary $\mathcal{D}$, and two words **v** and **w** from the dictionary.
>
> **Output:** $d_{LC}(\mathbf{v}, \mathbf{w})$

Problem 6.3

Find an algorithm to solve a generalization of the Lewis Carroll problem when insertions, deletions, and substitutions are allowed (rather than only substitutions).

Problem 6.4

Modify DPCHANGE to return not only the smallest *number* of coins but also the correct combination of coins.

Problem 6.5

Let $s(\mathbf{v}, \mathbf{w})$ be the length of a longest common subsequence of the strings $\mathbf{v}$ and $\mathbf{w}$ and $d(\mathbf{v}, \mathbf{w})$ be the edit distance between $\mathbf{v}$ and $\mathbf{w}$ under the assumption that insertions and deletions are the only allowed operations. Prove that $d(\mathbf{v}, \mathbf{w}) = n + m - 2s(\mathbf{v}, \mathbf{w})$, where n is the length of $\mathbf{v}$ and m is the length of $\mathbf{w}$.

Problem 6.6

Find the number of different paths from *source* $(0,0)$ to *sink* (n, m) in an $n \times m$ rectangular grid.

Problem 6.7

Can you find an approximation ratio of the greedy algorithm for the Manhattan Tourist problem?

Problem 6.8

Let $\mathbf{v} = v_1 v_2 \cdots v_n$ be a string, and let P be a $4 \times m$ profile. Generalize the sequence alignment algorithm for aligning a sequence against a profile. Write the corresponding recurrence (in lieu of pseudocode), and estimate the amount of time that your algorithm will take with respect to n and m.

Problem 6.9

There are only two buttons inside an elevator in a building with 50 floors. The elevator goes 11 floors up if the first button is pressed and 6 floors down if the second button is pressed. Is it possible to get from floor 32 to floor 33? What is the minimum number of buttons one has to press to do so? What is the shortest time one needs to get from floor 32 to floor 33 (time is proportional to the number of floors that are passed on the way)?

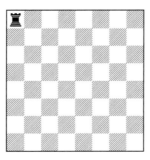

Problem 6.10

A rook stands on the upper left square of a chessboard. Two players make turns moving the rook either horizontally to the right or vertically downward (as many squares as they want). The player who can place the rook on the lower right square of the chessboard wins. Who will win? Describe the winning strategy.

Problem 6.11

A queen stands on the third square of the uppermost row of a chessboard. Two players take turns moving the queen either horizontally to the right or vertically downward or diagonally in the southeast direction (as many squares as they want). The player who can place the queen on the lower right square of the chessboard wins. Who will win? Describe the winning strategy.

Problem 6.12

Two players play the following game with two "chromosomes" of length n and m nucleotides. At every turn a player can destroy one of the chromosomes and break another one into two nonempty parts. For example, the first player can destroy a chromosome of length n and break another chromosome into two chromosomes of length $\frac{m}{3}$ and $m - \frac{m}{3}$. The player left with two single-nucleotide chromosomes loses. Who will win? Describe the winning strategy for each n and m.

Problem 6.13

Two players play the following game with two sequences of length n and m nucleotides. At every turn a player can either delete an arbitrary number of nucleotides from one sequence or an equal (but still arbitrary) number of nucleotides from both sequences. The player who deletes the last nucleotide wins. Who will win? Describe the winning strategy for each n and m.

Problem 6.14

Two players play the following game with two sequences of length n and m nucleotides. At every turn a player must delete two nucleotides from one sequence (either the first or the second) and one nucleotide from the other. The player who cannot move loses. Who will win? Describe the winning strategy for each n and m.

Problem 6.15

Two players play the following game with a nucleotide sequence of length n. At every turn a player may delete either one or two nucleotides from the sequence. The player who deletes the last letter wins. Who will win? Describe the winning strategy for each n.

Problem 6.16

Two players play the following game with a nucleotide sequence of length $n = n_A + n_T + n_C + n_G$, where n_A, n_T, n_C, and n_G are the number of A,T,C, and G in the sequence. At every turn a player may delete either one or two nucleotides from the sequence. The player who is left with a uni-nucleotide sequence of an arbitrary length (i.e., the sequence containing only one of 4 possible nucleotides) loses. Who will win? Describe the winning strategy for each n_A, n_T, n_C, and n_G.

Problem 6.17

What is the optimal global alignment for APPLE and HAPPE? Show all optimal alignments and the corresponding paths under the match premium $+1$, mismatch penalty -1, and indel penalty -1.

Problem 6.18

What is the optimal global alignment for MOAT and BOAST? Show all optimal alignments and the corresponding paths under the scoring matrix below and indel penalty -1.

	A	B	M	O	S	T
A	1	-1	-1	-2	-2	-3
B		1	-1	-1	-2	-2
M			2	-1	-1	-2
O				1	-1	-1
S					1	-1
T						2

Problem 6.19

Fill the global alignment dynamic programming matrix for strings AT and AAGT with affine scoring function defined by match premium 0, mismatch penalty -1, gap opening penalty -1, and gap extension penalty -1. Find all optimal global alignments.

Problem 6.20

Consider the sequences $\mathbf{v}$ = TACGGGTAT and $\mathbf{w}$ = GGACGTACG. Assume that the match premium is $+1$ and that the mismatch and indel penalties are -1.

- Fill out the dynamic programming table for a global alignment between $\mathbf{v}$ and $\mathbf{w}$. Draw arrows in the cells to store the backtrack information. What is the score of the optimal global alignment and what alignment does this score correspond to?

- Fill out the dynamic programming table for a local alignment between $\mathbf{v}$ and $\mathbf{w}$. Draw arrows in the cells to store the backtrack information. What is the score of the optimal local alignment in this case and what alignment achieves this score?

- Suppose we use an affine gap penalty where it costs -20 to open a gap, and -1 to extend it. Scores of matches and mismatches are unchanged. What is the optimal global alignment in this case and what score does it achieve?

Problem 6.21

For a pair of strings $\mathbf{v} = v_1 \ldots v_n$ and $\mathbf{w} = w_1 \ldots w_m$, define $M(\mathbf{v}, \mathbf{w})$ to be the matrix whose (i, j)th entry is the score of the optimal global alignment which aligns the character v_i with the character w_j. Give an $O(nm)$ algorithm which computes $M(\mathbf{v}, \mathbf{w})$.

Define an *overlap alignment* between two sequences $\mathbf{v} = v_1 \ldots v_n$ and $\mathbf{w} = w_1 \ldots w_m$ to be an alignment between a suffix of $\mathbf{v}$ and a prefix of $\mathbf{w}$. For example, if $\mathbf{v} = $ TATATA and $\mathbf{w} = $ AAATTT, then a (not necessarily optimal) overlap alignment between $\mathbf{v}$ and $\mathbf{w}$ is

ATA
AAA

Optimal overlap alignment is an alignment that maximizes the global alignment score between $v_i, \ldots, v_n$ and $w_1, \ldots w_j$, where the maximum is taken over all suffixes $v_i, \ldots, v_n$ of $\mathbf{v}$ and all prefixes $w_1, \ldots w_j$ of $\mathbf{w}$.

Problem 6.22

Give an algorithm which computes the optimal overlap alignment, and runs in time $O(nm)$.

Suppose that we have sequences $\mathbf{v} = v_1 \ldots v_n$ and $\mathbf{w} = w_1 \ldots w_m$, where $\mathbf{v}$ is longer than $\mathbf{w}$. We wish to find a substring of $\mathbf{v}$ which best matches *all* of $\mathbf{w}$. Global alignment won't work because it would try to align all of $\mathbf{v}$. Local alignment won't work because it may not align all of $\mathbf{w}$. Therefore this is a distinct problem which we call the *Fitting problem*. *Fitting* a sequence $\mathbf{w}$ into a sequence $\mathbf{v}$ is a problem of finding a substring $\mathbf{v}'$ of $\mathbf{v}$ that maximizes the score of alignment $s(\mathbf{v}', \mathbf{w})$ among all substrings of $\mathbf{v}$. For example, if $\mathbf{v} = $ GTAGGCTTAAGGTTA and $\mathbf{w} = $ TAGATA, the best alignments might be

	global	local	fitting
$\mathbf{v}$	GTAGGCTTAAGGTTA	TAG	TAGGCTTA
$\mathbf{w}$	-TAG----A---T-A	TAG	TAGA--TA
score	-3	3	2

The scores are computed as 1 for match, -1 for mismatch or indel. Note that the optimal local alignment is not a valid fitting alignment. On the other hand, the optimal global alignment contains a valid fitting alignment, but it achieves a suboptimal score among all fitting alignments.

Problem 6.23

Give an algorithm which computes the optimal fitting alignment. Explain how to fill in the first row and column of the dynamic programming table and give a recurrence to fill in the rest of the table. Give a method to find the best alignment once the table is filled in. The algorithm should run in time $O(nm)$.

We have studied two approaches to sequence alignment: global and local alignment. There is a middle ground: an approach known as *semiglobal* alignment. In semiglobal alignment, the entire sequences are aligned (as in global alignment). What makes it semiglobal is that the "internal gaps" of the alignment are counted, but the "gaps on the end" are not. For example, consider the following two alternative alignments:

```
Sequence 1:   CAGCA-CTTGGATTCTCGG
Sequence 2:   ---CAGCGTGG--------
```

```
Sequence 1:   CAGCACTTGGATTCTCGG
Sequence 2:   CAGC-----G-T----GG
```

The first alignment has 6 matches, 1 mismatch, and 12 gaps. The second alignment has 8 matches, no mismatches, and 10 gaps. Using the simplest scoring scheme (+1 match, −1 mismatch, −1 gap), the score for the first alignment is −7, and the score for the second alignment is −2, so we would prefer the second alignment. However, the first alignment is more biologically realistic. To get an algorithm which prefers the first alignment to the second, we can not count the gaps "on the ends."

Under this new ("semiglobal") approach, the first alignment would have 6 matches, 1 mismatch, and 1 gap, while the second alignment would still have 8 matches, no mismatches, and 10 gaps. Now the first alignment would have a score of 4, and the second alignment would have a score of −2, so the first alignment would have a better score.

Note the similarities and the differences between the Fitting problem and the Semiglobal Alignment problem as illustrated by the semiglobal—but not fitting—alignment of ACGTCAT against TCATGCA:

```
Sequence 1:   ACGTCAT---
Sequence 2:   ---TCATGCA
```

Problem 6.24

Devise an efficient algorithm for the Semiglobal Alignment problem and illustrate its work on the sequences **ACAGATA** and **AGT**. For scoring, use the match premium +1, mismatch penalty −1, and indel penalty −1.

Define a *NoDeletion* global alignment to be an alignment between two sequences $\mathbf{v} = v_1 v_2 \ldots v_n$ and $\mathbf{w} = w_1 w_2 \ldots w_m$, where only matches, mismatches, and insertions are allowed. That is, there can be no deletions from $\mathbf{v}$ to $\mathbf{w}$ (i.e., all letters of $\mathbf{w}$ occur in the alignment with no spaces). Clearly we must have $m \geq n$ and let $k = m - n$.

Problem 6.25

Give an $O(nk)$ algorithm to find the optimal *NoDeletion* global alignment (note the improvement over the $O(nm)$ algorithm when k is small).

Problem 6.26

Substrings $v_i, \ldots, v_{i+k}$ and $v_{i'}, \ldots, v_{i'+k}$ of the string $v_1, \ldots, v_n$ form a *substring pair* if $i' - i + k > MinGap$, where $MinGap$ is a parameter. Define the substring pair score as the (global) alignment score of $v_i, \ldots, v_{i+k}$ and $v_{i'}, \ldots, v_{i'+k}$. Design an algorithm that finds a substring pair with maximum score.

Problem 6.27

For a parameter k, compute the global alignment between two strings, subject to the constraint that the alignment contains at most k gaps (blocks of consecutive indels).

Nucleotide sequences are sometimes written in an alphabet with five characters: A, T, G, C, and N, where N stands for an unspecified nucleotide (in essence, a wild-card). Biologists may use N when sequencing does not allow one to unambiguously infer the identity of a nucleotide at a specific position. A sequence with an N is referred to as a *degenerate* string; for example, ATTNG may correspond to four different interpretations: ATTAG, ATTTG, ATTGG, and ATTCG. In general, a sequence with k unspecified nucleotides N will have 4^k different interpretations.

Problem 6.28

Given a non-degenerate string, **v**, and a degenerate string **w** that contains k Ns, devise a method to find the best interpretation of **w** according to **v**. That is, out of all 4^k possible interpretations of **w**, find **w**$'$ with the minimum alignment score $s(\mathbf{w}', \mathbf{v})$.

Problem 6.29

Given a non-degenerate string, **v**, and a degenerate string **w** that contains k Ns, devise a method to find the worst interpretation of **w** according to **v**. That is, out of all 4^k possible interpretations of **w**, find **w**$'$ with the minimum alignment score $s(\mathbf{w}', \mathbf{v})$.

Problem 6.30

Given two strings $\mathbf{v}_1$ and $\mathbf{v}_2$, explain how to construct a string **w** minimizing

$$|d(\mathbf{v}_1, \mathbf{w}) - d(\mathbf{v}_2, \mathbf{w})|$$

such that

$$d(\mathbf{v}_1, \mathbf{w}) + d(\mathbf{v}_2, \mathbf{w}) = d(\mathbf{v}_1, \mathbf{v_2}).$$

$d(\cdot, \cdot)$ is the edit distance between two strings.

Problem 6.31

Given two strings $\mathbf{v}_1$ and $\mathbf{v}_2$ and a text **w**, find whether there is an occurrence of $\mathbf{v}_1$ and $\mathbf{v}_2$ interwoven (without spaces) in **w**. For example, the strings **abac** and bbc occur interwoven in cabbabccdw. Give an efficient algorithm for this problem.

A string **x** is called a *supersequence* of a string **v** if **v** is a subsequence of **x**. For example, ABLUE is a supersequence for BLUE and ABLE.

Problem 6.32

Given strings **v** and **w**, devise an algorithm to find the shortest supersequence for both **v** and **w**.

A *tandem repeat* P^k of a pattern $P = p_1 \ldots p_n$ is a pattern of length $n \cdot k$ formed by concatenation of k copies of P. Let P be a pattern and T be a text of length m. The *Tandem Repeat* problem is to find a best local alignment of T with some tandem repeat of P. This amounts to aligning P^k against T and the standard local alignment algorithm solves this problem in $O(km^2)$ time.

Problem 6.33

Devise a faster algorithm for solving the tandem repeat problem.

An alignment of circular strings is defined as an alignment of linear strings formed by cutting (linearizing) these circular strings at arbitrary positions. The following problem asks to find the cut points of two circular strings that maximize the alignment of the resulting linear strings.

Problem 6.34

Devise an efficient algorithm to find an optimal alignment (local and global) of circular strings.

The early graphical method for comparing nucleotide sequences—dot matrices—still yields one of the best visual representations of sequence similarities. The axes in a dot matrix correspond to the two sequences $\mathbf{v} = v_1 \ldots v_n$ and $\mathbf{w} = w_1 \ldots w_m$. A dot is placed at coordinates (i, j) if the substrings $s_i \ldots s_{i+k}$ and $t_j \ldots t_{j+k}$ are sufficiently similar. Two such substrings are considered to be sufficiently similar if the Hamming distance between them is at most d.

When the sequences are very long, it is not necessary to show exact coordinates; figure 6.29 is based on the sequences corresponding to the β-globin gene in human and mouse. In these plots each axis is on the order of 1000 base pairs long, $k = 10$ and $d = 2$.

Problem 6.35

Use figure 6.29 to answer the following questions:

- How many exons are in the human β-globulin gene?

- The dot matrix in figure 6.29 (top) is between the mouse and human genes (i.e., all introns and exons are present). Do you think the number of exons in the β-globulin gene is different in the human genome as compared to the mouse genome?

- Label segments of the axes of the human and mouse genes in figure 6.29 to show where the introns and exons would be located.

A local alignment between two different strings $\mathbf{v}$ and $\mathbf{w}$ finds a pair of substrings, one in $\mathbf{v}$ and the other in $\mathbf{w}$, with maximum similarity. Suppose that we want to find a pair of (nonoverlapping) substrings *within* string $\mathbf{v}$ with maximum similarity (*Optimal Inexact Repeat problem*). Computing an optimal local alignment between $\mathbf{v}$ and $\mathbf{v}$ does not solve the problem, since the resulting alignment may correspond to overlapping substrings.

Problem 6.36

Devise an algorithm for the Optimal Inexact Repeat problem.

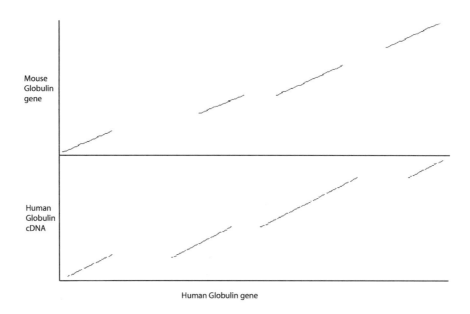

Figure 6.29   Human β-globulin cDNA vs. the gene sequence in two organisms.

In the *chimeric alignment* problem, a string $\mathbf{v}$ and a set of strings $\{\mathbf{w}_1, \ldots, \mathbf{w}_N\}$ are given, and the problem is to find $\max_{1 \leq i,j \leq N} s(\mathbf{v}, \mathbf{w}_i \circ \mathbf{w}_j)$ where $\mathbf{w}_i \circ \mathbf{w}_j$ is the concatenation of $\mathbf{w}_i$ and $\mathbf{w}_j$ ($s(\cdot, \cdot)$ stand for the score of optimal global alignment).

Problem 6.37

Devise an efficient algorithm for the chimeric alignment problem.

A virus infects a bacterium, and modifies a replication process in the bacterium by inserting

at every A, a polyA of length 1 to 5.
at every C, a polyC of length 1 to 10.
at every G, a polyG of arbitrary length ≥ 1.
at every T, a polyT of arbitrary length ≥ 1.

No gaps or other insertions are allowed in the virally modified DNA. For example, the sequence AAATAAAGGGGCCCCCTTTTTTTTCC is an infected version of ATAGCTC.

Problem 6.38

Given sequences $\mathbf{v}$ and $\mathbf{w}$, describe an efficient algorithm that will determine if $\mathbf{v}$ could be an infected version of $\mathbf{w}$.

Problem 6.39

Now assume that for each nucleotide (A, C, G, T) the virus will either delete a letter or insert a run of the letter of arbitrary length. Give an efficient algorithm to detect if **v** could be an infected version of **w** under these circumstances.

Problem 6.40

Define homodeletion as an operation of deleting a run of the same nucleotide and homoinsertion as an operation of inserting a run of the same nucleotide. For example, ACAAAAAAGCTTTTA is obtained from ACGCTTTTA by a homoinsertions of a run of six A, while ACGCTA is obtained from ACGCTTTTA by homodeletion of a run of three T. The homo-edit distance between two sequences is defined as the minimum number of homodeletions and homoinsertions to transform one sequence into another. Give an efficient algorithm to compute the homoedit distance between two arbitrary strings.

Problem 6.41

Suppose we wish to find an optimal global alignment using a scoring scheme with an affine *mismatch* penalty. That is, the premium for a match is $+1$, the penalty for an indel is $-\rho$, and the penalty for x consecutive mismatches is $-(\rho + \sigma x)$. Give an $O(nm)$ algorithm to align two sequences of length n and m with an affine mismatch penalty. Explain how to construct an appropriate "Manhattan" graph and estimate the running time of your algorithm.

Problem 6.42

Define a *NoDiagonal* global alignment to be an alignment where we disallow matches and mismatches. That is, only indels are allowed. Give a $\Theta(nm)$ algorithm to determine the number of *NoDiagonal* alignments between a sequence of length n and a sequence of length m. Give a closed-form formula for the number of *NoDiagonal* global alignments (e.g., something of the form $f(n, m) = n^2 m - \sqrt{n!} + \pi n^m$).

Problem 6.43

Estimate the number of different (not necessarily optimal) global alignments between two n-letter sequences.

Problem 6.44

Devise an algorithm to compute the number of distinct optimal global alignments (optimal paths in edit graph) between a pair of strings.

Problem 6.45

Estimate the number of different (not necessarily optimal) local alignments between two n-letter sequences.

Problem 6.46

Devise an algorithm to compute the number of distinct optimal local alignments (optimal paths in local alignment edit graph) between a pair of strings.

Problem 6.47

Let $s_{i,j}$ be a dynamic programming matrix computed for the LCS problem. Prove that for any i and j, the difference between $s_{i+1,j}$ and $s_{i,j}$ is at most 1.

Let $i_1, \ldots, i_n$ be a sequence of numbers. A subsequence of $i_1, \ldots, i_n$ is called an *increasing subsequence* if elements of this subsequence go in increasing order. Decreasing subsequences are defined similarly. For example, elements 2, 6, 7, 9 of the sequence 8, 2, 1, 6, 5, 7, 4, 3, 9 form an increasing subsequence, while elements 8, 7, 4, 3 form a decreasing subsequence.

Problem 6.48

Devise an efficient algorithm for finding longest increasing and decreasing subsequences in a permutation of integers.

Problem 6.49

Show that in any permutation of n distinct integers, there is either an increasing subsequence of length at least $\sqrt{n}$ or a decreasing subsequence of length at least $\sqrt{n}$.

A subsequence σ of permutation π is *2-increasing* if, as a set, it can be written as

$$\sigma = \sigma_1 \cup \sigma_2$$

where σ_1 and σ_2 are increasing subsequences of π. For example, 1, 5, 7, 9 and 2, 6 are increasing subsequences of $\pi = 821657439$ forming a 2-increasing subsequence 2, 1, 6, 5, 7, 9 consisting of six elements.

Problem 6.50

Devise an algorithm to find a longest 2-increasing subsequence.

RNAs adopt complex three-dimensional structures that are important for many biological functions. Pairs of positions in RNA with complementary nucleotides can form *bonds*. Bonds (i, j) and (i', j') are interleaving if $i < i' < j < j'$ and noninterleaving otherwise (fig. 6.30). Every set of noninterleaving bonds corresponds to a potential RNA structure. In a very naive formulation of the RNA folding problem, one tries to find a maximum set of noninterleaving bonds. The more adequate model, attempting to find a fold with the minimum energy, is much more difficult.

Problem 6.51

Develop a dynamic programming algorithm for finding the largest set of noninterleaving bonds given an RNA sequences.

The human genome can be viewed as a string of n ($\approx$ 3 billion) nucleotides, partitioned into substrings representing chromosomes. However, for many decades, biologists used a different *band* representation of the genome that is obtained via traditional light microscopy. Figure 6.31 shows 48 bands (as seen on chromosme 4) out of 862 observable bands for the entire human genome. Although several factors (e.g., local G/C frequency) have been postulated to govern the formation of these banding patterns, the mechanism behind their formation remains poorly understood. A mapping between the human genomic sequence (which itself only became available in 2001) and the banding pattern representation would be useful to leverage sequence level

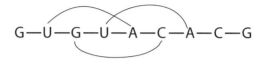

(a) Interleaving bonds

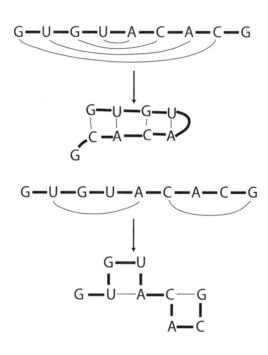

(b) Non-interleaving bonds

Figure 6.30   Interleaving and noninterleaving bonds in RNA folding.

Figure 6.31   Band patterns on human chromosome 4.

gene information against diseases that have been associated with certain band positions. However, until recently, no mapping between these two representations of the genome has been known.

The Band Positioning problem is to find the starting and ending nucleotide positions for each band in the genome (for simplicity we assume that all chromosomes are concatenated to form a single coordinate system). In other words, the Band Positioning problem is to find an increasing array $start(b)$, that contains the starting nucleotide position for each band b in the genome. Each band b begins at the nucleotide given by $start(b)$ and ends at $start(b+1) - 1$.[31]

A naive approach to this problem would be to use observed band width data to compute the nucleotide positions. However, this solution is inaccurate because it assumes that band width is perfectly correlated with its length in nucleotides. In reality, this correlation is often quite poor and a different approach is needed.

In the last decade biologists have performed a large number of *FISH (fluorescent in situ hybridization)* experiments that can help to solve the Band Positioning problem. FISH data consist of pairs (x, b), where x is a position in the genome, and b is the index of the band that contains x. FISH data are often subject to experimental error, so some FISH data points may contradict each other.

Given a solution $start(b)$ $(1 \leq b \leq 862)$ of the Band Positioning problem, we define its *FISH quality* as the number of FISH experiments that it supports, that is, the number of FISH experiments (x, b) such that $start(b) \leq x < start(b+1)$.

Problem 6.52

Find a solution to the Band Positioning problem that maximizes its FISH quality.

The FISH quality parameter ignores the width of the bands. A more adequate formulation is to find an optimal solution of the Band Positioning problem that is consistent with band width data, that is, the solution that minimizes

$$\sum_{b=1}^{862} |width(i) - (start(b+1) - start(b))|$$

, where $width(i)$ is the estimated width of the ith band.

Problem 6.53

Find an optimal solution of the Band Positioning problem that minimizes

$$\sum_{b=1}^{862} |width(i) - (start(b+1) - start(b))|$$

Problem 6.54

Describe recurrence relations for multiple alignment of 4 sequences under the SP (sum-of-pairs) scoring rule.

---

31. For simplicity we assume that $start(863) = n + 1$ thus implying that the last 862th band starts at the nucleotide $start(862)$ and ends at n.

Problem 6.55

Develop a likelihood ratio approach and design an algorithm that utilizes codon usage arrays for gene prediction.

Problem 6.56

Consider the Exon Chaining problem in the case when all intervals have the same weight. Design a greedy algorithm that finds an optimal solution for this limited case of the problem.

Problem 6.57

Estimate the running time of the spliced alignment algorithm. Improve the running time by transforming the spliced alignment graph into a graph with a smaller number of edges. This transformation is hinted at in figure 6.28.

Introns are spliced out of pre-mRNA during mRNA processing and biologists can perform *cDNA sequencing* that provides the nucleotide sequence complementary to the mRNA. The cDNA, therefore, represents the concatenation of exons of a gene. Consequently the exon-intron structure can be determined by aligning the cDNA against the genomic DNA with the aligned regions representing the exons and the large gaps representing the introns. This alignment can be aided by the knowledge of the conserved donor and acceptor splice site sequences (GT at the 5' splice site and AG at the 3' splice site).

While a spliced alignment can be used to solve this *cDNA Alignment* problem there exists a faster algorithm to align cDNA against genomic sequence. One approach is to introduce gap penalties that would adequately account for gaps in the cDNA Alignment problem. When aligning cDNA against genomic sequences we want to allow long internal gaps in the cDNA sequence. In addition, long gaps that respect the consensus sequences at the intron-exon junctions are favored over gaps that do not satisfy this property. Such gaps that exceed a given length threshold and respect the donor and acceptor sites should be assigned a constant penalty. This penalty is lower than the affine penalty for long gaps that do not respect the splice site consensus. The input to the cDNA Alignment problem is genomic sequence $\mathbf{v}$, cDNA sequence $\mathbf{w}$, match, mismatch, gap opening and gap extension parameters, as well as L (minimum intron length) and δ_L (fixed penalty for gaps longer than L that respect the consensus sequences). The output is an alignment of $\mathbf{v}$ and $\mathbf{w}$ where aligned regions represent putative exons and gaps in $\mathbf{v}$ represent putative introns.

Problem 6.58

Devise an efficient algorithm for the cDNA Alignment problem.

The spliced alignment algorithm finds exons in genomic DNA by using a related protein as a template. What if a template is not a protein but another (uninterpreted) genomic DNA sequence? Or, in other words, can (unannotated) mouse genomic DNA be used to predict human genes?

Problem 6.59

Generalize the spliced alignment algorithm for alignment of one genomic sequence against another.

Problem 6.60

For simplicity, the Spliced Alignment problem assumes that the genomic sequence and the target protein sequence are both written in the same alphabet. Modify the recurrence relations to handle the case when they are written in different alphabets (specifically, proteins are written in a twenty letter alphabet, and DNA is written in a four letter alphabet).

7 *Divide-and-Conquer Algorithms*

As the name implies, a divide-and-conquer algorithm proceeds in two distinct phases: a divide phase in which the algorithm splits a problem instance into smaller problem instances and solves them; and a conquer phase in which it stitches the solutions to the smaller problems into a solution to the bigger one. This strategy often works when a solution to a large problem can be built from the solutions of smaller problem instances. Divide-and-conquer algorithms are often used to improve the efficiency of a polynomial algorithm, for example, by solving a problem in $O(n \log n)$ time that would otherwise require quadratic time.

7.1 Divide-and-Conquer Approach to Sorting

In chapter 2 we introduced the Sorting problem, and developed an algorithm that required $O(n^2)$ time to sort a list of integers. The divide-and-conquer approach gives us a faster sorting algorithm.

Suppose that instead of a single list of n integers in an arbitrary order, we have two lists, **a** of length n_1 and **b** of length n_2, each with approximately $n/2$ elements, but these two lists are both sorted. How could we make a sorted list of n elements from these? A reasonable approach is to traverse each list simultaneously as if each were a sorted stack of cards, picking the smaller element on the top of either pile. The MERGE algorithm below combines two sorted lists into a single sorted list in $O(n_1 + n_2)$ time.

MERGE($\mathbf{a}, \mathbf{b}$)
1   $n1 \leftarrow$ size of $\mathbf{a}$
2   $n2 \leftarrow$ size of $\mathbf{b}$
3   $a_{n1+1} \leftarrow \infty$
4   $b_{n2+1} \leftarrow \infty$
5   $i \leftarrow 1$
6   $j \leftarrow 1$
7   **for** $k \leftarrow 1$ **to** $n1 + n2$
8   **if** $a_i < b_j$
9           $c_k \leftarrow a_i$
10          $i \leftarrow i + 1$
11   **else**
12          $c_k \leftarrow b_j$
13          $j \leftarrow j + 1$
14 **return c**

In order to use MERGE to sort an arbitrary list, we made an inductive leap: somehow we were presented with two half-size lists that were already sorted. It would seem to be impossible to get this input without actually solving the Sorting problem to begin with. However, the MERGE algorithm is easily applied if we have a list **c** with only two elements: break **c** into two lists, each list with one element. Since those sublists are sorted—a list of one element is always sorted—then we can merge them into a sorted 2-element list. If **c** has four elements, we can still break it into two lists, each with two elements, sort each of the two element lists, and merge the resulting sorted lists afterward. In fact, the same general idea applies to an arbitrary list and gives rise to the MERGESORT algorithm.

MERGESORT($\mathbf{c}$)
1   $n \leftarrow$ size of $\mathbf{c}$
2   **if** $n = 1$
3   **return c**
4   **left** $\leftarrow$ list of first $n/2$ elements of **c**
5   **right** $\leftarrow$ list of last $n - n/2$ elements of **c**
6   **sortedLeft** $\leftarrow$ MERGESORT(**left**)
7   **sortedRight** $\leftarrow$ MERGESORT(**right**)
8   **sortedList** $\leftarrow$ MERGE(**sortedLeft**, **sortedRight**)
9   **return sortedList**

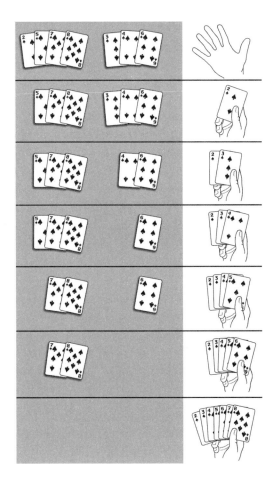

MERGESORT comprises two distinct phases: a divide phase in lines 4–7 where its input is split into parts and sorted; and a conquer phase in line 8, where the sorted sublists are then combined into a sorted version of the input list. In order to calculate the efficiency of this algorithm we need to account for the time spent by MERGESORT in each phase.

We will use $T(n)$ to represent the amount of time spent in a call to MERGE-SORT for a list with n elements; this involves two calls to MERGESORT on lists of size $n/2$, as well as a single call to MERGE. If MERGE is called on two lists each of size $n/2$, it will require $O(n/2 + n/2) = O(n)$ time to merge them. This leads to the following recurrence relation, where c is used to denote a positive constant:

$$T(n) = 2T(n/2) + cn$$
$$T(1) = 1$$

The solution to this recurrence relation is $T(n) = O(n \log n)$, a fact that can be verified through mathematical induction. Another way to establish the $O(n \log n)$ running time of MERGESORT is to construct the recursion tree (fig. 7.1) and to notice that it consists of $\log n$ levels.[1] At the top level, you have to merge two lists each with $\frac{n}{2}$ elements, requiring $O(\frac{n}{2} + \frac{n}{2}) = O(n)$ time. At the second level, there are four lists, each with $\frac{n}{4}$ elements, requiring $O(\frac{n}{4} + \frac{n}{4} + \frac{n}{4} + \frac{n}{4}) = O(n)$ time. At the ith level, there are 2^i lists, each with $\frac{n}{2^i}$ elements, again requiring $O(n)$ time. Therefore, merging requires overall $O(n \log n)$ time since there are $\log n$ levels in the recursion tree.

7.2   Space-Efficient Sequence Alignment

As another illustration of divide-and-conquer algorithms, we revisit the Sequence Alignment problem from chapter 6.

When comparing long DNA fragments, the limiting resource is usually not the running time of the algorithm, but the space required to store the dynamic programming table. In 1975 Daniel Hirschberg proposed a divide-and-conquer approach that performs alignment in linear space, at the expense of doubling the computational time.

The time complexity of the dynamic programming algorithm for aligning sequences of lengths n and m respectively is proportional to the number of *edges* in the edit graph, or $O(nm)$. On the other hand, the space complexity is proportional to the number of *vertices* in the edit graph, which is also $O(nm)$. However, if we only want to compute the score of the alignment rather than the alignment itself, then the space can be reduced to just twice the number of vertices in a single column of the edit graph, that is, $O(n)$. This reduction comes from the observation that the only values needed to compute the alignment scores in column j are the alignment scores in column $j - 1$ (fig. 7.2). Therefore, the alignment scores in the columns before $j - 1$ can be discarded while computing the alignment scores for columns $j, j + 1, \ldots, m$. Unfortunately, to find the longest path in the edit graph re-

---

1. How many times do you need to divide an array in half before you get to single-element sets?

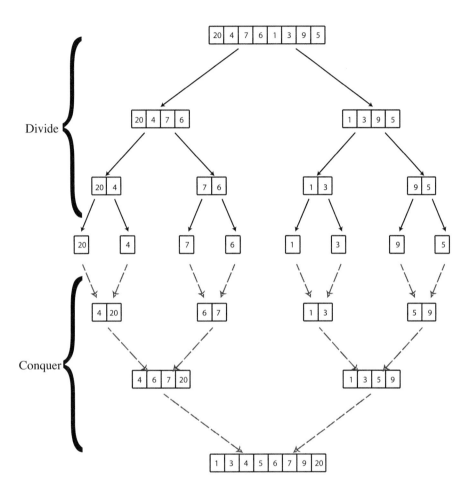

Figure 7.1   The recursion tree for MERGESORT. The divide (upper) part consists of
$\log 8 = 3$ levels (not counting the root) where the input is split into pieces. The
conquer (lower) part consists of the same number of levels where the split pieces are
merged back together.

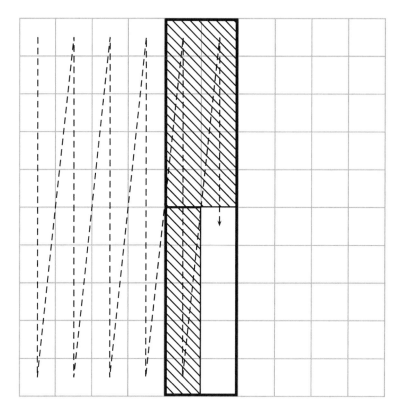

Figure 7.2 Calculating an alignment *score* requires no more than $2n$ space for an $n \times n$ alignment problem. Computing the alignment scores in each column requires only the scores in the preceding column. We show here the dynamic programming array–the data structure that holds the score at each vertex—instead of the graph.

quires backtracking pointers for the entire edit graph. Therefore, the entire backtracking matrix $\mathbf{b} = (b_{i,j})$ needs to be stored, causing the $O(nm)$ space requirement. However, we can finesse this to require only $O(n)$ space.

The longest path in the edit graph connects the *source* vertex $(0,0)$ with the *sink* vertex (n,m) and passes through some (unknown) *middle* vertex $(mid, \frac{m}{2})$, that is, the vertex somewhere on the middle column $(\frac{m}{2})$ of the graph (fig. 7.3). The key observation is that we can find this middle vertex without actually knowing the longest path in the edit graph. We define

$length(i)$ as the length of the longest path from $(0,0)$ to (n,m) that passes through the vertex $(i, \frac{m}{2})$. In other words, out of all paths from $(0,0)$ to (n,m), $length(i)$ is the length of the longest of the ones that pass through $(i, \frac{m}{2})$. Since the middle vertex, $(mid, \frac{m}{2})$, lies on the longest path from the source to the sink, $length(mid) = \max_{0 \leq i \leq n} length(i)$. Below we show that $length(i)$ can be efficiently computed without knowing the longest path. We will assume for simplicity that m is even and concentrate on finding only the middle vertex rather than the entire longest path.

Vertex $(i, \frac{m}{2})$ splits the $length(i)$-long path into two subpaths, which we will call *prefix* and *suffix*. The prefix subpath runs from the source to $(i, \frac{m}{2})$, and has length $prefix(i)$. The suffix subpath runs from $(i, \frac{m}{2})$ to the sink, and has length $suffix(i)$. It can be seen that $length(i) = prefix(i) + suffix(i)$, and an important observation is that $prefix(i)$ and $suffix(i)$ are actually very easy to compute in linear space. Indeed, $prefix(i)$ is simply the length of the longest path from $(0,0)$ to $(i, \frac{m}{2})$ and is given by $s_{i,\frac{m}{2}}$. Also, $suffix(i)$ is the length of the longest path from $(i, \frac{m}{2})$ to (n,m), or, equivalently, the length of the longest path from the sink (n,m) to $(i, \frac{m}{2})$ in the graph with all edges reversed. Therefore, $suffix(i)$ can be computed as a longest path in this "reversed" edit graph.

Computing $length(i)$ for $0 \leq i \leq n$ can be done in linear space by computing the scores $s_{i,\frac{m}{2}}$ (lengths of the prefix paths from $(0,0)$ to $(i, \frac{m}{2})$ for $0 \leq i \leq n$) and the scores of the paths from $(i, \frac{m}{2})$ to (n,m), which can be computed as the score $s_{i,\frac{m}{2}}^{reverse}$ of the path from (n,m) to $(i, \frac{m}{2})$ in the reversed edit graph. The value $length(i) = prefix(i) + suffix(i) = s_{i,\frac{m}{2}} + s_{i,\frac{m}{2}}^{reverse}$ is the length of the longest path from $(0,0)$ to (n,m) passing through the vertex $(i, \frac{m}{2})$. Therefore, $\max_{0 \leq i \leq n} length(i)$ computes the length of the longest path and determines mid.

Computing all $length(i)$ values requires time equal to the area of the left rectangle (from column 1 to $\frac{m}{2}$) plus the area of the right rectangle (from column $\frac{m}{2}+1$ to m) and the space $O(n)$, as shown in figure 7.3. After the middle vertex $(mid, \frac{m}{2})$ is found, the problem of finding the longest path from $(0,0)$ to (n,m) can be partitioned into two subproblems: finding the longest path from $(0,0)$ to the middle vertex $(mid, \frac{m}{2})$ and finding the longest path from the middle vertex $(mid, \frac{m}{2})$ to (n,m). Instead of trying to find these paths, we first try to find the middle vertices in the corresponding smaller rectangles (fig. 7.3). This can be done in the time equal to the area of these rectangles, which is half as large as the area of the original rectangle. Proceeding in this way, we will find the middle vertices of all rectangles in time proportional to $area + \frac{area}{2} + \frac{area}{4} + \ldots \leq 2 \times area$, and therefore compute the longest path in time $O(nm)$ and space $O(n)$:

PATH($source, sink$)
1   **if** $source$ **and** $sink$ are in consecutive columns
2   **output** longest path from $source$ to $sink$
3   **else**
4       $mid \leftarrow$ middle vertex $(i, \frac{m}{2})$ with largest score $length(i)$
5   PATH($source, mid$)
6   PATH($mid, sink$)

7.3   Block Alignment and the Four-Russians Speedup

We began our analysis of sorting with the quadratic SELECTIONSORT algorithm and later developed the MERGESORT algorithm with $O(n \log n)$ running time. A natural question to ask is whether one could design an even faster sorting algorithm, perhaps a linear one. Alas, for the Sorting problem there exists a lower bound for the complexity of *any* sorting algorithm, essentially stating that it will require at least $\Omega(n \log n)$ operations.[2] Therefore, it makes no sense to improve upon MERGESORT with the expectation of improving the worst-case running time (though improving the practical running time is worth the effort).

Similarly, we began our analysis of the Global Alignment problem from the dynamic programming algorithm that requires $O(n^2)$ time to align two n-nucleotide sequences, but never asked whether an even faster alignment algorithm existed. Could it be possible to reduce the running time of the alignment algorithm from $O(n^2)$ to $O(n \log n)$? Nobody has an answer to this question because nontrivial lower bounds for the Global Alignment problem remain unknown.[3] An $O(n \log n)$ alignment algorithm would revolutionize bioinformatics and would likely be the demise of the popular BLAST algorithm. Although nobody knows how to design an $O(n \log n)$ algorithm for global alignment, there exists a subquadratic $O(\frac{n^2}{\log n})$ algorithm for a similar Longest Common Subsequence (LCS) problem.

---

2. This result relies on certain assumptions about the nature of computation, which are not really germane to this book. As an example, if you had an unlimited supply of computers sorting a list in parallel, you could perhaps sort faster.

3. One cannot simply argue that the *problem* requires $O(n^2)$ time since one has to traverse the entire dynamic programming table, because the problem might be solved by some ingenious technique that does not rely on a dynamic programming recurrence.

Linear-Space Sequence Alignment

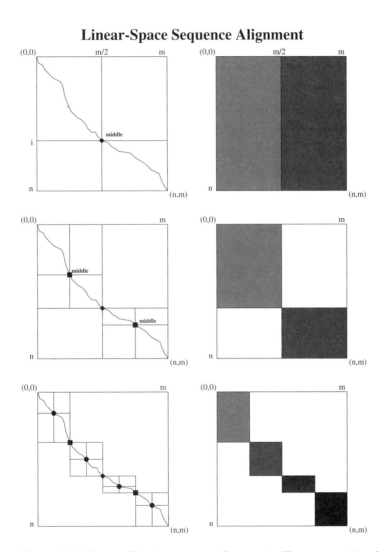

Figure 7.3   Space-efficient sequence alignment. The computational time (i.e., the area of the solid rectangles) decreases by a factor of 2 at every iteration.

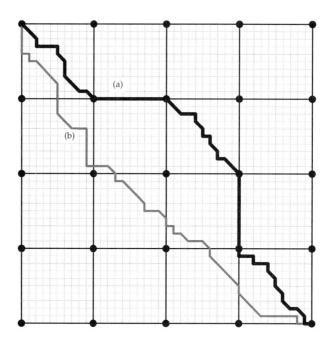

Figure 7.4 Two paths in a 40×40 grid partitioned into 16 subgrids of size 10×10. The black path (a) is a block path, while the gray path (b) is not.

Let $\mathbf{u} = u_1 \dots u_n$ and $\mathbf{v} = v_1 \dots v_n$ be two DNA sequences partitioned into *blocks* of length t, that is, $\mathbf{u} = |u_1 \dots u_t| \, |u_{t+1} \dots u_{2t}| \, \dots \, |u_{n-t+1} \dots u_n|$ and $\mathbf{v} = |v_1 \dots v_t| \, |v_{t+1} \dots v_{2t}| \, \dots \, |v_{n-t+1} \dots v_n|$. For simplicity we assume that $\mathbf{u}$ and $\mathbf{v}$ have the same length and that it is divisible by t. For example, if t were 3 one could view $\mathbf{u}$ and $\mathbf{v}$ as DNA sequences of genes partitioned into codons. The *block alignment* of $\mathbf{u}$ and $\mathbf{v}$ is an alignment in which every block in one sequence is either aligned against an *entire* block in the other sequence or is inserted or deleted as a whole. To be sure, the alignment path within a block can be completely arbitrary—it simply needs to enter and leave the block through vertices on the corners.

Figure 7.4 shows an $n \times n$ grid partitioned into $t \times t$ subgrids. A path in this edit graph is called a *block* path if it traverses every $t \times t$ square through its corners (i.e., enters and leaves every block at bold vertices). An equivalent statement of this definition is that a block path contains at least two bold vertices from every square that it passes through—it cannot, for example, lop off a corner of a square. Block alignments correspond to block paths in the edit graph and the *Block Alignment problem* is to find the highest-scoring, or longest, block path through this graph. We will see below that when t is on the order of the logarithm of the overall sequence length—neither too small nor too large–we can solve this problem in less than quadratic n^2 time.

Block Alignment Problem:
Find the longest block path through an edit graph.

 Input: Two sequences, **u** and **v** partitioned into blocks of
 size t.

 Output: The block alignment of **u** and **v** with the maximum
 score (i.e., the longest block path through the edit graph).

One can consider $\frac{n}{t} \times \frac{n}{t}$ pairs of blocks (each pair defines a square in the edit graph) and compute the alignment score $\beta_{i,j}$ for each pair of blocks $|u_{(i-1)\cdot t+1} \ldots u_{i \cdot t}|$ and $|v_{(j-1)\cdot t+1} \ldots v_{j \cdot t}|$. This amounts to solving $\frac{n}{t} \times \frac{n}{t}$ mini alignment problems of size $t \times t$ each and takes $O(\frac{n}{t} \cdot \frac{n}{t} \cdot t \cdot t) = O(n^2)$ time. If $s_{i,j}$ denotes the optimal block alignment score between the first i blocks of **u** and the first j blocks of **v**, then

$$s_{i,j} = \max \begin{cases} s_{i-1,j} - \sigma_{block} \\ s_{i,j-1} - \sigma_{block} \\ s_{i-1,j-1} + \beta_{i,j} \end{cases},$$

where σ_{block} is the penalty for inserting or deleting the entire block.[4] The indices i and j in this recurrence vary from 0 to $\frac{n}{t}$ and therefore, the running time of this algorithm is $O(\frac{n^2}{t^2})$ if we do not count time to precompute $\beta_{i,j}$ for $0 \leq i,j \leq \frac{n}{t}$. This approach allows one to solve the Block Alignment problem for any value of t, but as we saw before, precomputing all $\beta_{i,j}$ takes the same $O(n^2)$ time that the dynamic programming algorithm takes.

The speed reduction we promised is achieved by the Four-Russians technique when t is roughly $\log n$.[5] Instead of constructing $\frac{n}{t} \times \frac{n}{t}$ minialignments for all pairs of blocks from **u** and **v** we will construct $4^t \times 4^t$ minialignments

---

4. In the simplest case $\sigma_{block} = \sigma t$, where σ is the penalty for the insertion or deletion of a nucleotide.
5. Since the Block Alignment problem takes a partitioned grid as input, the algorithm does not get to make a choice for the value of t.

for all pairs of t-nucleotide strings and store their alignment scores in a large lookup table. At first glance this looks counterproductive, but if $t = \frac{\log n}{4}$ then $4^t \times 4^t = n^{\frac{1}{2}} \times n^{\frac{1}{2}} = n$, which is much smaller than $\frac{n}{t} \times \frac{n}{t}$.

The resulting lookup table, which we will call $Score$, has only $4^t \times 4^t = n$ entries. Computing each of the entries takes $O(\log n \cdot \log n)$ time, so the overall running time to compute all entries of this table is only $O(n \cdot (\log n)^2)$. We emphasize that the resulting two-dimensional lookup table $Score$ is indexed by a pair of t-nucleotide strings, thus leading to a slightly different recurrence:

$$ s_{i,j} = \max \begin{cases} s_{i-1,j} - \sigma_{block} \\ s_{i,j-1} - \sigma_{block} \\ s_{i-1,j-1} + Score(i\text{th block of } \mathbf{v}, \, j\text{th block of } \mathbf{u}) \end{cases} $$

Since the time to precompute the lookup table $Score$ in this case is relatively small, the overall running time is dominated by the dynamic programming step, for example, by the $\frac{n}{t} \times \frac{n}{t}$ accesses it makes to the lookup table. Since each access takes $O(\log n)$ time, the overall running time of this algorithm is $O(\frac{n^2}{\log n})$.

7.4   Constructing Alignments in Subquadratic Time

So now we have an algorithm for the Block Alignment problem that is subquadratic for convenient values of one of its input parameters, but it is not clear whether similar ideas could be used to solve any of the problems from chapter 6. In this section we show how to design a $O(\frac{n^2}{\log n})$ algorithm for finding the longest common subsequence of two strings, again using the Four-Russians speedup.

Unlike the block path in a partitioned edit graph, the path corresponding to a longest common subsequence can traverse the edit graph arbitrarily and does not have to pass through the bold vertices of figure 7.4. Therefore, precomputing the length of paths between the upper left corner and lower right corner of every $t \times t$ subsequence is not going to help.

Instead we will select all vertices at the borders of the squares (shown by bold vertices in figure 7.5) rather than just the vertices at the corners, as in figure 7.4. This results in a significantly larger number of bold vertices than in the case of block alignments but we can keep the number subquadratic. Taken together, these vertices form $\frac{n}{t}$ whole rows and $\frac{n}{t}$ whole columns in the edit graph; the total number of bold vertices is $O(\frac{n^2}{t})$. We will perform

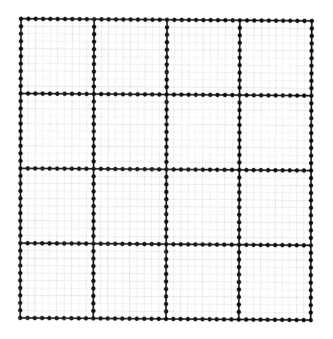

Figure 7.5   The partitioned edit graph for the LCS problem.

dynamic programming on only the $O(\frac{n^2}{t})$ bold vertices, effectively ignoring
the internal vertices in the edit graph.

In essence we are interested in the following problem: given the alignment
scores $s_{i,*}$ in the first row and the alignment scores $s_{*,j}$ in the first column of
a $t \times t$ minisquare, compute the alignment scores in the last row and column
of the minisquare. The values of s in the last row and last column depend
entirely on four variables: the values $s_{i,*}$ in the first row of the square, the
values $s_{*,j}$ in the first column of the square, and the two t-long substrings
corresponding to the rows and columns of the square. Of course, we could
use this information to fill in the entire dynamic programming matrix for a
$t \times t$ square but we cannot afford doing this (timewise) if we want to have a
subquadratic algorithm.

To use the Four-Russians technique, we again rely on the brute force men-
tality and build a lookup table on all possible values of the four variables:
all pairs of t-nucleotide sequences *and* all pairs of possible scores for the first

row $s_{i,*}$ and the first column $s_{*,j}$. For each such quadruple, we store the precomputed scores for the last row and last column. However, this will be an enormous table since there may be a large number of possible scores for the first row and first column. Therefore, we perform some trickery. A careful analysis of the LCS problem shows that the possible alignment scores in the first row (or first column) are not entirely arbitrary: $0, 1, 2, 2, 2, 3, 4$ is a possible sequence of scores, but $0, 1, 2, 4, 5, 8, 9, 10$ is not. Not only does the progression of scores have to be monotonically increasing, but adjacent elements cannot differ by more than 1 (see problem 6.47). We can encode this as a binary vector of *differences*; the above example $0, 1, 2, 2, 2, 3, 4$ would be encoded as $1, 1, 0, 0, 1, 1$.[6] Thus, since there are 2^t possible scores and 4^t possible strings, the entire lookup table will require $2^t \cdot 2^t \cdot 4^t \cdot 4^t = 2^{6t}$ space. Again, we set $t = \frac{\log n}{4}$ to make the size of the table collapse down to $2^{6\frac{\log n}{4}} = n^{1.5}$. Alas, this allows the precomputation step to be subquadratic, and the running time of the algorithm is dominated by the process of filling in the scores for the bold vertices in figure 7.5 which takes $O\left(\frac{n^2}{\log n}\right)$ time.

7.5   Notes

MergeSort was invented in 1945 by the legendary John von Neumann while he was designing EDVAC, the world's first stored-program electronic computer. The idea of using a divide-and-conquer approach for sequence comparison was proposed first by Daniel Hirschberg in 1975 for the LCS problem (49), and then in 1988 by Eugene Myers and Webb Miller for the Local Alignment problem (77). The Four-Russians speedup was proposed by Vladimir Arlazarov, Efim Dinic, Mikhail Kronrod, and Igor Faradzev in 1970 (6) and first applied to sequence comparison by William Masek and Michael Paterson (73).

---

6. $(1, 1, 0, 0, 1, 1)$ is $(1 - 0, 2 - 1, 2 - 2, 2 - 2, 3 - 2, 4 - 3)$.

Webb Miller (born 1943 in Washington State) is professor in the Departments of Biology and of Computer Science and Engineering at Pennsylvania State University. He holds a PhD in mathematics from the University of Washington. He is a pioneer and a leader in the area of DNA and protein sequence comparison, and in comparing whole genomes in particular.

For a number of years Miller worked on computational techniques for understanding the behavior of computer programs that use floating-point arithmetic. In 1987 he completely changed his research focus, after picking bioinformatics as his new field. He says:

> My reason for wanting a complete change was simply to bring more adventure and excitement into my life. Bioinformatics was attractive because I had no idea what the field was all about, and because neither did anyone else at that time.

The catalyst was his friendship with Gene Myers, who was already working in the new area. It wasn't even called "bioinformatics" then; Miller was switching to a field without a name. He loved the frontier spirit of the emerging discipline and the possibility of doing something useful for mankind.

> The change was difficult for me because I was completely ignorant of biology and statistics. It took a number of years before I really started to understand biology. I'm now on the faculty of a biology department, so in some sense I successfully made the transition. (Unfortunately, I'm still basically ignorant of statistics.) In another respect, the change was easy because there was so little already known about the field. I read a few papers by Mike Waterman and David Sankoff, and was off and running.

Miller came to the new field armed with two skills that proved very useful, and with a couple of ideas that helped focus his research initially. The skills were his mathematical training and his experience writing computer programs. The first idea that he brought to the field was that an optimal alignment between two sequences can be computed in space proportional to the length of the longer sequence. It is straightforward to compute the score of an optimal alignment in that amount of space, but it is much less obvious how to produce an alignment with that score. A very clever linear-space alignment algorithm had been discovered by Dan Hirschberg around 1975. The other idea was that when two sequences are very similar and when alignments are scored rather simply, an optimal alignment can be computed much more quickly than by dynamic programming, using a greedy algorithm. That idea was discovered independently by Gene Myers (with some prodding from Miller) and Esko Ukkonen in the mid-1980s. Miller hoped that these two ideas, or variants of them, would get him started in the new field; he had "solutions in search of biological problems" rather than "biological problems in search of solutions." Indeed, this is a common mode of entry into bioinformatics for scientists trained in a quantitative field.

During his first decade in bioinformatics, Miller coauthored a few papers about linear-space alignment methods. Finding a niche for greedy algorithms took longer, but for comparing very similar DNA sequences, particularly when the difference between them is due to sequencing errors rather than evolutionary mutations, they are quite useful; they deserve wider recognition in the bioinformatics community than they now have.

The most successful of Miller's bioinformatics projects have involved ideas other than the ones he brought with him to the field. His most widely known project was the collaboration to develop the **BLAST** program, where it was David Lipman's insights that drove the project in the right direction. However, it is Miller's work on comparison methods for long DNA sequences that brought him closer to biology and made Miller's algorithms a household name among teams of scientists analyzing mammalian and other whole-genome sequences. Miller picked this theme as his Holy Grail around 1989, and he has stuck with it ever since. When he started, there were only two people in the world brave—or foolish—enough to publicly advocate sequencing the mouse genome and comparing it with the human genome: Miller and his long-term collaborator, the biologist Ross Hardison. They occasionally went so far as to tout the sequencing of several additional mammals. Nowadays, it looks to everyone like the genome sequencing of mouse, rat, chimpanzee, dog, and so on, was inevitable, but perhaps Miller's many years of working on programs to compare genome sequences made the inevitable happen sooner.

What worked best for Miller was to envision an advance in bioinformatics that would foster new biological discoveries – namely, that development of methods to compare complete mammalian genome sequences would lead to a better understanding of evolution and of gene regulation – and to do everything he could think of to make it happen. This included developing algorithms that would easily align the longest sequences he could find, and helping Ross Hardison to verify experimentally that these alignments are useful for studying gene regulation. When Miller and Hardison decided to show how alignments and data from biological experiments could be linked through a database, they learned about databases. When they wanted to set up a network server to align DNA sequences, they learned about network servers. When nobody in his lab was available to write software that they needed, Miller wrote it himself. When inventing and analyzing a new algorithm seemed important, he worked on it. The methods changed but the biological motivation remained constant.

Miller has been more successful pursuing "a biological problem in search of solutions than the other way around. His colleague, David Haussler, has had somewhat the same experience; his considerable achievements bringing hidden Markov models and other machine learning techniques to bioinformatics have recently been eclipsed by his monumental success with the Human Genome Browser, which has directly helped a far wider community of scientists.

> The most exciting point so far in my career is today, with a new vertebrate genome sequence coming my way every year. Some day, I hope to look back with pride at my best achievement in bioinformatics, but perhaps it hasn't happened yet.

7.6   Problems

Problem 7.1

Construct the recursion tree for MERGESORT on the input $(2, 5, 7, 4, 3, 6, 1, 8)$.

Problem 7.2

How much memory does MERGESORT need overall? Modify the algorithm to use as little as possible.

Problem 7.3

Suppose that you are given an array A of n words sorted in lexicographic order and want to search this list for some arbitrary word, perhaps w (we write the number of characters in w as $|w|$). Design three algorithms to determine if w is in the list: one should have $O(n\,|w|)$ running time; another should have $O(|w| \log n)$ running time but use no space (except for A and w); and the third should have $O(|w|)$ running time but can use as much additional space as needed.

Problem 7.4

We normally consider multiplication to be a very fast operation on a computer. However, if the numbers that we are multiplying are very large (say, 1000 digits), then multiplication by the naive grade-school algorithm will take a long time. How long does it take? Write a faster divide-and-conquer algorithm for multiplication.

Problem 7.5

Develop a linear-space version of the local alignment algorithm.

Problem 7.6

Develop a linear-space version of global sequence alignment with affine gap penalties.

In the space-efficient approach to sequence alignment, the original problem of size $n \times n$ is reduced to two subproblems of sizes $i \times \frac{n}{2}$ and $(n - i) \times \frac{n}{2}$ (for the sake of simplicity, we assume that both sequences have the same length). In a fast parallel implementation of sequence alignment, it is desirable to have a *balanced partitioning* that breaks the original problem into subproblems of equal sizes.

Problem 7.7

Design a space-efficient alignment algorithm with balanced partitioning.

Problem 7.8

Design a divide-and-conquer algorithm for the Motif Finding problem and estimate its running time. Have you improved the running time of the exhaustive search algorithm?

Problem 7.9

Explore the possibilities of using a divide-and-conquer approach for the Median String problem. Can you split the problem into subproblems? Can you combine the solutions of the subproblem into a solution to the main problem?

Problem 7.10

Devise a space-efficient dynamic programming algorithm for multiple alignment of three sequences. Write the corresponding recurrence relations. For three n-nucleotide sequences your algorithm should use at most quadratic $O(n^2)$ memory. Write the recursive algorithm that outputs the resulting alignment.

Problem 7.11

Design a linear-space algorithm for the Block Alignment problem.

Problem 7.12

Write a pseudocode for constructing the LCS in subquadratic time.

8 *Graph Algorithms*

Many bioinformatics algorithms may be formulated in the language of *graph theory*. The use of the word "graph" here is different than in many physical science contexts: we do not mean a chart of data in a Cartesian coordinate system. In order to work with graphs, we will need to define a few concepts that may not appear at first to be particularly well motivated by biological examples, but after introducing some of the mathematical theory we will show how powerful they can be in such bioinformatics applications as DNA sequencing and protein identification.

8.1 Graphs

Figure 8.1 (a) shows two white and two black knights on a 3×3 chessboard. Can they move, using the usual chess knight's moves,[1] to occupy the positions shown in figure 8.1 (b)? Needless to say, two knights cannot occupy the same square while they are moving.

Figure 8.2b represents the chessboard as a set of nine points. Two points are connected by a line if moving from one point to another is a valid knight move. Figure 8.2c shows an equivalent representation of the resulting diagram that reveals that knights move around a "cycle" formed by points 1, 6, 7, 2, 9, 4, 3, and 8. Every knight's move on the chessboard corresponds to moving to a neighboring point in the diagram, in either a clockwise or counterclockwise direction. Therefore the white-white-black-black knight arrangement cannot be transformed into the alternating white-black-white-black arrangement.

---

1. In the game of chess, knights (the "horses") can move two steps in any of four directions (left, right, up, and down) followed by one step in a perpendicular direction, as shown in figure 8.1 (c).

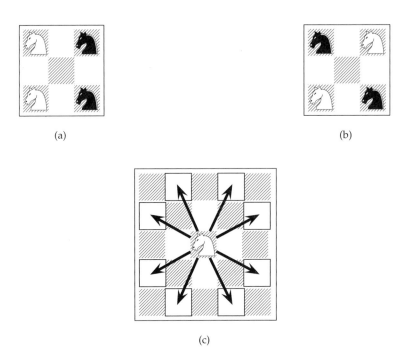

(a)   (b)

(c)

Figure 8.1   Two configurations of four knights on a chessboard. Can you use valid knight moves to turn the configuration in (a) into the configuration in (b)? Valid knight moves are shown in (c).

Diagrams with collections of points connected by lines are examples of *graphs*. The points are called *vertices* and lines are called *edges*. A simple graph shown in figure 8.3, consists of five vertices and six edges. We denote a graph by $G = G(V, E)$ and describe it by its set of vertices V and set of edges E (every edge can be written as a pair of vertices). The graph in figure 8.3 is described by the vertex set $V = \{a, b, c, d, e\}$ and the edge set $E = \{(a, b), (a, c), (b, c), (b, d), (c, d), (c, e)\}$.

The way the graph is actually drawn is irrelevant; two graphs with the same vertex and edge sets are equivalent, even if the particular pictures that represent the graph appear different (see figure 8.3). The only important feature of a graph is which vertices are connected and which are not.

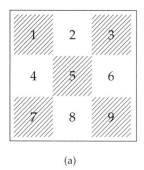

(a)

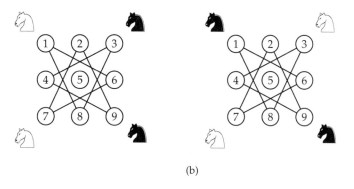

(b)

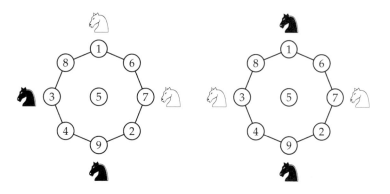

(c)

Figure 8.2   A graph representation of a chessboard. A knight sitting on some square can reach any of the squares attached to that square by an edge.

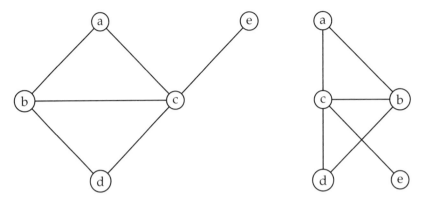

Figure 8.3   Two equivalent representations of a simple graph with five vertices and six edges.

Figure 8.4 represents another chessboard obtained from a 4×4 chessboard by removing the four corner squares. Can a knight travel around this board, pass through each square exactly once, and return to the same square it started on? Figure 8.4 (b) shows a rather complex graph with twelve vertices and sixteen edges revealing all possible knight moves. However, rearranging the vertices (fig. 8.4c) reveals the cycle that describes the correct sequence of moves.

The number of edges incident to a given vertex v is called the *degree* of the vertex and is denoted $d(v)$. For example, vertex 2 in figure 8.4 (c) has degree 3 while vertex 4 has degree 2. The sum of degrees of all 12 vertices is, in this case, 32 (8 vertices of degree 3 and 4 vertices of degree 2), twice the number of edges in the graph. This is not a coincidence: for every graph G with vertex set V and edge set E, $\sum_{v \in V} d(v) = 2 \cdot |E|$. Indeed, an edge connecting vertices v and w is counted in the sum $\sum_{v \in V} d(v)$ twice: first in the term $d(v)$ and again in the term $d(w)$. The equality $\sum_{v \in V} d(v) = 2 \cdot |E|$ explains why you cannot connect fifteen phones such that each is connected to exactly seven others, and why a country with exactly three roads out of every city cannot have precisely 1000 roads.

Many bioinformatics problems make use of *directed* graphs, in which every edge is directed from one vertex to another, as shown by the arrows in figure 8.5. Every vertex v in a directed graph is characterized by *indegree(v)* (the number of incoming edges) and *outdegree(v)* (the number of outgoing

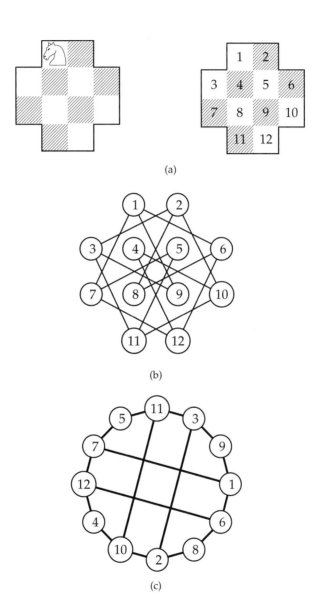

Figure 8.4 The knight's tour through the twelve squares in part (a) can be seen by constructing a graph (b) and rearranging its vertices in a clever way (c).

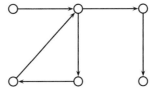

Figure 8.5   A directed graph.

edges). For every directed graph $G(V, E)$,

$$\sum_{v \in V} indegree(v) = \sum_{v \in V} outdegree(v),$$

since every edge is counted once on the right-hand side of the equation and once on the left-hand side.

A graph is called *connected* if all pairs of vertices can be connected by a *path*, which is a continuous sequence of edges, where each successive edge begins where the previous one left off. Paths that start and end at the same vertex are referred to as *cycles*. For example, the paths (3-2-10-11-3) and (3-2-8-6-12-7-5-11-3) in figure 8.4 (c) are cycles.

Graphs that are not connected are *disconnected* (fig. 8.6). Disconnected graphs can be partitioned into *connected components*. One can think of a graph as a map showing cities (vertices) and the freeways (edges) that connect them. Not all cities are connected by freeways: for example, you cannot drive from Miami to Honolulu. These two cities belong to two different connected components of the graph. A graph is called *complete* if there is an edge between every two vertices.

Graph theory was born in the eighteenth century when Leonhard Euler solved the famous Königsberg Bridge problem. Königsberg is located on the banks of the Pregel River, with a small island in the middle. The various parts of the city are connected by bridges (fig. 8.7) and Euler was interested in whether he could arrange a tour of the city in such a way that the tour visits each bridge exactly once. For Königsberg this turned out to be impossible, but Euler basically invented an algorithm to solve this problem for any city.

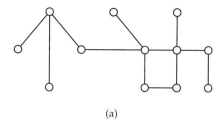

(a)

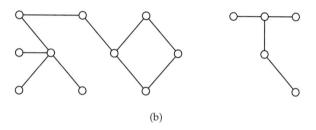

(b)

Figure 8.6   A connected (a) and a disconnected (b) graph.

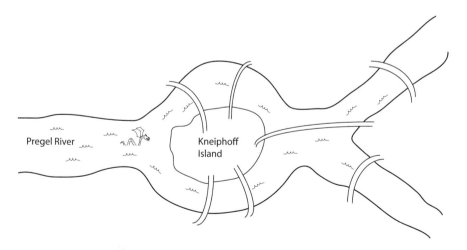

Figure 8.7   Bridges of Königsberg.

Bridge Obsession Problem:
Find a tour through a city (located on n islands connected by m bridges) that starts on one of the islands, visits every bridge exactly once, and returns to the originating island.

    Input: A map of the city with n islands and m bridges.

    Output: A tour through the city that visits every bridge exactly once and returns to the starting island.

Figure 8.8 shows a city map with ten islands and sixteen bridges, as well as the transformation of the map into a graph with ten vertices and sixteen edges (every island corresponds to a vertex and every bridge corresponds to an edge). After this transformation, the Bridge Obsession problem turns into the Eulerian Cycle problem that was solved by Euler and later found thousands of applications in different areas of science and engineering:

Eulerian Cycle Problem:
Find a cycle in a graph that visits every edge exactly once.

    Input: A graph G.

    Output: A cycle in G that visits every edge exactly once.

After the Königsberg Bridge problem was solved, graph theory was forgotten for a century before it was rediscovered by Arthur Cayley who studied the chemical structures of (noncyclic) saturated hydrocarbons C_nH_{2n+2} (fig. 8.9). Structures of this type of hydrocarbon are examples of *trees*, which are simply connected graphs with no cycles. It is not hard to show that every tree has at least one vertex with degree 1, or *leaf*.[2] This observation immediately implies that every tree on n vertices has $n - 1$ edges, regardless of the structure of the tree. Indeed, since every tree has a leaf, we can remove it and its attached edge, resulting in another tree. So far we have removed one edge and one vertex. In this smaller tree there exists a leaf that we, again, remove. So far, we have removed two vertices and two edges. We keep this up until we are left with a graph with a single vertex and no edges. Since we have removed $n - 1$ vertices and $n - 1$ edges, the number of edges in every tree is $n - 1$ (fig. 8.10).

---

2. Actually, every tree has at least two leaves, except for the trivial single-vertex tree.

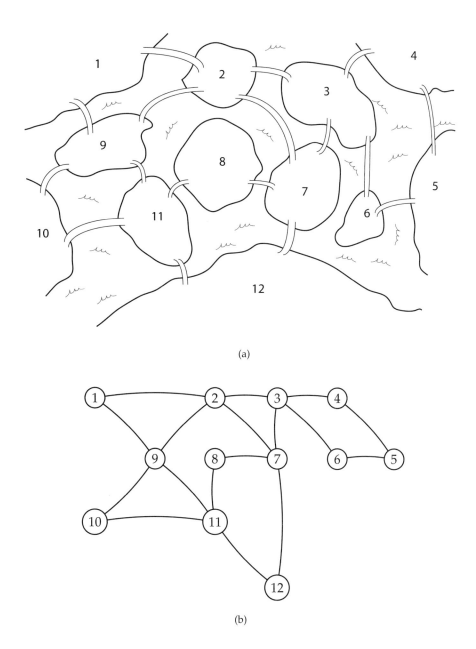

(a)

(b)

Figure 8.8   A more complicated version of Königsberg (a). To solve the Bridge Obsession problem, Euler transformed the map of Königsberg into a graph (b) and found an *Eulerian cycle*. The path that runs through vertices 1-2-3-4-5-6-3-7-2-9-11-8-7-12-11-10-9-1 is an Eulerian cycle.

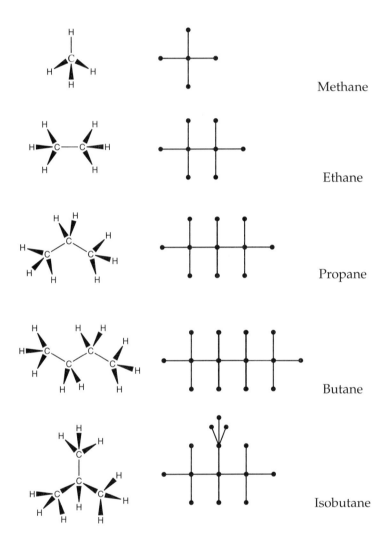

Methane

Ethane

Propane

Butane

Isobutane

Figure 8.9 Hydrocarbons (the saturated, nonaromatic variety) as chemists see them (left), and their graph representation (right). Two different molecules with the same number of the same types of atoms are called *structural isomers*.

Figure 8.10 Proving that a tree with n vertices has $n - 1$ edges.

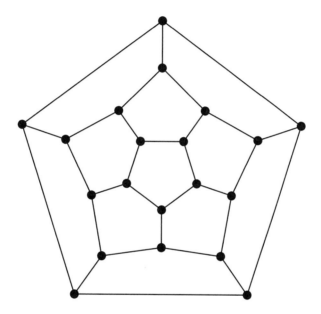

Figure 8.11 Can you travel from any one of the vertices in this graph, visit every other vertex exactly once, and end up at the original vertex?

Shortly after Cayley's work on tree enumeration, Sir William Hamilton invented a game corresponding to a graph whose twenty vertices were labeled with the names of twenty famous cities (fig. 8.11). The goal is to visit all twenty cities in such a way that every city is visited exactly once before returning back to the city where the tour started. As the story goes, Hamilton sold the game for 25 pounds to a London game dealer and it failed miserably. Despite the commercial failure of a great idea, the more general problem of finding "Hamiltonian cycles" in arbitrary graphs is of critical importance to many scientific and engineering disciplines. The problem of finding Hamiltonian cycles looks deceivingly simple and somehow similar to the Eulerian Cycle problem. However, it turns out to be $\mathcal{NP}$-complete while the Eulerian Cycle problem can be solved in linear time.[3]

Hamiltonian Cycle Problem:

Find a cycle in a graph that visits every vertex exactly once.

Input: A graph G.

Output: A cycle in G that visits every vertex exactly once (if such a cycle exists).

Graphs, like many freeway maps, often give some sort of weight to every edge, as in the Manhattan Tourist problem in chapter 6. The weight of an edge may reflect, depending on the context, different attributes. For example, the length of a freeway segment connecting two cities, the number of tourist attractions along a city block, and the alignment score between two amino acids are all natural weighting schemes. Weighted graphs are often formally represented as an ordered triple, $G = (V, E, w)$, where V is the set of vertices in the graph, E is the set of edges, and w is a weight function defined for every edge e in E (i.e., $w(e)$ is a number reflecting the weight of edge e). Given a weighted graph, one may be interested in finding some shortest path between two vertices (e.g., a shortest path between San Diego and New York). Though this problem may sound difficult if you were given a complicated road map, it turns out that there exist fast algorithms to answer this question.

Shortest Path Problem:

Given a weighted graph and two vertices, find the shortest distance between them.

Input: A weighted graph, $G = (V, E, w)$, and two distinguished vertices s and t.

Output: The shortest path between s and t in graph G.

---

3. The Hamiltonian Cycle problem is equivalent in complexity to the Traveling Salesman problem mentioned in chapter 2 and is therefore $\mathcal{NP}$-complete.

(a)

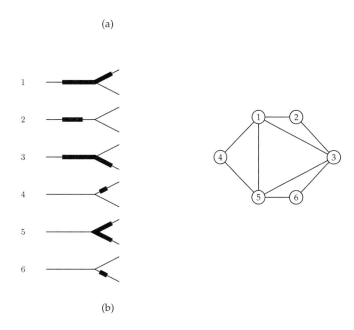

(b)

Figure 8.12 Two different hypothetical structures of a gene. (a) Linear structures exhibit very different interval graphs than the graphs exhibited by (b) branched structures. It is impossible to construct a linear sequence of overlapping intervals that gives rise to the graph in (b).

We emphasize that this problem is different—and somewhat more complicated—than the Longest Path in a Directed Acyclic Graph (DAG) problem that we considered in chapter 6.

Graphs are powerful tools in many applied studies and their role goes well beyond analysis of maps of freeways. For example, at first glance, the following "basketball" problem has nothing to do with graph theory:

> Fifteen teams play a basketball tournament in which each team plays with each other team. Prove that the teams can always be numbered from 1 to 15 in such a way that team 1 defeated team 2, team 2 defeated team 3, ... , team 14 defeated team 15.

A careful analysis reduces this problem to finding a Hamiltonian path in a directed graph on fifteen vertices.

8.2   Graphs and Genetics

Conceived by Euler, Cayley, and Hamilton, graph theory flourished in the twentieth century to become a critical component of discrete mathematics. In the 1950s, Seymour Benzer applied graph theory to show that genes are linear.

At that time, it was known that genes behaved as functional units of DNA, much like pearls on a necklace, but the chemical structure and organization of the genes was not clear. Were genes broken into still smaller components? If so, how were they organized? Prior to Watson and Crick's elucidation of the DNA double helix, it seemed a reasonable hypothesis that the DNA content of genes was branched, as in figure 8.12, or even looped, rather than linear. These two organizations have very different topological implications, which Benzer exploited in an ingenious experiment.

Benzer studied a large number of bacteriophage[4] T4 mutants, which happened to have a continuous interval deleted from an important gene. In their "normal" state, nonmutant T4 phages will kill a bacterium. The mutant T4 phages that were missing a segment of their genome could not kill the bacterium. Different mutants had different intervals deleted from the gene, and Benzer had to determine which interval had been deleted in each mutant. Though Benzer did not know exactly where in the gene the mutant's deletion was, he had a way to test whether two deletions (i.e., their intervals) overlapped, relying on how two phages with two different deletions behave

---

4. A bacteriophage is a virus that attacks bacteria.

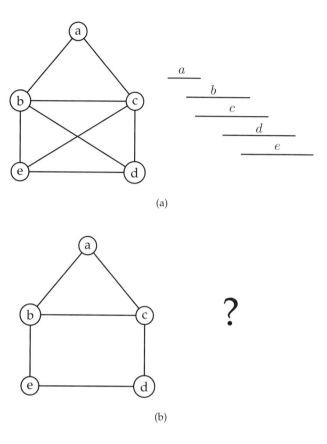

(a)

(b)

Figure 8.13   An interval graph (a) and a graph that is not an interval graph (b).

inside a bacterial host cell. When two mutant phages with two different deletions infect a bacterium, two outcomes are possible depending on whether
the deleted intervals overlap or not. If they do not overlap, then two phages
combined have all the genetic material of one normal phage. However, if two
deleted intervals overlap, some genetic material is absent in both phages.
The Benzer experiment was based on the observation that in the former case
(all genetic material of a normal phage is present), two mutants are able to
kill the bacterium; and in the latter case (where some genetic material is removed in both phages) the bacterium survives.

Benzer infected bacteria with each *pair* of mutant strains from his T4 phage
and simply noted which pairs killed the bacterial host. Pairs which were

lethal to the host were mutants whose deleted intervals did not overlap. Benzer constructed a graph[5] where each strain was a vertex and two vertices were connected when a double infection of a bacterial host was nonlethal. He reasoned that if genes were linear [fig. 8.12 (a)], he would probably see one type of graph, but if the genes were branched [fig. 8.12 (b)] he would see another.

Given a set of n intervals on a line segment, the *interval graph* is defined as a graph with n vertices that correspond to the intervals. There is an edge between vertices v and w if and only if the intervals v and w overlap. Interval graphs have several important properties that make them easy to recognize— for example, the graph in figure 8.13 (a) is an interval graph, whereas the "house" graph in figure 8.13 (b) is not. Benzer's problem was equivalent to deciding whether the graph obtained from his bacteriophage experiment represented an interval graph. Had the experiment resulted in a graph like the "house" graph, then the genes could not have been organized as linear structures. As it turned out, the graph was indeed an interval graph, indicating that genes were composed of linearly organized functional units.

8.3   DNA Sequencing

Imagine several copies of a magazine cut into millions of pieces. Each copy is cut in a different way, so a piece from one copy may overlap pieces from another. Assuming that some large number of pieces are just sort of lost, and the remaining pieces are splashed with ink, can you recover the original text? This, essentially, is the problem of fragment assembly in DNA sequencing. Classic DNA sequencing technology allows one to read short 500- to 700-nucleotide sequences per experiment, each fragment corresponding to one of the many magazine pieces. Assembling the entire genome from these short fragments is like reassembling the magazine from the millions of tiny slips of paper.[6] Both problems are complicated by unavoidable experimental errors—ink splashes on the magazine, and mistakes in reading nucleotides. Furthermore, the data are frequently incomplete—some magazine pieces get lost, while some DNA fragments never make it into the sequencing machine. Nevertheless, efforts to determine the DNA sequence of organisms have been remarkably successful, even in the face of these difficulties.

---

5. It is not clear that he actually knew anything about graph theory at the time, but graph theorists eventually noticed his work.
6. We emphasize that biologists reading these 500- to 700-nucleotide sequences have no idea where they are located within the entire DNA string.

Two DNA sequencing methods were invented independently and simultaneously in Cambridge, England by Fred Sanger and in Cambridge, Massachusetts by Walter Gilbert. Sanger's method takes advantage of how cells make copies of DNA. Cells copy a strand of DNA nucleotide by nucleotide in a reaction that adds one base at a time. Sanger realized that he could make copies of DNA fragments of different lengths if he starved the reaction of one of the four bases: a cell can only copy its DNA while it has all of the bases in supply. For a sequence ACGTAAGCTA, starving at T would produce a mixture of the fragments ACG and ACGTAAGC. By running one starvation experiment for each of A, T, G, and C and then separating the resulting DNA fragments by length, one can read the DNA sequence.[7] Each of four starvation experiments produces a *ladder* of fragments of varying lengths called the *Sanger ladder*.[8] This approach culminated in the sequencing of a 5386-nucleotide virus in 1977 and a Nobel Prize shortly thereafter. Since then the amount of DNA sequence data has been increasing exponentially, particularly after the launch of the Human Genome Project in 1989. By 2001, it had produced the roughly 3 billion-nucleotide sequence of the human genome.

Within the past twenty years, DNA sequencing technology has been developed to the point where modern sequencing machines can sequence 500- to 700-nucleotide DNA fragments, called *sequencing reads*. These reads then have to be assembled into a continuous genome, which turns out to be a very hard problem. Even though the DNA reading process has become quite automated, these machines are not microscope-like devices that simply scan 500 nucleotides as if they were a sentence in a book. The DNA sequencing machines measure the lengths of DNA fragments in the Sanger ladder, but even this task is difficult; we cannot measure a *single* DNA fragment, but must measure billions of identical fragments.

Shotgun sequencing starts with a large sample of genomic DNA. The sample is *sonicated*, a process which randomly partitions each piece of DNA in the sample into *inserts*; the inserts that are smaller than 500 nucleotides are removed from further consideration. Before the inserts can be read, each one must be multiplied billions of times so that it is possible to read the ladders produced by Sanger's technique. To amplify the inserts, a sample is cloned into a *vector*, and this vector used to infect a bacterial host. As the bacterium reproduces, it creates a colony that contains billions of copies of the vector

---

7. Later Sanger found chemicals—so-called dideoxynucleotides—that could be inserted in place of A, T, G, or C, and cause a growing DNA chain to end.
8. The Sanger ladder for T shows the lengths of all sub-fragments ending at T and therefore reveals the set of positions where T occurs.

and its associated insert. As a result, the cloning process results in the production of a large sample of one particular insert that can then be sequenced by the Sanger method. Usually, only the first 500 to 700 nucleotides of the insert can be interpreted from this experiment. DNA sequencing is therefore a two stage process including both experimental (reading 500–700 nucleotide sequences form different inserts) and computational (assembling these reads into a single long sequence) components.

8.4   Shortest Superstring Problem

Since every string, or *read*, that we sequence came from the much longer genomic string, we are interested in a *superstring* of the reads—that is, we want a long string that "explains" all the reads we generated. However, there are many possible superstrings to choose from—for example, we could concatenate all the reads together to get a (not very helpful) superstring. We choose to be most interested in the shortest one, which turns out to be a reasonable first approximation to the unknown genomic DNA sequence. With this in mind, the simplest approximation of DNA sequencing corresponds to the following problem.

Shortest Superstring Problem:
Given a set of strings, find a shortest string that contains all of them.

Input: Strings $s_1, s_2, \ldots, s_n$.

Output: A string s that contains all strings $s_1, s_2, \ldots, s_n$ as substrings, such that the length of s is as small as possible.

Figure 8.14 presents two superstrings for the set of all eight three-letter strings in a 0–1 alphabet. The first (trivial) superstring is obtained by the concatenation of all eight strings, while the second one is a shortest superstring.

Define $overlap(s_i, s_j)$ to be the length of the longest prefix of s_j that matches a suffix of s_i. The Shortest Superstring problem can be cast as a Traveling Salesman problem in a complete directed graph with n vertices corresponding to strings $s_1, \ldots, s_n$ and edges of length $-overlap(s_i, s_j)$ (fig. 8.15). This reduction, of course, does not lead to an efficient algorithm since the TSP is $\mathcal{NP}$-complete. Moreover it is known that the Shortest Superstring problem is itself $\mathcal{NP}$-complete, so that a polynomial algorithm for this problem is un-

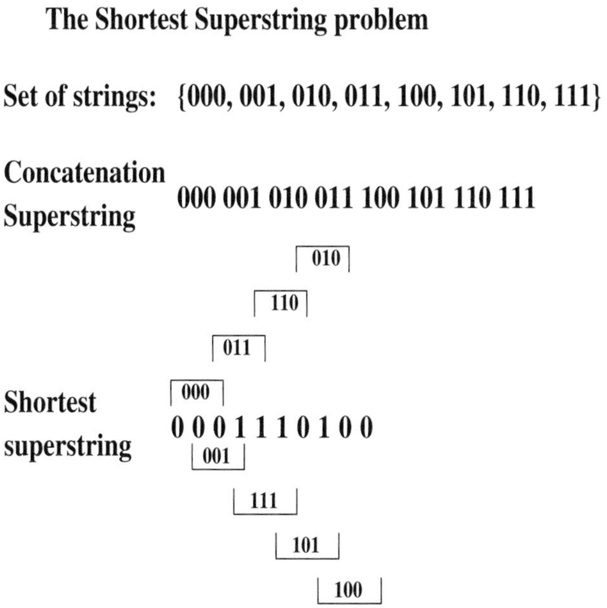

The Shortest Superstring problem

Set of strings: {000, 001, 010, 011, 100, 101, 110, 111}

Concatenation
                    000 001 010 011 100 101 110 111
Superstring

$$\boxed{010}$$
$$\boxed{110}$$
$$\boxed{011}$$

Shortest $\boxed{000}$
                    0 0 0 1 1 1 0 1 0 0
superstring $\boxed{001}$
                         $\boxed{111}$
                              $\boxed{101}$
                                   $\boxed{100}$

Figure 8.14   Superstrings for the set of eight three-letter strings in a 0–1 alphabet. Concatenating all eight strings results in a 24-letter superstring, while the shortest superstring contains only 10 letters. The shortest superstring in this case represents a solution of the Clever Thief problem—it is the minimum string of tests a thief has to conduct to try all possible k-letter passwords for a combination lock.

likely. The early DNA sequencing algorithms used a simple greedy strategy: repeatedly merge a pair of strings with maximum overlap until only one string remains. It has been conjectured, but not yet proved, that this greedy algorithm has performance guarantee 2.

8.5   DNA Arrays as an Alternative Sequencing Technique

When the Human Genome Project started, DNA sequencing was a routine but time-consuming and hard-to-automate procedure. In 1988 four groups of biologists independently and simultaneously suggested a different sequencing technique called *Sequencing by Hybridization*. SBH involves building a miniature *DNA array*, also known as a *DNA chip*, that contains thousands of short DNA fragments called *probes*. Each of these short fragments reveals

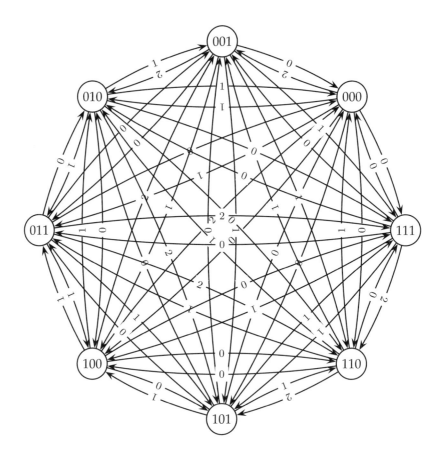

Figure 8.15 The overlap graph for the eight strings in figure 8.14.

whether or not a known—but short—sequence occurs in the unknown DNA sequence; all these pieces of information together should reveal the identity of the target DNA sequence.

Given a short probe (an 8- to 30-nucleotide single-stranded synthetic DNA fragment) and a single-stranded target DNA fragment, the target will *hybridize* with the probe if the probe is a substring of the target's Watson-Crick *complement*. When the probe and the target are mixed together, they form a weak chemical bond and stick together. For example, a probe ACCGTGGA will hybridize to a target CCCTGGCACCTA since it is complementary to the substring **TGGCACCT** of the target.

In 1988 almost nobody believed that the idea of using DNA probes to sequence long genomes would work, because both the biochemical problem of synthesizing thousands of short DNA fragments and the combinatorial problem of sequence reconstruction appeared too complicated. Shortly after the first paper describing DNA arrays was published, the journal *Science* wrote that given the amount of work required to synthesize a DNA array, "[using DNA arrays for sequencing] would simply be substituting one horrendous task for another." A major breakthrough in DNA array technology was made by Steve Fodor and colleagues in 1991. Their approach to array manufacturing relies on *light-directed polymer synthesis*,[9] which has many similarities to computer chip manufacturing (fig. 8.16). Using this technique, building an array with all 4^l probes of length l requires just $4 \cdot l$ separate reactions, rather than the presumed 4^l reactions. With this method, a California-based biotechnology company, Affymetrix, built the first 64-kb DNA array in 1994. Today, building 1-Mb or larger arrays is routine, and the use of DNA arrays has become one of the most widespread new biotechnologies.

SBH relies on the hybridization of the target DNA fragment against a very large array of short probes. In this manner, probes can be used to test the unknown target DNA to determine its *l-mer composition*.[10] The *universal* DNA array contains all 4^l probes of length l and is applied as follows (fig. 8.17):

- Attach all possible probes of length l (l=8 in the first SBH papers) to a flat surface, each probe at a distinct and known location. This set of probes is called the *DNA array*.

- Apply a solution containing fluorescently labeled DNA fragment to the array.

- The DNA fragment hybridizes with those probes that are complementary to substrings of length l of the fragment.

- Using a spectroscopic detector, determine which probes hybridize to the DNA fragment to obtain the *l*-mer composition of the target DNA fragment.

- Apply the combinatorial algorithm described below to reconstruct the sequence of the target DNA fragment from the *l*-mer composition.

---

9. Light as in photons, not light as in "not heavy."
10. The *l*-mer composition of a string is simply the set of all *l*-mers present in the string.   For example, the 8-mer composition of **CCCTGGCACCTA** is {**CCCTGGCA, CCTGGCAC, CTGGCACC, TGGCACCT, GGCACCTA**}.

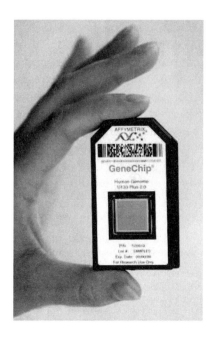

Figure 8.16   A GeneChip, produced by Affymetrix. (Picture courtesy of Affymetrix, Inc.)

8.6   Sequencing by Hybridization

Given an unknown DNA sequence, an array provides information about all strings of length l that the sequence contains, but does not provide information about their positions in the sequence. For a string s of length n, the l-mer composition, or *spectrum*, of s, is the multiset of $n - l + 1$ l-mers in s and is written $Spectrum(s, l)$. If $l = 3$ and $s = $ TATGGTGC, then $Spectrum(s, l) = \{$TAT, ATG, TGG, GGT, GTG, TGC$\}$.[11] We can now formulate the problem of sequencing a target DNA fragment from its DNA array data.

---

11. The l-mers in this spectrum are listed in the order of their appearance in s creating the impression that we know which nucleotide occur at each position. We emphasize that the order of these l-mers in s is unknown and it is probably more appropriate to list them in lexicographic order, like ATG, GGT, GTG, TAT, TGC, TGG.

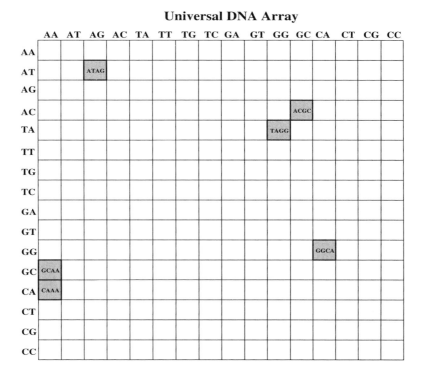

Universal DNA Array

DNA target TATCCGTTT (complement of ATAGGCAAA)

hybridizes to the array of all 4-mers:

```
A T A G G C A A A
A T A G
  T A G G
    A G G C
      G G C A
        G C A A
          C A A A
```

Figure 8.17 Hybridization of TATCCGTTT with the universal DNA array consisting of all 4^4 4-mers.

Sequencing by Hybridization (SBH) Problem:
Reconstruct a string from its l-mer composition.

Input: A set, $\mathcal{S}$, representing all l-mers from an (unknown)
string s.

Output: String s such that $Spectrum(s, l) = \mathcal{S}$.

Although conventional DNA sequencing and SBH are very different experimental approaches, you can see that the corresponding computational
problems are quite similar. In fact, SBH is a particular case of the Shortest
Superstring problem when the strings $s_1, \ldots, s_n$ represent the set of all substrings of s of fixed size. However, in contrast to the Shortest Superstring
problem, there exists a simple linear-time algorithm for the SBH problem.
Notice that it is not a contradiction that the Shortest Superstring problem
is $\mathcal{NP}$-complete, yet we claim to have a linear-time algorithm for the SBH
problem, since the Shortest Superstring problem is more general than the
SBH problem.

Although DNA arrays were originally proposed as an alternative to conventional DNA sequencing, de novo sequencing with DNA arrays remains
an unsolved problem in practice. The primary obstacle to applying DNA
arrays for sequencing is the inaccuracy in interpreting hybridization data to
distinguish between perfect matches [i.e., l-mers present in $Spectrum(s, l)$]
and highly stable mismatches (i.e, l-mers not present in $Spectrum(s, l)$, but
with sufficient chemical bonding potential to generate a strong hybridization signal). This is a particularly difficult problem for the short probes used
in universal arrays. As a result, DNA arrays have become more popular in
gene expression analysis[12] and studies of genetic variations—both of which
are done with longer probes—than in *de novo* sequencing. In contrast to SBH
where the target DNA sequence is unknown, these approaches assume that
the DNA sequence is either known or "almost" known (i.e., known up to a
small number of mutations). For example, to detect genetic variations one
can design twenty- to thirty-nucleotide probes to reliably detect mutations,
bypassing the still unsolved problem of distinguishing perfect matches from
highly stable mismatches in the case of short probes. To detect mutations in

---

12. In gene expression analysis, a solution containing mRNA (rather than DNA) is applied to
the array with the goal of figuring out whether a given gene is switched on or switched off. In
this case, absence of a hybridization signal indicates that a gene is not being transcribed into an
mRNA, and is therefore switched off.

Sequence reconstruction (Hamiltonian path approach)

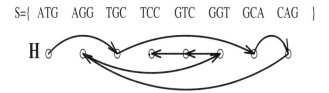

S={ ATG   AGG   TGC   TCC   GTC   GGT   GCA   CAG }

Vertices: l-tuples from the spectrum S.   Edges: overlapping l-tuples.

Path visiting ALL VERTICES corresponds to sequence reconstruction   ATGCAGGTCC

Figure 8.18   SBH and the Hamiltonian path problem.

the (known) sequence S, an array should contain all 25-mers from S, as well as selected mutated versions of these 25-mers.[13]

8.7   SBH as a Hamiltonian Path Problem

Two l-mers p and q *overlap* if $overlap(p, q) = l - 1$, that is, the last $l - 1$ letters of p coincide with the first $l - 1$ letters of q. Given the measured spectrum $Spectrum(s, l)$ of a DNA fragment s, construct a directed graph, H, by introducing a vertex for every l-mer in $Spectrum(s, l)$, and connect every two vertices p and q by the directed edge (p, q) if p and q overlap. There is a one-to-one correspondence between paths that visit each vertex of H exactly once and DNA fragments with the spectrum $Spectrum(s, l)$. The spectrum presented in figure 8.18 corresponds to the sequence reconstruction ATGCAGGTCC, which is the only path visiting all vertices of H:

ATG → TGC → GCA → CAG → AGG → GGT → GTC → TCC

The spectrum shown in figure 8.19 yields a more complicated graph with two Hamiltonian paths, each path corresponding to two possible reconstructions: ATGCGTGGCA and ATGGCGTGCA. As the overlap graph becomes

13. In practice, the mutated versions of an l-mer are often limited to the 3 l-mers with mutations at the middle position. Arrays constructed in this manner are called *tiling arrays*.

Multiple sequence reconstructions (Hamiltonian path approach)

S={ ATG   TGG   TGC   GTG   GGC   GCA   GCG   CGT   }

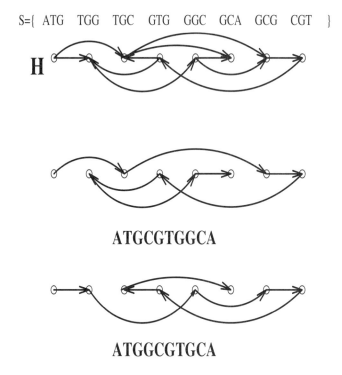

ATGCGTGGCA

ATGGCGTGCA

Figure 8.19   Spectrum S yields two possible reconstructions corresponding to distinct Hamiltonian paths.

larger, this approach ceases to be practically useful since the Hamiltonian Path problem is $\mathcal{NP}$-complete.

8.8   SBH as an Eulerian Path Problem

As we have seen, reducing the SBH problem to a Hamiltonian Path problem does not lead to an efficient algorithm. Fortunately, reducing SBH to the *Eulerian Path* problem in a directed graph,[14] which leads to the simple linear-time algorithm for sequence reconstruction mentioned earlier.

---

14. A directed path is a path $v_1 \rightarrow v_2 \rightarrow \ldots \rightarrow v_n$ from vertex v_1 to vertex v_n in which every edge (v_i, v_{i+1}) is directed from v_i to v_{i+1}.

The reduction of the SBH problem to an Eulerian Path problem is to construct a graph whose edges—rather than vertices—correspond to l-mers from $Spectrum(s, l)$, and then to find a path in this graph visiting every edge exactly once. In this approach we build a graph G on the set of all $(l - 1)$-mers, rather than on the set of all l-mers as in the previous section. An $(l - 1)$-mer v is joined by a directed edge with an $(l - 1)$-mer w if the spectrum contains an l-mer for which the first $l - 1$ nucleotides coincide with v and the last $l - 1$ nucleotides coincide with w (fig. 8.20). Each l-mer from the spectrum corresponds to a directed edge in G rather than to a vertex as it does in H; compare figures 8.19 and 8.20. Therefore, finding a DNA fragment containing all l-mers from the spectrum corresponds to finding a path visiting all *edges* of G, which is the problem of finding an Eulerian path. Superficially, finding an Eulerian path looks just as hard as finding a Hamiltonian path, but, as we show below, finding Eulerian paths turns out to be simple.

We will first consider *Eulerian cycles*, that is, Eulerian paths in which the first and the last vertices are the same. A directed graph G is *Eulerian* if it contains an Eulerian cycle. A vertex v in a graph is *balanced* if the number of edges entering v equals the number of edges leaving v, that is, if $indegree(v) = outdegree(v)$. For any given vertex v in an Eulerian graph, the number of times the Eulerian cycle enters v is exactly the same as the number of times it leaves v. Thus, $indegree(v) = outdegree(v)$ for every vertex v in an Eulerian graph, motivating the following theorem characterizing Eulerian graphs.

Theorem 8.1 *A connected graph is Eulerian if and only if each of its vertices is balanced.*

Proof First, it is easy to see that if a graph is Eulerian, then each vertex must be balanced. We show that if each vertex in a connected graph is balanced, then the graph is Eulerian.

To construct an Eulerian cycle, we start from an arbitrary vertex v and form any arbitrary path by traversing edges that have not already been used. We stop the path when we encounter a vertex with no way out, that is, a vertex whose outgoing edges have already been used in the path. In a balanced graph, the only vertex where this can happen is the starting vertex v since for any other vertex, the balance condition ensures that for every incoming edge there is an outgoing edge that has not yet been used. Therefore, the resulting path will end at the same vertex where it started, and with some luck will be Eulerian. However, if the path is not Eulerian, it must contain a vertex w that still has some number of untraversed edges. The cycle we just constructed

Multiple sequence reconstructions (the Eulerian path approach)

S={ATG, TGG, TGC, GTG, GGC, GCA, GCG , CGT}

Vertices correspond to (l-1)-tuples.

Edges correspond to l-tuples from the spectrum

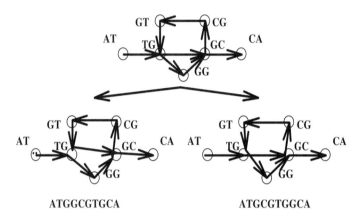

ATGGCGTGCA   ATGCGTGGCA

Paths visiting ALL EDGES correspond to sequence reconstructions

Figure 8.20   SBH and the Eulerian path problem.

forms a balanced subgraph.[15] Since the original graph was balanced, then the edges that were *not* traversed in the first cycle also form a balanced subgraph. Since all vertices in the graph with untraversed edges are balanced there must exist some other path starting and ending at w, containing only untraversed edges. This process is shown in figures 8.21 and 8.22.

One can now combine the two paths into a single one as follows. Traverse the first path from v to w, then traverse the second path from w back to itself, and then traverse the remainder of the first path from w back to v. Repeating this until there are no more vertices with unused edges will eventually yield an Eulerian cycle. This algorithm can be implemented in time linear in the number of edges in the graph.   □

───────────

15. A subgraph is a graph obtained by removing some edges from the original graph.

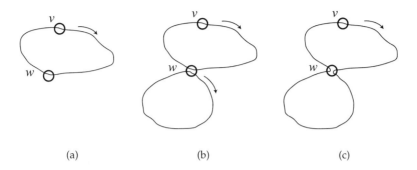

Figure 8.21   Constructing an Eulerian cycle in an Eulerian graph.

Notice that we have described Eulerian graphs as containing an Eulerian *cycle*, rather than an Eulerian path, but we have said that the SBH problem reduces to that of finding an Eulerian path. A vertex v in a graph is called *semibalanced* if $|indegree(v) - outdegree(v)| = 1$. If a graph has an Eulerian path starting at vertex s and ending at vertex t, then all its vertices are balanced, with the possible exception of s and t, which may be semibalanced. The Eulerian path problem can be reduced to the Eulerian cycle problem by adding an edge between two semibalanced vertices. This transformation balances all vertices in the graph and therefore guarantees the existence of an Eulerian cycle in the graph with the added edge. Removing the added edge from the Eulerian cycle transforms it into an Eulerian path. The following theorem characterizes all graphs that contain Eulerian paths.

Theorem 8.2   *A connected graph has an Eulerian path if and only if it contains at most two semibalanced vertices and all other vertices are balanced.*

8.9   Fragment Assembly in DNA Sequencing

As we mentioned previously, after the short 500- to 700-bp *DNA reads* are sequenced, biologists need to assemble them together to reconstruct the entire genomic DNA sequence. This is known as fragment assembly. The Shortest Superstring problem described above is an overly simplified abstraction that does not adequately capture the essence of the fragment assembly problem,

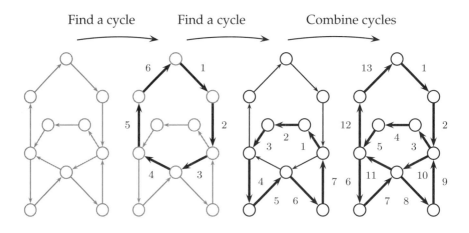

Figure 8.22   Constructing an Eulerian cycle. Untraversed edges are shown in gray.

since it assumes error-free reads. The error rate in DNA reads produced by modern sequencing machines varies from 1% to 3%. A further complication in fragment assembly is the fact that one does not know a priori which of two DNA strands a read came from. DNA is double-stranded, and which of the two strands was sequenced by a read depends on how the insert was oriented in the vector. Since this is essentially arbitrary, one never knows whether a read came from a target strand DNA sequence or from its Watson-Crick complement.

However, sequencing errors and assignments of reads to one of two strands are just minor annoyances compared to the major problem in fragment assembly: repeats in DNA. The human genome contains many sequences that repeat themselves throughout the genome a surprisingly large number of times. For example, the roughly 300 nucleotide *Alu* sequence is repeated more than a million times throughout the genome, with only 5% to 15% sequence variation. Even more troublesome for fragment assembly algorithms is the fact that repeats occur at several scales. The human T-cell receptor locus contains five closely located repeats of the trypsinogen gene, which is 4 kb long and varies only by 3% to 5% between copies. These long repeats are particularly difficult to assemble, since there are no reads with unique portions flanking the repeat region. The human genome contains more than 1 million Alu repeats ($\approx$ 300 bp) and 200,000 $LINE$ repeats ($\approx$ 1000 bp), not to mention that an estimated 25% of genes in the human genome have duplicated copies. A little arithmetic shows that these repeats and duplicated genes represent about half the human genome.[16]

If one models the genome as a 3 billion-letter sequence produced by a random number generator, then assembling it from 500-letter reads is actually relatively simple. However, because of the large number of repeats, it is theoretically impossible to uniquely assemble real reads as long as some repeats are longer than the typical read length. Increasing the length of the reads (to make them longer than most repeats) would solve the problem, but the sequencing technology has not significantly improved the read length yet.

Figure 8.23 (upper) presents a puzzle that looks deceivingly simple and has only sixteen triangular pieces. People usually assemble puzzles by connecting matching pieces. In this case, for every triangle in the puzzle, there is a variety of potentially matching triangles (every frog in the puzzle is repeated several times). As a result, you cannot know which of the potentially matching triangles is the correct one to use at any step. If you proceed without some sort of guidance, you are likely to end up in the situation shown in figure 8.23 (lower). Fourteen of the pieces have been placed completely consistently, but the two remaining pieces are impossible to place. It is difficult to design a strategy that can avoid such dead ends for this particular puzzle, and it is even more difficult to design strategies for the linear puzzle presented by repeats in a genome.

Since repeats present such a challenge in assembling long genomes, the original strategy for sequencing the human genome was first to clone it into BACs,[17] each BAC carrying an approximately $150,000$ bp long insert. After constructing a library of overlapping BACs that covers the entire human genome (which requires approximately 30,000 BACs), each one can be sequenced as if it were a separate minigenome. This BAC-by-BAC sequencing strategy significantly simplifies the computational assembly problem (by virtue of the fact that the number of repeats present within a BAC is 30,000 times smaller than the number of repeats in the entire genome) but makes the sequencing project substantially more cumbersome. Although the Human Genome project demonstrated that this BAC-by-BAC strategy can be successful, recent large-scale sequencing projects (including the 2002 mouse

---

16. Fortunately, as different copies of these repeats have evolved differently over time, they are not exact repeats.
17. Bacterial Artificial Chromosomes.

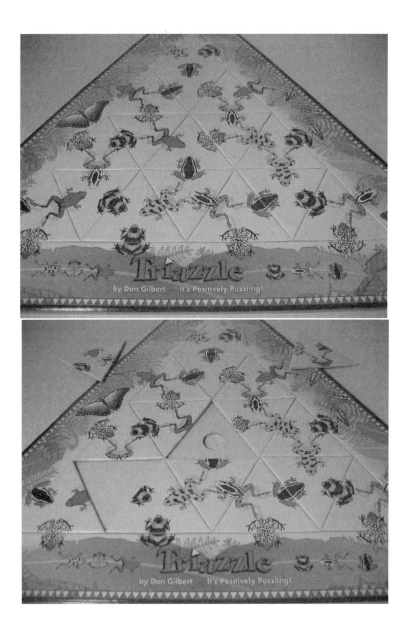

Figure 8.23   Repeats are not just a problem in DNA sequence assembly. This puzzle has deceptively few pieces but is harder than many jigsaw puzzles that have thousands of pieces. (With permission of Dan Gilbert Art Group, Inc.)

genome assembly) mainly follow the whole-genome assembly paradigm advocated by James Weber and Gene Myers in 1997. Myers led the fragment assembly efforts at Celera Genomics, the company that announced the completion of a (draft) human genomic sequence in 2001.[18] Weber and Myers suggested a virtual increase in the length of a read, by pairing reads that were separated by a fixed-size gap. This suggestion resulted in the so-called *mate-pair reads* sequencing technique. In this method, inserts of length approximately L (where L is much longer than the length of a read) are selected, and *both* ends of the insert are sequenced. This produces a pair of reads called *mates* at a known (approximate) distance, L, from each other. The insert length L is chosen in such a way that it is larger than the length of most repeats in the human genome. The advantage of mate-pair reads is that it is unlikely that both reads of the mate-pair will lie in a large-scale DNA repeat. Thus, the read that lies in a unique portion of DNA determines which copy of a repeat its mate is in.

Most fragment assembly algorithms consist of the following three steps:

- *Overlap:* Finding potentially overlapping reads

- *Layout:* Finding the order of reads along DNA

- *Consensus:* Deriving the DNA sequence from the layout

The overlap problem is to find the best match between the suffix of one read and the prefix of another. In the absence of sequencing errors, we could simply find the longest suffix of one string that exactly matches the prefix of another string. However, sequencing errors force us to use a variation of the dynamic programming algorithm for sequence alignment. Since errors are small (1% to 3%), the common practice is to filter out pairs of fragments that do not share a significantly long common substring, an idea we will return to in chapter 9 when we discuss combinatorial pattern matching.

Constructing the layout is the hardest step in fragment assembly. The difficulty is in deciding whether two fragments really overlap (i.e., their differences are caused by sequencing errors) or actually come from two different copies of a repeat. Repeats represent a major challenge for whole-genome shotgun sequencing and make the layout problem very difficult.

The final consensus step of fragment assembly amounts to correcting errors in sequence reads. The simplest way to build the consensus is to report

---

18. The human genome sequence was sequenced in 2001 by both the publicly-funded Human Genome Consortium and the privately-financed Celera Genomics. As a result, there exist two slightly different versions of the human genome.

the most frequent character in the layout constructed in the layout step. This assumes that each position in the genome was represented by a sufficiently large number of reads to ensure that experimental errors are reduced to minor noise.

8.10   Protein Sequencing and Identification

Few people remember that before DNA sequencing had even been seriously suggested, scientists routinely sequenced proteins. Frederick Sanger was awarded his first (of two) Nobel prize for determining the amino acid sequence of insulin, the protein needed by people suffering from diabetes. Sequencing the 52-amino acid bovine insulin in the late 1940s seemed more challenging than sequencing an entire genome seems today. The computational problem facing protein sequencing at that time was similar to that facing modern DNA sequencing; the main difference was in the length of the sequenced fragments. In the late 1940s, biologists discovered how to apply the *Edman degradation reaction* to chop off one terminal amino acid at a time from the end of a protein and read it. Unfortunately, this only works for a few terminal amino acids before the results become impossible to interpret. To get around this problem, Sanger digested insulin with *proteases* (enzymes that cleave proteins) into *peptides* (short protein fragments) and sequenced each of the resulting fragments independently. He then used these overlapping fragments to reconstruct the entire sequence, exactly like the DNA sequencing "break—read the fragments—assemble" method today, as shown in figure 8.24.

The Edman degradation reaction became the predominant protein sequencing method for the next twenty years, and by the late 1960s protein sequencing machines were on the market. Despite these advances, protein sequencing ceased to be of central interest in the field as DNA sequencing technology underwent rapid improvements in the late 1970s. In DNA sequencing, obtaining reads is relatively easy; it is the assembly that is difficult. In protein sequencing, obtaining reads is the primary problem, while assembly is easy.[19]

Having DNA sequence data for a cell is critical to understanding the molecular processes that the cell goes through. However, it is not the only impor-

---

19. Modern protein sequencing machines are capable of reading more than fifty residues from a peptide fragment. However, these machines work best when the protein is perfectly purified, which is hard to achieve in biological experiments.

```
GIVE
GIVEECCA
GIVEECCASV
GIVEECCASVC
GIVEECCASVCSL
GIVEECCASVCSLY
          SVC
           SLY
           SLYELEDYC
            YE
            YEL
            YELE
             ELEDY
             ELEDYCD
              LE
               LEDYCD
                EDYCD
                 DYCD
                  CD


              FVDEHLCG
              FVDEHLCGSHL
                  HLCGSHL
                    SHLVEA
                      VEAL
                      VEALY
                       AL
                       ALY
                        YLVCG
                         LVCGERGF
                         LVCGERGFF
                           GERG
                            GF
                            GFFYTPK
                             YTPKA
                             TPKA
```

Figure 8.24 The peptide fragments that Frederick Sanger obtained from insulin through a variety of methods. The protein is split into two parts, the A-chain (shown on the left) and the B-chain (shown on the right) as a result of an enzymatic digestion process. Sanger's further elucidation of the disulfide bridges linking the various cystein residues was the result of years of painstaking laboratory work. The sequence was published in three parts: the A-chain, the B-chain, and then the disulfide linkages. Insulin is not a particularly large protein, so better techniques would be useful.

tant component: one also needs to know what proteins the cell produces and what they do. On the one hand, we do not yet know the full set of proteins that cells produce, so we need a way to discover the sequence of previously unknown proteins. On the other hand, it is important to identify which specific proteins interact in a biological system (e.g., those proteins involved in DNA replication). Lastly, different cells in an organism have different repertoires of expressed proteins. Brain cells need different proteins to function than liver cells do, and an important problem is to identify proteins that are present or absent in each biological tissue under different conditions.

There are two types of computational problems motivated by protein sequencing. *De novo protein sequencing* is the elucidation of a protein's sequence in the case when a biological sample contains a protein that is either not present in a database or differs from a canonical version present in a database (e.g., mutated proteins or proteins with biochemical modifications). The other problem is the identification of a protein that is present in a database; this is usually referred to as *protein identification*. The main difference between protein sequencing algorithms and protein identification algorithms is the difficulty of the underlying computational problems.

Perhaps the easiest way to illustrate the distinction between protein identification and sequencing is with a gedanken experiment. Suppose a biologist wants to determine which proteins form the DNA polymerase complex in rats. Having the complete rat genome sequence and knowing the location of all the rat genes does not yet allow a biologist to determine what chemical reactions occur during the DNA replication process. However, isolating a rat's DNA polymerase complex, breaking it apart, and sequencing the proteins that form parts of the complex will yield a fairly direct answer to the researcher's question. Of course, if we presume that the biologist has the complete rat genome sequence and all of its gene *products*, he may not actually have to sequence every amino acid in every protein in the DNA polymerase complex—just enough to figure out which proteins are present. This is protein identification. On the other hand, if the researcher decides to study an organism for which complete genome data are not available (perhaps an obscure species of ant), then the researcher will need to perform de novo protein sequencing.[20]

---

20. As usual with gedanken experiments, reality is more complicated. Even if the complete genomic sequence of a species is known and annotated, the repertoire of all possible proteins usually is not, due to the myriad alternative splicings (different ways of constructing mRNA from the gene's transcript) and post-translational modifications that occur in a living cell.

For many problems, protein sequencing and identification remain the only ways to probe a biological process. For example, gene splicing (see chapter 6) is a complex process performed by the large molecular complex called the *spliceosome*, which consists of over 100 different proteins complexed with some functional RNA. Biologists want to determine the "parts list" of the spliceosome, that is, the identity of proteins that form the complex. DNA sequencing is not capable of solving this problem directly: even if all the proteins in the genome were known, it is not clear which of them are parts of the spliceosome. Protein sequencing and identification, on the other hand, are very helpful in discovering this parts list. Recently, Matthias Mann and colleagues purified the spliceosome complex and used protein sequencing and protein identification techniques to find a detailed parts list for it.

Another application of these technologies is the study of proteins involved in *programmed cell death*, or *apoptosis*. In the development of many organisms cells must die at specific times. A cell dies if it fails to acquire certain survival factors, and the death process can be initiated by the expression of certain genes. In a developing nematode, for example, the death of individual cells in the nervous system may be prevented by mutations in several genes that are the subject of active investigation. DNA sequence data alone are not sufficient to find the genes involved in programmed cell death, and until recently, nobody knew the identity of these proteins. Protein analysis by *mass spectrometry* allowed the sequencing of proteins involved in programmed cell death, and the discovery of some proteins involved in the death-inducing signaling complex.

The exceptional sensitivity of mass spectrometry has opened up new experimental and computational possibilities for protein studies. A protein can be digested into peptides by proteases like trypsin. In a matter of seconds, a *tandem mass spectrometer* breaks a peptide into even smaller fragments and measures the mass of each. The *mass spectrum* of a peptide is a collection of masses of these fragments. The protein sequencing problem is to derive the sequence of a peptide given its mass spectrum. For an ideal fragmentation process where every fragment of a peptide is generated, and in an ideal mass spectrometer, the peptide sequencing problem is simple. However, the fragmentation process is not ideal, and mass spectrometers measure mass with some imprecision. These details make peptide sequencing difficult.

A mass spectrometer works like a charged sieve. A large molecule (peptide) gets broken into smaller fragments that have an electrical charge. These fragments are then spun around and accelerated in a magnetic field until they hit a detector. Because large fragments are harder to spin than small ones, one can distinguish between fragments with different masses based on the amount of energy required to fling the different fragments around. It happens that most molecules can be broken in several places, generating several different *ion types*. The problem is to reconstruct the amino acid sequence of the peptide from the masses of these broken pieces.

8.11   The Peptide Sequencing Problem

Let $A = \{a_1, a_2, \ldots, a_{20}\}$ be the set of amino acids, each with molecular masses $m(a_i)$. A *peptide* $P = p_1 \cdots p_n$ is a sequence of amino acids, with *parent* mass $m(P) = \sum_{i=1}^{n} m(p_i)$. We will denote the partial *N-terminal* peptide $p_1, \ldots, p_i$ of mass $m_i = \sum_{j=1}^{i} m(p_j)$ as P_i and the partial *C-terminal* peptide $p_{i+1}, \ldots, p_n$ of mass $m(P) - m_i$ as P_i^-, for $1 \leq i \leq n$. Mass spectra obtained by tandem mass spectrometry (MS/MS) consist predominantly of partial N-terminal peptides and C-terminal peptides.[21]

A mass spectrometer typically breaks a peptide $p_1 p_2 \cdots p_n$ at different peptide bonds and detects the masses of the resulting partial N-terminal and C-terminal peptides.[22] For example, the peptide GPFNA may be broken into the N-terminal peptides G, GP, GPF, GPFN, and C-terminal peptides PFNA, FNA, NA, A. Moreover, while breaking GPFNA into GP and FNA, it may lose some small parts of GP and FNA, resulting in fragments of a lower mass. For example, the peptide GP might lose a water (H_2O), and the peptide FNA might lose an ammonia (NH_3). The resulting masses detected by the spectrometer will be equal to the mass of GP minus the mass of water (water happens to weigh $1 + 1 + 16 = 18$ daltons), and the mass of FNA minus the mass of ammonia ($1 + 1 + 1 + 14 = 17$ daltons). Peptides missing water and ammonia are two different *ion types* that can occur in fragmenting a peptide in a mass spectrometer.[23]

---

21. Every protein is a linear chain of amino acids, connected by a peptide bond. The peptide bond starts with a nitrogen (N) and ends with a carbon (C); therefore, every protein begins with an "unstarted" peptide bond that begins with N and another "unfinished" peptide bond that ends with C. An *N-terminal* peptide is a fragment of a protein that includes the "leftmost" end (i.e., the N-terminus). A *C-terminal* peptide is a fragment of a protein that includes the "rightmost" end (i.e., the C-terminus).

22. Biologists typically work with billions of identical peptides in a solution. A mass spectrometry machine breaks different peptide molecules at different peptide bonds (some peptide bonds are more prone to breakage than others). As a result, many N-terminal and C-terminal peptide may be detected by a mass spectrometer.

23. This is a simplified description of the complex and messy fragmentation process. In this section we intentionally hide many of the technical details and focus only on the computational challenges.

Peptide fragmentation in a tandem mass spectrometer can be character-ized by a set of numbers $\Delta = \{\delta_1, \ldots, \delta_k\}$ representing the different types of ions that correspond to the removal of a certain chemical group from a pep-tide fragment. We will call Δ the set of ion types. A *δ-ion* of an N-terminal partial peptide P_i is a modification of P_i that has mass $m_i - \delta$, correspond-ing to the loss of a (typically small) chemical group of mass δ when P was fragmented into P_i. The δ-ion of C-terminal peptides is defined similarly. The most frequent N-terminal ions are called *b-ions* (ion b_i corresponds to P_i with $\delta = -1$) and the most frequent C-terminal ions are called *y-ions* (ion y_i corresponds to P_i^- with $\delta = 19$), shown in figure 8.25 (a). Examples of other frequent N-terminal ions are represented by b-H_2O (a b-fragment that loses a water) or y-NH_3 and some others like b-H_2O-NH_3.

For tandem mass spectrometry, the *theoretical* spectrum $T(P)$ of peptide P can be calculated by subtracting all possible ion types $\delta_1, \ldots, \delta_k$ from the masses of all partial peptides of P, such that every partial peptide generates k masses in the theoretical spectrum, as in figure 8.25 (b).[24]

An *experimental* spectrum $S = \{s_1, \ldots, s_q\}$ is a set of numbers obtained in a mass spectrometry experiment that includes masses of some fragment ions as well as chemical noise.[25] Note that the distinction between the theoretical spectrum $T(P)$ and the experimental spectrum S is that you mathematically generate $T(P)$ given the peptide sequence P, but you experimentally gener-ate S without knowing what the peptide sequence is that generated it.

The *match* between the experimentally measured spectrum S and peptide P is the number of masses in S that are equal to masses in $T(P)$. This is often referred to as the *shared peaks count*. In reality, peptide sequencing algorithms use more sophisticated objective functions than a simple shared peaks count, incorporating different weighting functions for the matching masses. We formulate the Peptide Sequencing problem as follows.

---

24. A theoretical spectrum of a peptide may contain as many as $2nk$ masses but it sometimes contains less since some of these masses are not unique.

25. In reality, a mass spectrometer detects *charged ions* and measures mass-to-charge ratios. As a result, an experimental spectrum contains the values $\frac{m}{z}$ where m is the mass and z is an integer (typically, 1 or 2) equal to the charge of a fragment ion. For simplicity we assume that $z = 1$ through the remainder of this chapter.

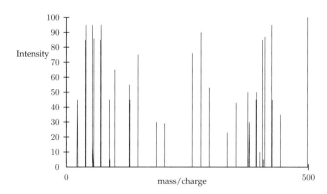

(a) The fragmentation pattern of the peptide GPFNA.

	Sequence	mass	less H_2O	less NH_3	less both
	GPFNA	498	480	481	463
b_1	G	58	40	41	23
y_4	PFNA	442	424	425	405
b_2	GP	149	131	132	114
y_3	FNA	351	333	334	316
b_3	GPF	296	278	279	261
y_2	NA	204	186	187	169
b_4	GPFN	410	392	393	375
y_l	A	90	72	73	55

(b) A theoretical mass spectrum of GPFNA.

(c) An "experimental" mass spectrum of GPFNA.

Figure 8.25   Tandem MS of the peptide $GPFNA$. Two different types of fragment ions, b-ions and y-ions are created (a) when the carbon-nitrogen bond breaks in the spectrometer. Each of these ion types can also lose H_2O or NH_3, or both, resulting in the masses presented in (b). Many other ion types are seen in typical experiments. If we were to measure the mass spectrum of this peptide, we would see a result similar to (c), where some peaks are missing and other noise peaks are present.

Peptide Sequencing Problem:
Find a peptide whose theoretical spectrum has a maximum match to a measured experimental spectrum.

    Input: Experimental spectrum S, the set of possible ion types Δ, and the parent mass m.

    Output: A peptide P of mass m whose theoretical spectrum matches S better than any other peptide of mass m.

In reality, mass spectrometers measure both mass and *intensity*, which reflects the number of fragment ions of a given mass are detected in the mass spectrometer. As a result, mass spectrometrists often represent spectra in two dimensions, as in figure 8.25 (c), and refer to the masses in the spectrum as "peaks."[26]

8.12   Spectrum Graphs

There are two main approaches to solving the Peptide Sequencing problem that researchers have tried: either through exhaustive search among all amino acid sequences of a certain length, or by analyzing the *spectrum graph* which we define below. The former approach involves the generation of all 20^l amino acid sequences of length l and their corresponding theoretical spectra, with the goal of finding a sequence with the best match between the experimental spectrum and the sequence's theoretical spectrum. Since the number of sequences grows exponentially with the length of the peptide, different branch-and-bound techniques have been designed to limit the combinatorial explosion in these methods. *Prefix pruning* restricts the computational space to sequences whose prefixes match the experimental spectrum well. The difficulty with the prefix pruning is that it frequently discards the correct sequence if its prefixes are poorly represented in the spectrum.

    The spectrum graph approach, on the other hand, does not involve generating all amino acid sequences, and leads to a fast algorithm for peptide sequencing. In this approach, we construct a graph from the experimental spectrum. Assume for simplicity that an experimental spectrum $S =$

---

26. A match of a theoretical spectrum against an experimental spectrum with varying intensity, then, needs to reflect the intensity of the fragment ions. While accounting for intensities is important for statistical analysis, it does not seriously affect the algorithmic details and we ignore intensities in the remainder of this chapter.

$\{s_1, \ldots, s_q\}$ consists of N-terminal ions and we will ignore the C-terminal ions for a while. Every mass $s \in S$ may have been created from a partial peptide by one of the k different ion types. Since we do not know which ion type from $\Delta = (\delta_1, \ldots, \delta_k)$ created the mass s in the experimental spectrum, we generate k different "guesses" for each of masses in the experimental spectrum. Every guess corresponds to the hypothesis that $s = x - \delta_j$, where x is the mass of some partial peptide and $1 \leq j \leq k$. Therefore, for every mass s in the experimental spectrum, there are k guesses for the mass x of some partial peptide: $s + \delta_1, s + \delta_2, \ldots, s + \delta_k$. As a result, each mass in the experimental spectrum is transformed into a set of k vertices in the spectrum graph, one for each possible ion type. The vertex for δ_i for the mass s is labeled with mass $s + \delta_i$. We connect any two vertices u and v in the graph by the directed edge (u, v) if the mass of v is larger than that of u by the mass of a single amino acid. If we add a vertex at 0 and a vertex at the parent mass m (connecting them to other vertices as before), then the Peptide Sequencing problem can be cast as finding a path from 0 to m in the resulting DAG.[27]

In summary, the vertex set of the resulting spectrum graph is a set of numbers $s_i + \delta_j$ representing potential masses of N-terminal peptides adjusted by the ion type δ_j. Every mass s_i of spectrum S generates k distinct vertices $V_i(s) = \{s_i + \delta_1, \ldots, s_i + \delta_k\}$, though the sets V_i and V_j may overlap if s_i and s_j are close. The set of vertices in a spectrum graph is therefore $\{s_{initial}\} \cup V_1 \cup \cdots \cup V_q \cup \{s_{final}\}$, where $s_{initial} = 0$ and $s_{final} = m$. The spectrum graph may have at most $qk + 2$ vertices. We label the edges of the spectrum graph by the amino acid whose mass is equal to difference between vertex masses. If we look at vertices as putative N-terminal peptides,[28] the edge from u to v implies that the N-terminal sequence corresponding to v may be obtained by extending the sequence at u by the amino acid that labels (u, v).

A spectrum S of a peptide $P = p_1 \ldots p_n$ is called *complete* if S contains at least one ion type corresponding to every N-terminal partial peptide P_i for every $1 \leq i \leq n$. The use of a spectrum graph is based on the observation that for a complete spectrum there exists a path of length $n + 1$ from $s_{initial}$ to s_{final} in the spectrum graph that is labeled by P. This observation casts the Peptide Sequencing problem as one of finding the "correct" path in the set of all paths between two vertices in a directed acyclic graph. If the spectrum

---

27. In addition to the experimental spectrum, every mass spectrometry experiment always produces the parent mass m of a peptide.

28. Although we ignored C-terminal ions in this simplified construction of the spectrum graph, these ions can be taken into account by combining the spectrum S with its "reversed" version.

is complete, then the correct path that we are looking for is often the path with the maximum number of edges, the familiar Longest Path in a DAG problem.

Unfortunately, experimental spectra are frequently incomplete. Moreover, even if the experimental spectrum is complete, there are often many paths in the spectrum graph to choose from that have the same (or even larger) length, preventing one from unambiguously reconstructing the peptide.

The problem with choosing a path with a maximum number of edges is that it does not adequately reflect the "importance" of different vertices. For example, a vertex in the spectrum graph obtained by a shift of $+1$ as $s_i + 1$ (corresponding to the most frequent b-ions) should be scored higher than a vertex obtained by a shift of the rare b-H_2O-NH_3 ion ($s_i + 1 - 18 - 17 = s_i - 34$). Further, whenever there are two peaks s_i and $s_{i'}$ such that $s_i + \delta_j = s_{i'} + \delta_{j'}$, the vertex corresponding to that mass should also get a higher score than a vertex obtained by a single shift.

In the probabilistic approach to peptide sequencing, each ion type δ_i has some probability of occurring, which we write as $p(\delta_i)$. Under the simplest assumption, the probability that δ_i occurs for some partial peptide is independent of whether δ_j also occurs for the same partial peptide. Under this assumption, any given partial peptide may contribute as many as k masses in the spectrum [this happens with probability $\prod_{i=1}^{k} p(\delta_i)$] and as few as 0 [this happens with probability $\prod_{i=1}^{k}(1 - p(\delta_i))$]. The probabilistic model below scores the vertices of the spectrum graph based on these simple assumptions.

Suppose that an N-terminal partial peptide P_i with mass m_i produces ions $\delta_1, \ldots, \delta_l$ ("present" ions of mass $m_i - \delta_1, m_i - \delta_2, \ldots, m_i - \delta_l$) but fails to produce ions $\delta_{l+1}, \ldots, \delta_k$ ("missing" ions) in the experimental spectrum. All l present ions will result in a vertex in the spectrum graph at mass m_i, corresponding to P_i. How should we score this vertex? A naive approach would be to reward P_i for every ion type that explains it, suggesting a score of $\prod_{i=1}^{l} p(\delta_i)$. However, this approach has the disadvantage of not considering the missing ions, so we combine those in by defining the score for the partial peptide to be

$$\left(\prod_{i=1}^{l} p(\delta_i) \right) \left(\prod_{i=l+1}^{k} (1 - p(\delta_i)) \right).$$

However, there is some inherent probability of chemical noise, that is, it can produce *any* mass (that has nothing to do with a peptide of interest) with certain probability p_R. Therefore, we adjust the probabilistic score as

$$\left(\prod_{i=1}^{l} \frac{p(\delta_i)}{p_R} \right) \left(\prod_{i=l+1}^{k} \frac{1 - p(\delta_i)}{1 - p_R} \right).$$

8.13   Protein Identification via Database Search

De novo protein sequencing algorithms are invaluable for identification of (both known and unknown) proteins, but they are most useful when working with complete or nearly complete high-quality spectra. Many spectra are far from complete, and de novo peptide sequencing algorithms often produce ambiguous solutions for such spectra. If we had access to a database of all proteins from a genome, then we would no longer need to consider all 20^l peptide sequences to interpret an MS/MS spectrum, but could instead limit our search to peptides present in this database.

Currently, most proteins are identified by database search— effectively looking the answer up in "the back of a book." Indeed, an experimental spectrum can be compared with the theoretical spectrum for each peptide in such a database, and the entry in the database that best matches the observed spectrum usually provides the sequence of the experimental peptide. This forms the basis of the popular **SEQUEST** algorithm developed by John Yates and colleagues. We formulate the Protein Identification problem as follows.

Protein Identification Problem:
Find a protein from a database that best matches the experimental spectrum.

   Input: A database of proteins, an experimental spectrum S, a set of ion types Δ, and a parent mass m.

   Output: A protein of mass m from the database with the best match to spectrum S.

Though the logic that **SEQUEST** uses to determine whether a database entry matches an experimental spectrum is somewhat involved, the basic approach of the algorithm is just a linear search through the database. One drawback to MS/MS database search algorithms like **SEQUEST** is that peptides in a cell are often slightly different from the "canonical" peptides present in databases. The synthesis of proteins on ribosomes is not the final step in a protein's life: many proteins are subject to further modifications that regulate protein activities and these modifications may be either permanent or reversible. For example, the enzymatic activity of some proteins is regulated by the addition or removal of a phosphate group at a specific residue.[29] *Phosphorylation* is a reversible process: *protein kinases* add phosphate groups while *phosphatases* remove them.

Proteins form complex systems necessary for cellular signaling and metabolic regulation, and are therefore often subject to a large number of biochemical modifications (e.g., phosphorylation and glycosylation). In fact, almost all protein sequences are modified after they have been constructed from their mRNA template, and as many as 200 distinct types of modifications of amino acid residues are known. Since we are unable to predict these *post-translational modifications* from DNA sequences, finding naturally occurring modifications remains an important open problem. Computationally, a chemical modification of the protein $p_1 p_2 \cdots p_i \cdots p_n$ at position i results in increased mass of the N-terminal peptides $P_i, P_{i+1}, \ldots, P_n$ and increased mass of the C-terminal peptides $P_1^-, P_2^-, \ldots, P_{i-1}^-$.

The computational analysis of modified peptides was also pioneered by John Yates, who suggested an exhaustive search approach that (implicitly) generates a virtual database of all modified peptides from a small set of potential modifications, and matches the experimental spectrum against this virtual database. It leads to a large combinatorial problem, even for a small set of modification types.

Modified Protein Identification Problem:

Find a peptide from the database that best matches the experimental spectrum with up to k modifications.

   Input: A database of proteins, an experimental spectrum S, a set of ion types Δ, a parent mass m, and a parameter k capping the number of modifications.

   Output: A protein of mass m with the best match to spectrum S that is at most k modifications away from an entry in the database.

The major difficulty with the Modified Protein Identification problem is

---

29. Phosphorylation uses serine, threonine, or tyrosine residues to add a phosphate group.

that very similar peptides P_1 and P_2 may have very different spectra S_1 and S_2.[30] Our goal is to define a notion of spectral similarity that correlates well with sequence similarity. In other words, if P_1 and P_2 are a few modifications apart, the spectral similarity between S_1 and S_2 should be high. The shared peaks count is, of course, an intuitive measure of spectral similarity. However, this measure diminishes very quickly as the number of mutations increases, thus leading to limitations in detecting similarities by database search. Moreover, there are many correlations between the spectra of related peptides, and only a small proportion of these correlations is captured by the shared peaks count.

The spectral convolution algorithm, below, reveals potential peptide modifications without an exhaustive search and therefore does not require generating a virtual database of modified peptides.

8.14   Spectral Convolution

Let S_1 and S_2 be two spectra. Define the *spectral convolution* to be the multiset $S_2 \ominus S_1 = \{s_2 - s_1 :\ s_1 \in S_1, s_2 \in S_2\}$ and let $(S_2 \ominus S_1)(x)$ be the multiplicity of element x in this multiset. In other words, $(S_2 \ominus S_1)(x)$ is the number of pairs $(s_1 \in S_1, s_2 \in S_2)$ such that $s_2 - s_1 = x$ (fig. 8.26).

The shared peak count that we introduced earlier in this chapter is the number of masses common to both S_1 and S_2, and is simply $(S_2 \ominus S_1)(0)$. MS/MS database search algorithms that maximize the shared peak count find a peptide in the database that maximizes $(S_2 \ominus S_1)(0)$, where S_2 is an experimental spectrum and S_1 is the theoretical spectrum of a peptide in the database. However, if S_1 and S_2 correspond to peptides that differ by k mutations or modifications, the value of $(S_2 \ominus S_1)(0)$ may be too small to determine that S_1 and S_2 really were generated by similar peptides. As a result, the power of the shared peak count to discern that two peptides are similar diminishes rapidly as the number of modifications increases—it is bad at $k = 1$, and nearly useless for $k > 1$.

The peaks in the spectral convolution allow us to detect mutations and modifications, even if the shared peak count is small. If peptides P_2 and P_1 (corresponding to spectra S_2 and S_1) differ by only one mutation ($k = 1$) with amino acid difference $\delta = m(P_2) - m(P_1)$, then $S_2 \ominus S_1$ is expected to have two approximately equal peaks at $x = 0$ and $x = \delta$. If the mutation

---

30. P_1 corresponds to a peptide from the database, while P_2 corresponds to the modified version of P_1, whose experimental spectrum is being used to search the database.

occurs at position t in the peptide, then the peak at $(S_2 \ominus S_1)(0)$ corresponds to N-terminal peptides P_i for $i < t$ and C-terminal peptides P_i^- for $i \geq t$. The peak at $(S_2 \ominus S_1)(\delta)$ corresponds to N-terminal peptides P_i for $i \geq t$ and C-terminal peptides P_i^- for $i < t$.

Now assume that P_2 and P_1 are two substitutions apart, one with mass difference δ' and another with mass difference $\delta - \delta'$, where δ denotes the difference between the parent masses of P_1 and P_2. These modifications generate two new peaks in the spectral convolution at $(S_2 \ominus S_1)(\delta')$ and at $(S_2 \ominus S_1)(\delta - \delta')$. It is therefore reasonable to define the similarity between spectra S_1 and S_2 as the overall height of the k highest peaks in $S_2 \ominus S_1$.

Although spectral convolution helps to identify modified peptides, it does have a limitation. Let

$$S = \{10, 20, 30, 40, 50, 60, 70, 80, 90, 100\}$$

be a spectrum of peptide P, and assume for simplicity that P produces only b-ions. Let

$$S' = \{10, 20, 30, 40, 50, 55, 65, 75, 85, 95\}$$

and

$$S'' = \{10, 15, 30, 35, 50, 55, 70, 75, 90, 95\}$$

be two theoretical spectra corresponding to peptides P' and P'' from the database. Which of the two peptides fits S better? The shared peaks count does not allow one to answer this question, since both S' and S'' have five peaks in common with S. Moreover, the spectral convolution also does not answer this question, since both $S \ominus S'$ and $S \ominus S''$ reveal strong peaks of the same height at 0 and 5. This suggests that both P' and P'' can be obtained from P by a single mutation with mass difference 5. However, a more careful analysis shows that although this mutation can be realized for P' by introducing a shift 5 after mass 50, it cannot be realized for P''. The major difference between S' and S'' is that the matching positions in S' come in clumps while the matching positions in S'' do not. Below we describe the spectral alignment approach, which addresses this problem.

8.15   Spectral Alignment

Let $A = \{a_1, \ldots, a_n\}$ be an ordered set of integers $a_1 < a_2 < \cdots < a_n$. A *shift* Δ_i transforms A into $\{a_1, \ldots a_{i-1}, a_i + \Delta_i, \ldots, a_n + \Delta_i\}$. That is, Δ_i alters all elements in the sequence except for the first $i - 1$ elements. We only

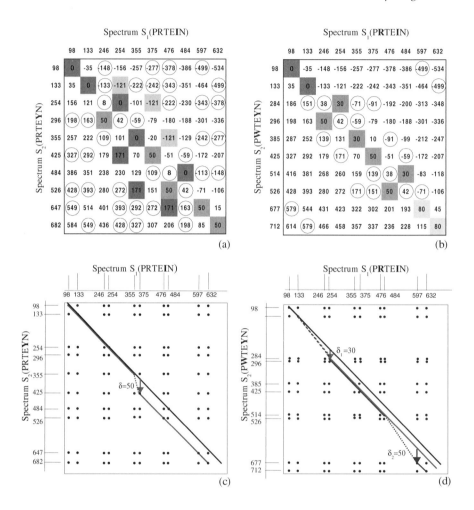

Figure 8.26 Detecting modifications of the peptide $PRTEIN$. (a) Elements of the spectral convolution $S_2 \ominus S_1$ represented as elements of a difference matrix. S_1 and S_2 are the theoretical spectra of the peptides $PRTEIN$ and $PRTEYN$, respectively. Elements in the spectral convolution that have multiplicity larger than 2 are shaded, while the elements with multiplicity exactly 2 are shown circled. The high multiplicity element 0 corresponds to all of the shared masses between the two spectra, while another high multiplicity element (50) corresponds to the shift of masses by $\delta = 50$, due to the mutation of I to Y in $PRTEIN$ (the difference in mass between Y and I is 50). In (b), two mutations have occurred in $PRTEIN$: $R \to W$ with $\delta' = 30$, and $I \to Y$ with $\delta'' = 50$. Spectral alignments for (a) and (b) are shown in (c) and (d), respectively. The main diagonals represent the paths for $k = 0$. The lines parallel to the main diagonals represent the paths for $k > 0$. Every jump between diagonals corresponds to an increase in k. Mutations and modifications to a peptide can be detected as jumps between the diagonals.

consider shifts that do not change the order of elements, that is, the shifts with $\Delta_i \geq a_{i-1} - a_i$. The *k-similarity*, $D(k)$, between sets A and B is defined as the maximum number of elements in common between these sets after k shifts. For example, a shift -5_6 transforms

$$S = \{10, 20, 30, 40, 50, 60, 70, 80, 90, 100\}$$

into

$$S' = \{10, 20, 30, 40, 50, \mathbf{55}, \mathbf{65}, \mathbf{75}, \mathbf{85}, \mathbf{95}\}.$$

Therefore $D(1) = 10$ for these sets. The set

$$S'' = \{10, 15, 30, 35, 50, 55, 70, 75, 90, 95\}$$

has five elements in common with S (the same as S') but there is no single shift transforming S into S'' ($D(1) = 6$). Below we analyze and solve the following Spectral Alignment problem:

Spectral Alignment Problem:
Find the k-similarity between two sets.

    Input: Sets A and B, which represent the two spectra, and a number k (number of shifts).

    Output: The *k-similarity*, $D(k)$, between sets A and B.

One can represent sets $A = \{a_1, \ldots, a_n\}$ and $B = \{b_1, \ldots, b_m\}$ as 0–1 arrays **a** and **b** of length a_n and b_m correspondingly. The array **a** will contain n ones (at positions $a_1, \ldots, a_n$) and $a_n - n$ zeros, while the array **b** will contain m ones (at positions $b_1, \ldots, b_m$) and $b_m - m$ zeros.[31] In such a model, a shift $\Delta_i < 0$ is simply a deletion of Δ_i zeros from **a**, while a shift $\Delta_i > 0$ is simply an insertion of Δ_i zeros in **a**. With this model in mind, the Spectral Alignment problem is simply to find the edit distance between **a** and **b** when the elementary operations are deletions and insertions of blocks of zeros. As we saw in chapter 6, these operations can be modeled by long horizontal and vertical edges in a Manhattan-like graph. The only differences between the traditional Edit Distance problem and the Spectral Alignment problem are a somewhat unusual alphabet and the scoring of paths in the resulting graph. The analogy between the Edit Distance problem and the Spectral Alignment

---

31. We remark that this is not a particularly dense encoding of the spectrum.

problem leads us to frame spectral alignment as a type of longest path problem.

Define a *spectral product* $A \otimes B$ to be the $a_n \times b_m$ two-dimensional matrix with nm ones corresponding to all pairs of indices (a_i, b_j) and all remaining elements zero. The number of ones on the main diagonal of this matrix describes the shared peaks count between spectra A and B, or in other words, 0-similarity between A and B. Figure 8.27 shows the spectral products $S \otimes S'$ and $S \otimes S''$ for the example from the previous section. In both cases the number of ones on the main diagonal is the same, and $D(0) = 5$. The δ-shifted peaks count is the number of ones on the diagonal that is δ away from the main diagonal. The limitation of the spectral convolution is that it considers diagonals separately without combining them into feasible mutation scenarios. The k-*Similarity* between spectra is defined as the maximum number of ones on a path through the spectral matrix that uses at most $k + 1$ diagonals, and the k-*optimal spectral alignment* is defined as the path that uses these $k + 1$ diagonals. For example, 1-similarity is defined by the maximum number of ones on a path through this matrix that uses at most two diagonals. Figure 8.27 demonstrates the notion that 1-similarity shows that S is closer to S' than to S''; in the first case the optimal two-diagonal path covers ten 1s (left matrix), versus six in the second case (right matrix). Figure 8.28 illustrates that the spectral alignment detects more and more subtle similarities between spectra, simply by increasing k [compare figures 8.26 (c) and (d)].[32] Below we describe a dynamic programming algorithm for spectral alignment.

Let A_i and B_j be the i-prefix of A and j-prefix of B, respectively. Define $D_{ij}(k)$ as the k-similarity between A_i and B_j such that the last elements of A_i and B_j are matched. In other words, $D_{ij}(k)$ is the maximum number of ones on a path to (a_i, b_j) that uses at most $k + 1$ different diagonals. We say that (i', j') and (i, j) are *codiagonal* if $a_i - a_{i'} = b_j - b_{j'}$ and that $(i', j') < (i, j)$ if $i' < i$ and $j' < j$. To take care of the initial conditions, we introduce a fictitious element $(0, 0)$ with $D_{0,0}(k) = 0$ and assume that $(0, 0)$ is codiagonal with any other (i, j). The dynamic programming recurrence for $D_{ij}(k)$ is then

$$D_{ij}(k) = \max_{(i',j')<(i,j)} \begin{cases} D_{i'j'}(k) + 1, & \text{if } (i', j') \text{ and } (i, j) \text{ are codiagonal} \\ D_{i'j'}(k - 1) + 1, & \text{otherwise.} \end{cases}$$

The k-similarity between A and B is given by $D(k) = \max_{ij} D_{ij}(k)$.

---

32. To a limit, of course. When k is too large, the spectral alignment is not very useful.

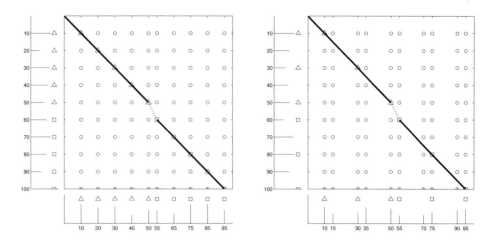

Figure 8.27   Spectral products $S \otimes S'$ (left) and $S \otimes S''$ (right), where $S = \{10, 20, 30, 40, 50, 60, 70, 80, 90, 100\}$, $S' = \{10, 20, 30, 40, 50, 55, 65, 75, 85, 95\}$, and $S'' = \{10, 15, 30, 35, 50, 55, 70, 75, 90, 95\}$. The matrices have dimensions 100×95, with ones shown by circles (zeros are too numerous to show). The spectrum S can be transformed into S' by a single shift and $D(1) = 10$. However, the spectrum S cannot be transformed into S'' by a single shift and $D(1) = 6$.

The above dynamic programming algorithm for spectral alignment is rather slow, with a running time of $O(n^4 k)$ for two n-element spectra, and below we describe an $O(n^2 k)$ algorithm for solving this problem. Define $diag(i, j)$ as the maximal codiagonal pair of (i, j) such that $diag(i, j) < (i, j)$. In other words, $diag(i, j)$ is the position of the previous "1" on the same diagonal as (a_i, b_j) or $(0, 0)$ if such a position does not exist. Define

$$M_{ij}(k) = max_{(i', j') \leq (i,j)} D_{i'j'}(k).$$

Then the recurrence for $D_{ij}(k)$ can be re-written as

$$D_{ij}(k) = max \begin{cases} D_{diag(i,j)}(k) + 1, \\ M_{i-1, j-1}(k-1) + 1. \end{cases}$$

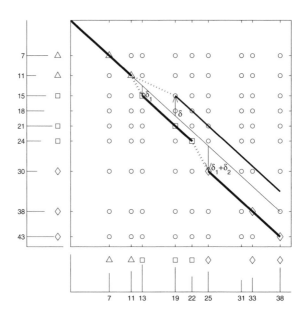

Figure 8.28   Aligning spectra. The shared peaks count reveals only $D(0) = 3$ matching peaks on the main diagonal, while spectral alignment reveals more hidden similarities between spectra ($D(1) = 5$ and $D(2) = 8$) and detects the corresponding mutations.

The recurrence for $M_{ij}(k)$ is given by

$$M_{ij}(k) = \max \begin{cases} D_{ij}(k), \\ M_{i-1,j}(k), \\ M_{i,j-1}(k). \end{cases}$$

The transformation of the dynamic programming graph can be achieved by introducing horizontal and vertical edges that provide the ability to switch between diagonals (fig. 8.29). The score of a path is the number of ones on this path, while k corresponds to the number of switches (number of used diagonals minus 1).

The simple dynamic programming algorithm outlined above hides many details that make the spectral alignment problem difficult. A spectrum can be thought of as a combination of *two* series of numbers, one increasing (the N-terminal ions) and the other decreasing (the C-terminal ions). These two

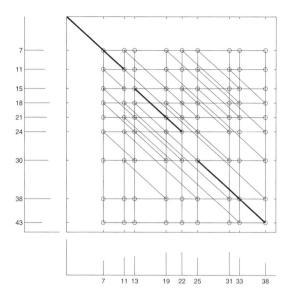

Figure 8.29   Modification of a dynamic programming graph leads to a fast spectral alignment algorithm.

series form diagonals in the spectral product $S \otimes S$, the main diagonal and the perpendicular diagonal. These correspond, respectively, to pairings of N-terminal and C-terminal ions. The algorithm we have described deals with the main diagonal only. Finding post-translationally modified proteins via mass spectrometry remains a difficult problem that nobody has yet solved, and significant efforts are underway to extend the Spectral Alignment algorithm to handle these complications and to develop new algorithmic ideas for protein identification.

8.16   Notes

The earliest paper on graph theory seems to be that of Leonhard Euler, who, in 1736, discussed whether or not it was possible to stroll around Königsberg crossing each of its bridges across the Pregel River exactly once. Euler remains one of the most prolific writers in mathematics: aside from graph theory, we owe him the notation $f(x)$ for a function, i for the square root of -1, and π for pi. He worked hard throughout his entire life only to become blind. He commented: "Now I will have fewer distractions," and proceeded to write hundreds of papers more.

Graph theory was forgotten for a century, but was revived in the second half of the nineteenth century by prominent scientists such as Sir William Hamilton (who, among many other things, invented quaternions) and Gustav Kirchhoff (who is responsible for Kirchhoff's laws).

DNA arrays were proposed simultaneously and independently in 1988 by Radoje Drmanac and colleagues in Yugoslavia (29), Andrey Mirzabenov and colleagues in Russia (69), and Ed Southern in the United Kingdom (100). The inventors of DNA arrays suggested using them for DNA sequencing, and the original name for this technology was sequencing by hybridization. A major breakthrough in DNA array technology was made by Steve Fodor and colleagues in 1991 (38) when they adapted photolithography (a process similar to computer chip manufacturing) to DNA synthesis. The Eulerian path approach to SBH was described in (83).

Sanger's approach to protein sequencing influenced work on RNA sequencing. Before biologists figured out how to sequence DNA, they routinely sequenced RNA. The first RNA sequencing project resulted in seventy-seven ribonucleotides and took seven years to complete, though in 1965 RNA sequencing used the same "break—read the fragments—assemble" approach that is used for DNA sequencing today. For many years, DNA sequencing was done by first transcribing DNA to RNA and then sequencing the RNA.

DNA sequencing methods were invented independently and simultaneously in 1977 by Frederick Sanger and colleagues (91) and Walter Gilbert and colleagues (74). The overlap-layout-consensus approach to DNA sequencing was first outlined in 1984 (82) and further developed by John Kececioglu and Eugene Myers in 1995 (55). DNA sequencing progressed to handle the entire 1800 kb *H. influenzae* bacterium genome in the mid-1990s. In 1997, inspired by this breakthrough, James Weber and Eugene Myers (110) proposed the whole-genome shotgun approach (first outlined by Jared Roach and colleagues in 1995 (87)) to sequence the entire human genome. The human genome was sequenced in 2001 by J. Craig Venter and his team at Celera Genomics (104) with the whole genome shotgun approach, and independently by Eric Lander and his colleagues at the Human Genome Consortium (62) using the BAC-by-BAC approach.

Early approaches to protein sequencing by mass spectrometry were based on manual peptide reconstruction and the assembly of those peptides into protein sequences (51). The description of the spectrum graph approach pre-

sented in this chapter is from Vlado Dancik and colleagues (25). Searching
a database for the purposes of protein identification in mass spectrometry
was pioneered by Matthias Mann and John Yates in 1994 (71; 34). In 1995
Yates (112) extended his original **SEQUEST** algorithm to search for modified
peptides based on a virtual database of all modified peptides. The spectral
alignment algorithm was introduced five years later (84).

8.17   Problems

Problem 8.1

Can 99 phones be connected by wires in such a way that each phone is connected with exactly 11 others?

Problem 8.2

Can a kingdom in which 7 roads lead out of each city have exactly 100 roads?

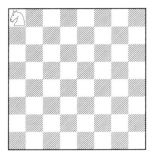

Problem 8.3

Can a knight travel around a chessboard pass through every square exactly once, and end on the same square it started on?

Problem 8.4

Can a knight travel around a chessboard, start at the upper left corner, pass through every square exactly once, and end on the lower right corner?

Problem 8.5

Can one use a 12-inch-long wire to form a cube (each of the 12 cube edges is 1-inch long). If not, what is the smallest number of cuts one must make to form this cube?

Problem 8.6

Find the shortest common superstring for eight 3-mers:

$$\{AGT, AAA, ACT, AAC, CTT, GTA, TTT, TAA\}$$

and solve the following two problems:

- Construct the graph with 8 vertices corresponding to these 3-mers (Hamiltonian path approach) and find a Hamiltonian path (7 edges) which visits each vertex exactly once. Does this path visit every edge of the graph? Write the superstring corresponding to this Hamiltonian path.
- Construct the graph with 8 edges corresponding to these 3-mers (Eulerian path approach) and find an Eulerian path (8 edges) which visits each edge exactly once. Does this path visit every vertex of the graph exactly once? Write the superstring corresponding to this Eulerian path.

Problem 8.7

Find the shortest common superstring for all 100 2-digit numbers from 00 to 99.

Problem 8.8

Find the shortest common superstring for all 16 3-digit binary numbers in 0-1 alphabet.

Problem 8.9

Use the Eulerian path approach to solve the SBH problem for the following spectrum:

$$S = \{\text{ATG, GGG, GGT, GTA, GTG, TAT, TGG}\}$$

Label edges and vertices of the graph, and give all possible sequences s such that $Spectrum(s, 3) = S$.

Problem 8.10

The SBH problem is to reconstruct a DNA sequence from its l-mer composition

- Suppose that instead of a single target DNA fragment, we have two target DNA fragments and we simultaneously analyze both of them with a universal DNA array. Give a precise formulation of the resulting problem (something like the formulation of the SBH problem).

- Give an approach to the above problem which resembles the Hamiltonian Path approach to SBH.

- Give an approach to the above problem which resembles the Eulerian Path approach to SBH.

Problem 8.11

Suppose we have k target DNA fragments, and that we are able to measure the overall multiplicity of each l-mer in these strings. Give an algorithm to reconstruct these k strings from the overall l-mer composition of these strings.

Problem 8.12

Prove that if n random reads of length L are chosen from a genome of length G, then the expected fraction of the genome represented in these reads is approximately $1 - e^c$, where $c = \frac{nL}{G}$ is the average *coverage* of the genome by reads.

The simplest heuristic for the Shortest Superstring problem is an obvious greedy algorithm: repeatedly merge a pair of strings with maximum overlap until only one string remains. The *compression* of an approximation algorithm for the Shortest Superstring problem is defined as the number of symbols saved by this algorithm compared to plainly concatenating all the strings.

Problem 8.13

Prove that the greedy algorithm achieves at least $\frac{1}{2}$ the compression of an optimal superstring, that is,

$$\frac{greedy\ compression}{optimal\ compression} \geq \frac{1}{2}$$

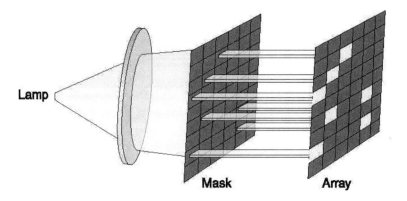

Figure 8.30   A mask used in the synthesis of a DNA array.

Let $\mathcal{P} = \{s_1, \ldots, s_m\}$ be a set of *positive* strings and $\mathcal{N} = \{t_1, \ldots, t_k\}$ be a set of *negative* strings. We assume that no negative string t_i is a substring of any positive string s_j. A *consistent superstring* is a string s such that each s_i is a substring of s and no t_i is a substring of s.

Problem 8.14

Design an approximation algorithm for the shortest consistent superstring problem.

Problem 8.15

DNA sequencing reads contain errors that lead to complications in fragment assembly. Fragment assembly with sequencing errors motivates the *Shortest k-Approximate Superstring problem*: Given a set of strings $\mathcal{S}$, find a shortest string s such that each string in $\mathcal{S}$ matches some substring of s with at most k errors. Design an approximation algorithm for this problem.

DNA arrays can be manufactured with the use of a *photolithographic* process that grows probes one nucleotide at a time through a series of chemical steps. Every nucleotide carries a photolabile protection group protecting the probe from further growth. This group can be removed by illuminating the probe. In each step, a predefined region of the array is illuminated, thus removing a photolabile protecting group from that region and "activating" it for further nucleotide growth. The entire array is then exposed to a particular nucleotide (which bears its own photolabile protecting group), but reactions only occur in the activated region. Each time the process is repeated, a new region is activated and a single nucleotide is appended to each probe in that region. By appending nucleotides to the proper regions in the appropriate sequence, it is possible to grow a complete set of l-mer probes in as few as $4 \cdot l$ steps. The light-directed synthesis allows random access to all positions of the array and can be used to make arrays with any probes at any site (fig. 8.30).

The proper regions are activated by illuminating the array through a series of masks, like those in figure 8.31. White areas of a mask correspond to the region of the array to be illuminated,

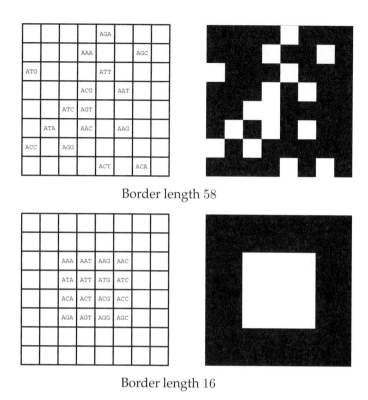

Border length 58

Border length 16

Figure 8.31 Two masks with different border lengths (only 3-mers starting with A are shown).

and black areas correspond to the region to be shadowed. Unfortunately, because of diffraction and light scattering, points that are close to the border between an illuminated region and a shadowed region are often subject to unintended illumination. In such a region, it is uncertain whether a nucleotide will be appended or not. This uncertainty gives rise to probes with unknown sequences and unknown lengths, that may hybridize to a target DNA strand, thus complicating interpretation of the experimental data. Methods are being sought to minimize the lengths of these borders so that the level of uncertainty is reduced.

Figure 8.31 presents two universal arrays with different arrangements of 3-mers and masks for synthesizing the first nucleotide A (only probes with first nucleotide A are shown). The *border length* of the mask at the bottom of figure 8.31 is significantly smaller than the border length of the mask at the top of figure 8.31. Companies producing DNA arrays try to arrange the 4^l probes on the universal array in such a way that the overall border length of *all* $4 \times l$ masks is minimal.

Problem 8.16

Find a lower bound on the overall border length of the universal array with 4^l l-mers.

For two l-mers x and y, let $d_H(x, y)$ be the Hamming distance between x and y, that is, the number of positions in which x and y differ. The overall border length of all masks equals $2 \sum d_H(x, y)$, where the sum is taken over all pairs of neighboring probes on the array. This observation establishes the connection between minimization of border length and Gray codes.

An l-bit *Gray code* is defined as a permutation of the binary numbers from $000 \cdots 0$ to $111 \cdots 1$, each containing l binary digits, such that neighboring numbers have exactly one differing bit, as do the first and last numbers. For example, the arrangement of sixteen 4-mers in a 4-bit Gray code is shown below:

0	1	2	3	4	5	6	7	8	9	10	11	12	13	14	15
0	0	0	0	0	0	0	0	1	1	1	1	1	1	1	1
0	0	0	0	1	1	1	1	1	1	1	1	0	0	0	0
0	0	1	1	1	1	0	0	0	0	1	1	1	1	0	0
0	1	1	0	0	1	1	0	0	1	1	0	0	1	1	0

This Gray code can be generated recursively, starting with the 1-bit Gray code

$$G_1 = \{0, 1\},$$

as follows. For an l-bit Gray code,

$$G_l = \{g_1, g_2, ..., g_{2^l-1}, g_{2^l}\},$$

define an $(l + 1)$-bit Gray code as follows:

$$G_{l+1} = \{0g_1, 0g_2, ..., 0g_{2^l-1}, 0g_{2^l}, 1g_{2^l}, 1g_{2^l-1}, ..., 1g_2, 1g_1\}.$$

The elements of G_l are simply copied with 0s added to the front, then reversed with 1s added to the front. Clearly, all elements in G_{l+1} are distinct, and consecutive elements in G_{l+1} differ by exactly one bit. For example, the 2-bit Gray code is $\{00, 01, 11, 10\}$, and the 3-bit Gray code is $\{000, 001, 011, 010, 110, 111, 101, 100\}$.

Problem 8.17

Design a Gray code for all 4-digit decimal numbers from 0000 to 9999.

We are interested in a *two-dimensional* Gray code composed of strings of length l over a four-letter alphabet. In other words, we would like to generate a 2^l-by-2^l matrix in which each of the 4^l l-mers in a universal array is present at some position, and each pair of adjacent l-mers (horizontally or vertically) differs in exactly one position. Constructing such a two-dimensional Gray code is equivalent to minimizing the border length of the universal array.

Problem 8.18

Find the arrangement of probes on a universal array minimizing the border length.

Accurate determination of the peptide parent mass is very important in de novo peptide sequencing. An error in parent mass leads to systematic errors in the construction of the spectrum graph when both N-terminal and C-terminal fragment ions are considered (wrong pairing of N-terminal and C-terminal fragment ions).

Problem 8.19

Given an MS/MS spectrum (without parent mass), devise an algorithm that estimates the parent mass.

Problem 8.20

Develop an algorithm for combining N-terminal and C-terminal series together in the spectral alignment algorithm.

Problem 8.21

Develop a version of the spectral alignment algorithm that is geared to mutations rather than modifications. In this case the jumps between diagonals are not arbitrary and one has to limit the possible shifts between diagonals to mass differences between amino acids participating in the mutation.

The *i-th prefix mass* of protein $P = p_1 \ldots p_n$ is $m_i = \sum_{j=1}^{i} m(p_j)$ where $m(p_j)$ is the mass of amino acid p_j. The *prefix spectrum* of protein P is the increasing sequence $m_1, \ldots, m_n$ of its prefix masses. For example, the prefix spectrum of protein CSE is $\{103, 103 + 87, 103 + 87 + 129\} = \{103, 190, 319\}$. A *peptide* is any substring $p_i \ldots p_j$ of P for $1 \leq i \leq j \leq n$. The prefix spectrum of a peptide $p_i \ldots p_j$ contains $j - i + 1$ masses, for example, the prefix spectrum of CS is $\{103, 190\}$, while the prefix spectrum of SE is $\{87, 216\}$. For simplicity we assume that every amino acid has an unique mass.

Problem 8.22

(Spectral Assembly problem) Given a set of prefix spectra from a set of (overlapping) peptides extracted from an (unknown) protein P, reconstruct the amino acid sequence of P.

The Spectral Assembly problem is equivalent to the classic *Shortest Common Superstring* problem. Assume that the protein P and the set of its peptides $\mathcal{P}$ satisfy the following conditions:

- $p_i p_{i+1} \neq p_j p_{j+1}$ for $1 \leq i < j \leq n - 1$, that is, every amino acid 2-mer (dipeptide) occurs at most once in protein P.

- For every two consecutive amino acids $p_i p_{i+1}$ in protein P there exists a peptide in $\mathcal{P}$ containing dipeptide $p_i p_{i+1}$.

For example, a protein STRAND and the set of peptides STR, TR, TRAN, RA, and AND satisfy these conditions.

Problem 8.23

Design an algorithm to solve the Spectral Assembly problem under the above conditions. Does the problem have a unique solution? If your answer is "no," then also provide an algorithm to find all possible protein reconstructions.

Problem 8.24

Extend this algorithm to solve the spectral assembly problem under the following conditions:

- $p_i p_{i+1} \ldots p_{i+l-1} \neq p_j p_{j+1} \ldots p_{j+l-1}$ for $1 \leq i < j \leq n-l+1$, that is, every amino acid l-mer (l-peptide) occurs at most once in protein P.
- For every l consecutive amino acids $p_i p_{i+1} \ldots p_{i+l-1}$ in protein P there exists a peptide in $\mathcal{P}$ containing l-peptide $p_i p_{i+1} \ldots p_{i+l-1}$.

Problem 8.25

Consider two proteins P_1 and P_2. The combined prefix spectrum of proteins P_1 and P_2 is defined as the union of their prefix spectra. Describe an algorithm for reconstructing P_1 and P_2 from their combined prefix spectrum. Give an example when such a reconstruction is non-unique. Generalize this algorithm for three and more proteins.

When analyzing a protein $p_1 \ldots p_n$, a mass spectrometer measures the masses of both the prefix peptides $p_1 \ldots p_i$ and of the suffix peptides $p_i \ldots p_n$, for $1 \leq i \leq n$. The *prefix-suffix* mass spectrum includes the masses of all prefix and suffix peptides. For example, CSE produces the following prefix-suffix spectrum $\{103, 129, 103 + 87, 129 + 87, 103 + 87 + 129\} = \{103, 129, 190, 216, 319\}$ and it remains unknown which masses in the prefix-suffix spectrum are derived from the prefix peptides and which are derived from the suffix peptides.

The prefix-suffix spectrum may contain as few as n masses (for palindromic peptides with every suffix mass matched by a prefix mass) and as many as $2n - 1$ masses (if the overall peptide mass is the only match between suffix and prefix masses).

Problem 8.26

Reconstruct a peptide given its prefix-suffix spectrum. Devise an efficient algorithm for this problem under the assumption that the prefix-suffix spectrum of a peptide of length n contains $2n - 1$ masses.

In 1993 David Schwartz and colleagues developed the *optical mapping* technique for construction of restriction maps. In optical mapping, single copies of DNA molecules are stretched and attached to a glass under a microscope. When restriction enzymes are activated, they cleave the DNA molecules at their restriction sites. The molecules remain attached to the surface, but the elasticity of the stretched DNA pulls back the molecule ends at the cleaved sites. These can be identified under the microscope as tiny gaps in the fluorescent line of the molecule. Thus a "photograph" of the DNA molecule with gaps at the positions of cleavage sites gives a snapshot of the restriction map.

Optical mapping bypasses the problem of reconstructing the order of restriction fragments, but raises new computational challenges. The problem is that not all sites are cleaved in each molecule and that some may incorrectly appear to be cut. In addition, inaccuracies in measuring the length of fragments and the unknown orientation of each molecule (left to right or vice versa) make the reconstruction difficult. In practice, data from many molecules are gathered to build a consensus restriction map.

The input to the optical mapping problem is a 0-1 $n \times m$ matrix $S = (s_{ij})$ where each row corresponds to a DNA molecule (straight or reversed), each column corresponds to a position in that molecule, and $s_{ij} = 1$ if (and only if) there is a cut in position j of molecule i. The goal is to reverse the orientation of a subset of molecules (subset of rows in S) and to declare a subset of the t columns "real cut sites" so that the number of ones in cut site columns is maximized.

A naive approach to this problem is to find t columns with a large proportion of ones and declare them potential cut sites. However, in this approach every real site will have a reversed twin (since each "photograph" corresponds to either straight or reversed DNA molecules with equal probabilities). Let $w(i,j)$ be the number of molecules with both cut sites i and j present (in either direct or reverse orientation). In a different approach, a graph on vertices $\{1, \ldots, m\}$ is constructed and two vertices are connected by an edge (i,j) of weight $w(i,j)$.

Problem 8.27

Establish a connection between optical mapping and finding paths in graphs.

9 *Combinatorial Pattern Matching*

In chapter 5, we considered the Motif Finding problem, which is to find some overrepresented pattern in a DNA sample. We are not given any particular pattern to search for; rather we must infer it from the sample. Combinatorial pattern matching, on the other hand, looks for exact or approximate occurrences of *given* patterns in a long text. Although pattern matching is in some ways simpler than motif finding since we actually know what we are looking for, the large size of genomes makes the problem, in practice, difficult. The alignment techniques of chapter 6 become impractical for whole genomes, particularly when one searches for approximate occurrences of many long patterns at one time. In this chapter we develop a number of ways to make pattern matching in a long string practical.

One recurring theme in this chapter is how to organize data into efficient *data structures*, often a crucial part of solving a problem. For example, suppose you were in a library with 2 million volumes in no discernible order and you needed to find one particular title. The only way to guarantee finding the title would be to check every book in the library. However, if the books in the library were sorted by title, then the check becomes very easy. A sorted list is only one of many types of data structures, and in this chapter we discuss significantly more sophisticated ways of organizing data.

9.1 Repeat Finding

Many genetic diseases are associated with deletions, duplications, and rearrangements of long chromosomal regions. These are dramatic events that affect the large-scale genomic architecture and may involve millions of nucleotides. For example, *DiGeorge syndrome*, which commonly results in an impaired immune system and heart defects, is associated with a large 3 Mb deletion on human chromosome 22. A deletion of this size is likely to remove important genes and lead to disease.

Such dramatic changes in genomic architecture often require—as in the case of DiGeorge syndrome—a pair of very similar sequences flanking the deleted segment. These similar sequences form a *repeat* in DNA, and it is important to find all repeats in a genome.

Repeats in DNA hold many evolutionary secrets. A striking and still unexplained phenomenon is the large number of repeats in many genomes: for example, repeats account for about 50% of the human genome. Algorithms that search for repeats in the human genome need to analyze a 3 billion nucleotide genome, and quadratic sequence alignment algorithms are too slow for this. The simplest approach to finding *exact* repeats is to construct a table that holds, for each l-mer, all the positions where the l-mer is located in the genomic DNA sequence. Such a table would contain 4^l bins, each bin containing some number between 0 and M of positions, where M is the frequency of occurrence of the most common l-mer in the genome. The average number of elements in each bin is $\frac{n}{4^l}$, where n is the length of the genome. In many applications, the parameter l varies from 10 to 13, so this table is not unmanageably large. Although this tabulation approach allows one to quickly find all repeats of length l, such short repeats are not very interesting for DNA analysis. Biologists instead are interested in long *maximal repeats*, that is, repeats that cannot be extended to the left or to the right. To find maximal repeats that are longer than a predefined parameter L, each exact repeat of length l must be extended to the left or to the right to see whether it is embedded in a repeat longer than L. Since typically $l << L$, there are many more repeats to be extended than there are maximal repeats to be found. For example, the bacterial *Escherichia coli* genome of 4.6 million nucleotides has millions of repeats of length $l = 12$ but only about 8000 maximal repeats of length $L = 20$ or longer.[1] Thus, most of the work in this approach to repeat finding is wasted trying to pointlessly extend short repeats. The popular **RE-Puter** algorithm gets around this by using a suffix tree, a data structure that we describe in a later section.

------

1. It may seem confusing at first that a genome of size 4.6 million base pairs has more than a million repeats. A *repeat* is defined as a *pair* of positions in the genome, and there are $\binom{n}{2}$ potential pairs of positions in a genome of size n. The most frequent 12-mer in *E. coli* (ACGCCGCATCCG) appears ninety-four times and corresponds to $(94 \cdot 93)/2 = 4376$ repeats.

9.2   Hash Tables

Consider the problem of listing all unique elements from a list.

Duplicate Removal Problem:
Find all unique entries in a list of integers.

Input: A list **a** of n integers.

Output: List **a** with all duplicates removed.

For example, the list $(8, 1, 5, 1, 0, 4, 5, 10, 1)$ has the elements 1 and 5 repeated multiple times, so the resulting list would be $(8, 1, 5, 0, 4, 10)$ or $(0, 1, 4, 5, 8, 10)$–the order of elements in the list does not matter. If the list was already sorted, one could simply traverse it and remove identical consecutive elements. This approach requires $O(n \log n)$ time to sort the list of size n, and then $O(n)$ time to traverse the sorted version.

Another approach is to use the elements in array **a** as addresses into another array **b**. That is, we can construct another array **b** consisting of 0s and 1s such that $b_i = 1$ if i is in **a** (any number of times), and $b_i = 0$ otherwise. For $\mathbf{a} = (8, 1, 5, 1, 0, 4, 5, 10, 1)$, elements b_0, b_1, b_4, b_5, b_8, and b_{10} are all equal to 1 and all other elements of **b** are equal to 0, as in $\mathbf{b} = (1, 1, 0, 0, 1, 1, 0, 0, 1, 0, 1)$. After **b** is constructed, simply traversing **b** solves the Duplicate Removal problem as in the DUPLICATEREMOVAL algorithm below.[2]

DUPLICATEREMOVAL($\mathbf{a}, n$)
1    $m \leftarrow$ largest element of **a**
2   **for** $i \leftarrow 0$ **to** m
3        $b_i \leftarrow 0$
4   **for** $i \leftarrow 0$ **to** n
5        $b_{a_i} \leftarrow 1$
6   **for** $i \leftarrow 0$ **to** m
7   **if** $b_i = 1$
8   **output** i
9   **return**

The array **b** can be created and traversed in time proportional to its length, so this algorithm is linear in the length of **b**.[3] Superficially, this sounds better

---

2. This assumes that all elements of **a** are non-negative.
3. We remark that DUPLICATEREMOVAL can be viewed as a sorting algorithm.

than an $O(n \log n)$ algorithm, which is not linear in n. However, one critical difficulty is that the entries in **a** can be arbitrarily large. If $\mathbf{a} = (1, 5, 10^{23})$, then **b** will have 10^{23} elements. Creation and traversing of a list with 10^{23} elements is certainly not efficient, and probably not even possible.

To work around this problem, we can introduce a function that operates on any integer and maps it to some integer in a small range (shown schematically in figure 9.1). For example, suppose we wanted the list **b** to contain fewer than 1000 entries. If we could take any integer (even 10^{23}) and map it to some number between 1 and 1000, then we could apply this "binning" strategy efficiently. The function that performs this mapping is often referred to as a *hash* function, and the list **b** is generally referred to as a *hash table*. The hash function, h, must be easy to calculate and integer-valued.[4] A fairly simple hash function is to take $h(x)$ to the integer remainder of $x / |\mathbf{b}|$, where $|\mathbf{b}|$ is the size of **b**. Usually the size of **b** is selected to fit into computer memory. For the Duplicate Remove problem, the hash table **b** is built according to the rule $b_i = 1$ if $h(a_j) = i$ for some element a_j from **a**, and $b_i = 0$ otherwise. If $|\mathbf{b}| = 10$, then $h(1) = 1$, $h(5) = 5$, and $h(10^{23}) = 0$, and $\mathbf{a} = (1, 5, 10^{23})$ results in the array $\mathbf{b} = (1, 1, 0, 0, 0, 1, 0, 0, 0, 0)$. However, this approach will not immediately work for duplicate removal since different elements, a_i and a_j, may collapse to the same bin, that is, $h(a_i) = h(a_j)$.

Ideally, we would like different integers in the array **a** to map to different integers in **b**, but this is clearly impossible when the number of unique values in **a** is larger than the length of **b**. Even when the number of unique values in **a** is smaller than the length of **b**, it can be particularly tricky to design a hash function h such that different values map to different bins. Furthermore, you would rather not have to redesign h for every different input. For example, if $h(x) = \frac{x}{1000}$, then elements 3, 1003, 2003, and so on, *collide* and fall into the same bin. A common technique to deal with collision is *chaining*. Elements collapsing to the same bin are often organized into a linked list[5] (fig. 9.2). Further analysis of the elements that collapsed to the same bin can reveal duplicate elements and leads to a fast algorithm for the Duplicate Removal problem.

While we have been concerned with removing duplicate integers from lists in this section, it is relatively easy to extend the concept of hashing to many other problems. For example, to find exact repeats of l-mers in a genome we treated the genome as a list of l-mers and relied on a hash table; it just hap-

---

4. Integer valued functions return only integers, regardless of their input.
5. For more about data structures, see (24).

x	$h(x)$
Penguin	1
Octopus	4
Turtle	3
Mouse	2
Snake	3
Heron	1
Tiger	2
Iguana	3
Ape	2
Cricket	4
Sparrow	1

Figure 9.1 An illustration of hashing: birds eat at Birdseed King [$h(x) = 1$]; mammals eat at Mammal Express [$h(x) = 2$]; reptiles, snakes, and turtles eat at McScales [$h(x) = 3$], and all other animals eat at Fred's [$h(x) = 4$].

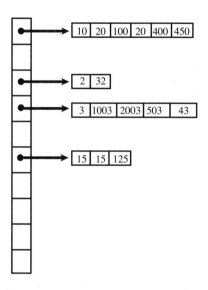

Figure 9.2   Chaining in a hash table.

pened that the hash table had exactly as many elements as we could possibly have seen, so we needed no strategy to deal with collision. On the other hand, if we wanted to find exact repeats for large values of l (say, $l = 40$), then we could not construct a hash table of 4^{40} entries. Instead, we could design a suitable hashing function and use a table of much smaller size.

9.3   Exact Pattern Matching

A common problem in bioinformatics is to search a database of sequences for a known sequence. Given a *pattern* string $\mathbf{p} = p_1 \cdots p_n$ and a longer *text* string $\mathbf{t} = t_1 \cdots t_m$, the Pattern Matching problem is to find any and all occurrences of pattern $\mathbf{p}$ in text $\mathbf{t}$.

Pattern Matching Problem:
Given a pattern and a text, find all occurrences of the pattern in the text.

Input: Pattern $\mathbf{p} = p_1 \ldots p_n$ and text $\mathbf{t} = t_1 \ldots t_m$.

Output: All positions $1 \leq i \leq m - n + 1$ such that the n-letter substring of $\mathbf{t}$ starting at i coincides with $\mathbf{p}$ (i.e, $t_i \ldots t_{i+n-1} = p_1 \ldots p_n$).

For example, if $\mathbf{t} = \mathsf{ATGGTCGGT}$ and $\mathbf{p} = \mathsf{GGT}$, then the result of an algorithm that solves the Pattern Matching problem would be positions 3 and 7.

We will use the notation $\mathbf{t}_i = t_i \ldots t_{i+n-1}$ to denote a substring of length n from $\mathbf{t}$ starting at position i. If $\mathbf{t}_i = \mathbf{p}$, then we have found an occurrence of the pattern in the text. By checking all possible values of i in increasing order, we are in effect scanning $\mathbf{t}$ by sliding a window of length n from left to right, and noting where the window is when the pattern $\mathbf{p}$ appears. A brute force algorithm for solving the Pattern Matching problem does exactly this.

PATTERNMATCHING($\mathbf{p}, \mathbf{t}$)
1   $n \leftarrow$ length of pattern $\mathbf{p}$
2   $m \leftarrow$ length of text $\mathbf{t}$
3   **for** $i \leftarrow 1$ **to** $m - n + 1$
4   **if** $\mathbf{t}_i = \mathbf{p}$
5   **output** i

At every position i, PATTERNMATCHING($\mathbf{p}, \mathbf{t}$) may need up to n operations to verify whether $\mathbf{p}$ is in the window by testing whether $t_i = p_1$, $t_{i+1} = p_2$, and so on. For typical instances, PATTERNMATCHING spends the bulk of its time discovering that the pattern *does not* appear at position i in the text. This test may take a single operation (if $t_i \neq p_1$), showing conclusively that $\mathbf{p}$ does not appear in $\mathbf{t}$ at position i; however, we may need as many as n operations to conduct this test. Therefore, the worst-case running time of PATTERNMATCHING can be estimated as $O(nm)$. Such a worst-case scenario can be seen by searching for $\mathbf{p} = \mathsf{AAAAT}$ in $\mathbf{t} = \mathsf{AAAA\ldots AAAAA}$. If m is 1 million, not only does the search take roughly 5 million operations, but it completes with no output, which is a bit disappointing.

One can evaluate the running time of PATTERNMATCHING in the case of finding a pattern in a "random" text. First, there is a good chance that the very first test (comparing t_1 with p_1) will be a mismatch, thus saving us the trouble of checking the remaining $n - 1$ letters of **p**. In general, the probability that the first letter of the pattern matches the text at position i is $\frac{1}{A}$, while the probability that it does not match is $\frac{A-1}{A}$. Similarly, the probability that the first two letters of the pattern match the text is $\frac{1}{A^2}$, while the probability that the first letter matches and the second letter does not match is $\frac{A-1}{A^2}$. The probability that PATTERNMATCHING matches exactly j out of n characters from **p** starting at t_i is $\frac{A-1}{A^j}$ for j between 1 and $n - 1$. Since this number shrinks rapidly as j increases, one can see that the probability that PATTERNMATCHING performs long tests gets to be quite small. Therefore, we would expect that, when presented with a text generated uniformly at random, the amount of work that PATTERNMATCHING will really need to perform when checking for **p** is somewhat closer to $O(m)$ than the rather pessimistic worst-case $O(nm)$ running time.

In 1973 Peter Weiner invented an ingenious data structure called a *suffix tree* that solves the Pattern Matching problem in *linear*-time $O(m)$ for any text and any pattern. The surprising thing about Weiner's result is that the size of the pattern does not seem to matter at all when it comes to the complexity of the Pattern Matching problem. Before we introduce suffix trees, we will first describe *keyword* trees used in a generalization of the Pattern Matching problem to multiple patterns.

9.4   Keyword Trees

Multiple Pattern Matching Problem:
Given a set of patterns and a text, find all occurrences of any of the patterns in the text.

Input: A set of k patterns $\mathbf{p^1}, \mathbf{p^2}, \ldots, \mathbf{p^k}$ and text $\mathbf{t} = t_1 \ldots t_m$.

Output: All positions $1 \leq i \leq m$ such that a substring of **t** starting at position i coincides with a pattern $\mathbf{p^j}$ for $1 \leq j \leq k$.

For example, if $\mathbf{t} = \mathsf{ATGGTCGGT}$ and $\mathbf{p^1} = \mathsf{GGT}$, $\mathbf{p^2} = \mathsf{GGG}$, $\mathbf{p^3} = \mathsf{ATG}$, and $\mathbf{p^4} = \mathsf{CG}$, then the result of an algorithm that solves the Multiple Pattern Matching problem would be positions 1, 3, 6, and 7.

Of course, the Multiple Pattern Matching problem with k patterns can be reduced to k individual Pattern Matching problems and solved in $O(knm)$ time, where n is the length of the longest of the k patterns, by k applications of the PATTERNMATCHING algorithm.[6] If one substitutes PATTERNMATCHING by a linear-time pattern matching algorithm, the Multiple Pattern Matching problem could be solved in just $O(km)$ time. However, there exists an even faster way to solve this problem in $O(N + m)$ time where N is the total length of patterns $\mathbf{p^1}, \mathbf{p^2}, \ldots, \mathbf{p^k}$.

In 1975 Alfred Aho and Margaret Corasick proposed using the *keyword tree* data structure to solve the Multiple Pattern Matching problem. The keyword tree for the set of patterns `apple`, `apropos`, `banana`, `bandana`, and `orange` is shown in figure 9.3. More formally, the keyword tree for a set of patterns $\mathbf{p^1}, \mathbf{p^2}, \ldots, \mathbf{p^k}$ is a rooted labeled tree satisfying the following conditions, assuming for simplicity that no pattern in the set is a prefix of another pattern:

- Each edge of the tree is labeled with a letter of the alphabet,

- Any two edges out of the same vertex have distinct labels,

- Every pattern $\mathbf{p^i}$ ($1 \leq i \leq k$) from the set of patterns is spelled on some path from the root to a leaf.

Clearly, the keyword tree has at most N vertices where N is the total length of all patterns, but may have fewer. One can construct the keyword tree in $O(N)$ time by progressively extending the keyword tree for the first j patterns into the keyword tree for $j + 1$ patterns.

The keyword tree can be used for finding whether there is a pattern in the set $\mathbf{p^1}, \mathbf{p^2}, \ldots, \mathbf{p^k}$ that matches the text starting at a fixed position i of the text. To do this, one should simply traverse the keyword tree using letters $t_i t_{i+1} t_{i+2} \ldots$ of the text to decide where to move at each step as in figure 9.4. This process either ends in a leaf, in which case there is a match to a pattern represented by the leaf, or interrupts before arriving at a leaf, in which case there is no match starting at position i. If the length of the longest pattern is n, then the Multiple Pattern Matching problem can be solved in $O(N + nm)$ time to construct the keyword tree and then use it to search through the text. The Aho-Corasick algorithm, which we do not give the details for, further reduces the running time for the Multiple Pattern Matching problem to $O(N + m)$.

---

6. More correctly, $O(Nm)$, where N is the total length of patterns $\mathbf{p^1}, \mathbf{p^2}, \ldots, \mathbf{p^k}$.

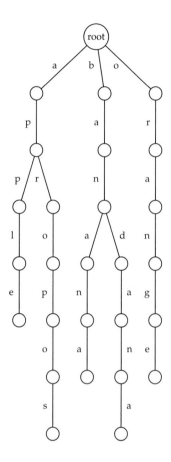

Figure 9.3   The keyword tree for apple, apropos, banana, bandana, and orange.

9.5   Suffix Trees

Suffix trees allow one to preprocess a text in such a way that for *any* pattern of length n, one can answer whether or not it occurs in the text, using only $O(n)$ time, regardless of how long the text is. That is, if you created one gigantic book out of all the issues of the journal *Science*, you could search for a pattern **p** in time proportional to the length of **p**. If you concatenated all the books ever written, it would take the same amount of time to search for **p**. Of course, the amount of time it takes to construct the suffix tree is different for the two texts.

Patterns: proud, perfect, muggle, ugly, rivet

t = "mr and mrs dursley of number 4 privet drive were proud to say that they were perfectly normal thank you very much"

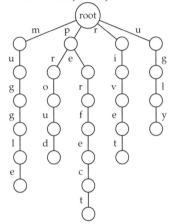

t = "mr and mrs dursley of number 4 privet drive were proud to say that they were perfectly normal thank you very much"

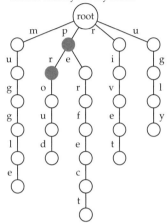

t = "mr and mrs dursley of number 4 privet drive were proud to say that they were perfectly normal thank you very much"

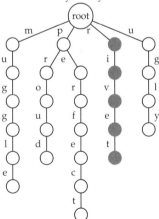

t = "mr and mrs dursley of number 4 privet drive were proud to say that they were perfectly normal thank you very much"

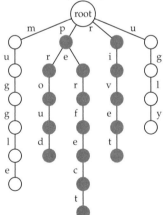

Figure 9.4 Searching for keywords proud, perfect, ugly, rivet, muggle in the text "mr and mrs dursley of number 4 privet drive were proud to say that they were perfectly normal thank you very much."

It turns out that a suffix tree for a text of length m can be constructed in $O(m)$ time, which leads immediately to a linear $O(n + m)$ algorithm for the Pattern Matching problem: construct the suffix tree for **t**, in $O(m)$ time, and then check whether **p** occurs in the tree, which requires $O(n)$ time.

Figure 9.5 (a) shows the keyword tree for all six suffixes of the string AT-CATG: G, TG, ATG, CATG, TCATG, and ATCATG. The suffix tree of a text can be obtained from the keyword tree of its suffixes by collapsing every path of nonbranching vertices into a single edge, as in figure 9.5 (b). Although one can use the keyword tree construction to build the suffix tree, it does not lend itself to an efficient algorithm for the suffix trees construction. Indeed, the keyword tree for k patterns can be constructed in $O(N)$ time where N is the total length of all these patterns. A text of length m has m suffixes varying in length from 1 to m and therefore the total length of these suffixes is $1 + 2 + \cdots + m = \frac{m(m+1)}{2}$, quadratic in the length of the text. Weiner's contribution to the construction of suffix trees in $O(m)$ time bypasses the keyword tree construction step altogether.[7]

The suffix tree for a text $\mathbf{t} = t_1 \ldots t_m$ is a rooted labeled tree with m leaves (numbered from 1 to m) satisfying the following conditions[8]:

- Each edge is labeled with a substring of the text,[9]

- Each internal vertex (except possibly the root) has at least 2 children,

- Any two edges out of the same vertex start with a different letter,

- Every suffix of text **t** is spelled out on a path from the root to some leaf.

Suffix trees immediately lead to a fast algorithm for pattern matching. We define the *threading* of a pattern **p** through a suffix tree T as the matching of characters from **p** along the unique path in T until either all characters of **p** are matched, which is a complete threading, or until no more matches are possible, which is an incomplete threading. If the threading of a pattern is complete, it ends at some vertex or edge of T and we define the **p**-*matching leaves* as all of the leaves that are descendants of that vertex or edge. Every **p**-matching leaf in the tree corresponds to an occurrence of **p** in **t**. An example of a complete threading, and the **p**-matching leaves is shown in figure 9.6.

---

7. The linear-time algorithm for suffix tree construction is rather complicated and is beyond the scope of this book. See (44) for an excellent review of suffix trees.

8. For simplicity we assume that the last letter of **t** does not appear anywhere in the text so that no suffix string is a prefix of another suffix string. If this is not the case, we can add an extra letter (like $) to the end of the text to satisfy this condition.

9. Note the difference between this and keyword trees.

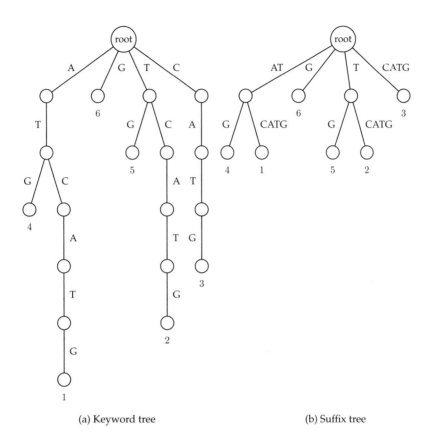

(a) Keyword tree   (b) Suffix tree

Figure 9.5   The difference between a keyword tree and a suffix tree for the string ATCATG. The suffix starting at position i corresponds to the leaf labeled by i.

SUFFIXTREEPATTERNMATCHING$(\mathbf{p}, \mathbf{t})$
1   Build the suffix tree for text $\mathbf{t}$
2   Thread pattern $\mathbf{p}$ through the suffix tree.
3   **if** threading is complete
4   **output** positions of every $\mathbf{p}$-matching leaf in the tree
5   **else**
6   **output** "pattern does not appear anywhere in the text"

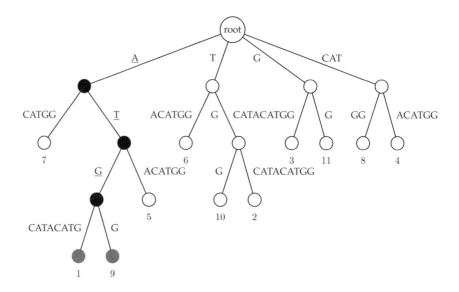

Figure 9.6   Threading the pattern ATG through the suffix tree for the text ATGCATA-CATGG. The suffixes ATGCATACATGG and ATGG both match, as noted by the gray vertices in the tree (the **p**-matching leaves). Each *p*-matching leaf corresponds to a position in the text where **p** occurs.

We stress that SUFFIXTREEPATTERNMATCHING searches for exact occurrences of **p** in **t**. Below we describe a few algorithms for inexact matching.

9.6   Heuristic Similarity Search Algorithms

Twenty years ago, it was surprising to find similarities between a newly sequenced cancer-causing gene and a gene involved in normal growth and development. Today, it would be even more surprising not to find a match for a newly sequenced gene, given the huge size of the GenBank database.[10] However, searching the GenBank database is not as easy as it was twenty years ago. The suffix tree algorithm above, while fast, can only find *exact*, rather than *approximate*, occurrences of a gene in a database. When we are trying to find an approximate match to a gene of length 10^3 in a database of size 10^{10}, the quadratic dynamic programming algorithms (like the Smith-Waterman local alignment algorithm) may be too slow. The situation gets

---

10. The number of nucleotides in GenBank already exceeds 10^{10}.

even worse when one tries to find all similarities between two mammalian genomes, each with about 3×10^9 nucleotides. Starting in the early 1990s biologists had no choice but to use fast heuristics[11] as an alternative to quadratic sequence alignment algorithms.

Many heuristics for fast database search in molecular biology use the same *filtration* idea. Filtration is based on the observation that a good alignment usually includes short identical or highly similar fragments. Thus one can search for short exact matches, for example, by using a hash table or a suffix tree, and use these short matches as seeds for further analysis. In 1973 Donald Knuth suggested a method for pattern matching with one mismatch based on the observation that strings differing by a single mismatch must match exactly in either the first or second half. For example, matching 9-mers with one allowable error can be reduced to exactly matching 4-mers, followed by extending the 4-mer exact matches into 9-mer approximate matches. This provides us with an opportunity to filter out positions that do not share common 4-mers, which is a large portion of all pairs of positions. In 1985 the idea of filtration in computational molecular biology was used by David Lipman and Bill Pearson, in their **FASTA** algorithm. It was further developed in **BLAST**, now the dominant database search tool in molecular biology.

Biologists frequently depict similarities between two sequences in the form of *dot matrices*. A dot matrix is simply a matrix with each entry either 0 or 1, where a 1 at position (i, j) indicates some similarity between the ith position of the first sequence and the jth position of the second sequence, as in figure 9.7. The similarity criteria may vary; for example, many dot matrices are based on matches of length t with at most k mismatches in an alignment of position i of the first sequence to position j of the second. However, no criterion is perfect in its ability to distinguish biologically relevant similarities from chance similarities. In biological applications, noise often disguises the real similarity, creating the problem of filtering the noise from dot matrices without filtering the biologically relevant similarity.

The positions of l-mers shared by a sequence and a database form an implicit dot matrix representation of the similarities between these strings. The popular **FASTA** protein database search tool uses exact matches of length l to construct a dot matrix. From this dot matrix, it selects some diagonal with a high concentration of 1s, and groups runs of 1s on this diagonal into longer runs.

---

11. A heuristic is an algorithm that will yield reasonable results, even if it is not provably optimal or lacks even a performance guarantee.

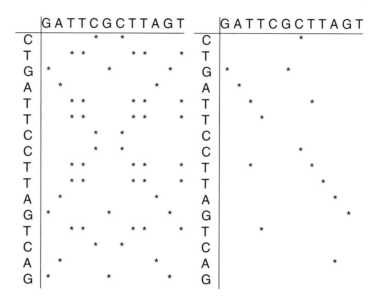

Figure 9.7 Two dot matrices for the strings CTGATTCCTTAGTCAG and GATTCGCTTAGT. A dot in position (i, j) of the first matrix indicates that the ith nucleotide of the first sequence matches the jth nucleotide of the second. A dot in position (i, j) of the second matrix indicates that the ith dinucleotide in the first sequence and jth dinucleotide in the second sequence are the same, i.e., the ith and $(i + 1)$-th symbols of the first sequence match the jth and $(j + 1)$-th symbols of the second sequence. The second matrix reveals a nearly diagonal run of 9 dots that points to an approximate match (with one insertion) between substring GATTCCT-TAGT in the first sequence and substring GATTCGCTTAG of the second one. The same diagonal run is present in the first matrix but disguised by many spurious dots.

9.7   Approximate Pattern Matching

Finding approximate matches of a pattern in a text (i.e., all substrings in the text with k or fewer mismatches from the pattern) is an important problem in computational molecular biology.

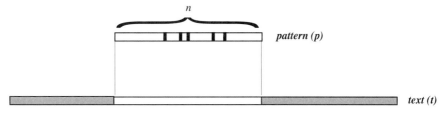

(a) Approximate Pattern Matching

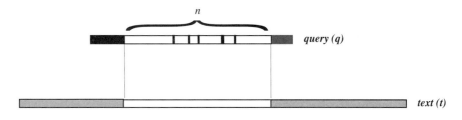

(b) Query Matching

Figure 9.8   The difference between the Approximate Pattern Matching problem and the Query Matching problem is that the Approximate Pattern Matching problem matches the *entire* pattern of length n against the text, while the Query Matching problem matches all *substrings* of length n in the query against the text.

Approximate Pattern Matching Problem:
Find all approximate occurrences of a pattern in a text.

Input: A pattern $\mathbf{p} = p_1 p_2 \ldots p_n$, text $\mathbf{t} = t_1 t_2 \ldots t_m$, and parameter k, the maximum number of mismatches.

Output: All positions $1 \leq i \leq m - n + 1$ such that $t_i t_{i+1} \ldots t_{i+n-1}$ and $p_1 p_2 \ldots p_n$ have at most k mismatches (i.e., $d_H(\mathbf{t}_i, \mathbf{p}) \leq k$).

The naive brute force algorithm for approximate pattern matching runs in $O(nm)$ time. The following algorithm will output all locations in $\mathbf{t}$ where $\mathbf{p}$ occurs with no more than k mismatches.

APPROXIMATEPATTERNMATCHING($\mathbf{p}, \mathbf{t}, k$)
1   $n \leftarrow$ length of pattern $\mathbf{p}$
2   $m \leftarrow$ length of text $\mathbf{t}$
3   **for** $i \leftarrow 1$ **to** $m - n + 1$
4       $dist \leftarrow 0$
5   **for** $j \leftarrow 1$ **to** n
6   **if** $t_{i+j-1} \neq p_j$
7               $dist \leftarrow dist + 1$
8   **if** $dist \leq k$
9   **output** i

In 1985 Gadi Landau and Udi Vishkin found an algorithm for approximate string matching with $O(km)$ worst-case running time. Although this algorithm yields the best known worst-case performance, it is not necessarily the best in practice. Consequently, several filtration-based approaches have better running times in practice, even though their worst-case running times are worse.

The Query Matching problem further generalizes the Approximate Pattern Matching problem. The difference between these two problems is illustrated in figure 9.8. *Query matching with k mismatches* involves a text $\mathbf{t}$, a query sequence $\mathbf{q}$, and a parameter k to limit the Hamming distance between some portion of the query and some portion of the text. We are also given a parameter n, which is the overall length of the match. Rather than looking for approximate matches between an *entire* string ($\mathbf{p}$) and all substrings of length n from $\mathbf{t}$, the Query Matching problem seeks approximate matches between *any* substrings of length n from the query and all other substrings of length n from the text.

Query Matching Problem:
Find all substrings of the query that approximately match the text.

Input: Query $\mathbf{q} = q_1 \ldots q_p$, text $\mathbf{t} = t_1 \ldots t_m$, and integers n and k.

Output: All pairs of positions (i, j) where $1 \leq i \leq p - n + 1$ and $1 \leq j \leq m - n + 1$ such that the n-letter substring of $\mathbf{q}$ starting at i approximately matches the n-letter substring of $\mathbf{t}$ starting at j, with at most k mismatches.

When $p = n$, the Query Matching problem becomes the Approximate

String Matching problem with k mismatches.

Using filtration algorithms for approximate query matching involves a two-stage process. The first stage preselects a set of positions in the text that are potentially similar to the query. The second stage verifies each potential position, rejecting potential matches with more than k mismatches. If the number of potential matches is small and potential match verification is not too slow, this method yields a fast query matching algorithm on most inputs that arise in practice.

The *l-mer filtration* technique is based on the simple observation that if an n-letter substring of a query approximately matches an n-letter substring of the text, then the two substrings share at least one l-mer for a sufficiently large value of l. All l-mers shared by the query and the text can be found easily by hashing. If the number of shared l-mers is relatively small, they can be verified, and all *real* matches with k mismatches can be located rapidly. The following theorem guides our choice of l, based on n and k:

Theorem 9.1 *If the strings $x_1 \ldots x_n$ and $y_1 \ldots y_n$ match with at most k mismatches, then they share an l-mer for $l = \lfloor \frac{n}{k+1} \rfloor$, that is, $x_{i+1} \ldots x_{i+l} = y_{i+1} \ldots y_{i+l}$ for some $1 \leq i \leq n - l + 1$.*

Proof: Partition the set of positions from 1 to n into $k + 1$ groups $\{1, \ldots, l\}$, $\{l + 1, \ldots, 2l\}$, $\{2l + 1, \ldots, 3l\}$, and so on, with $l = \lfloor \frac{n}{k+1} \rfloor$ positions in each group (the $k + 1$st group may have more than l positions). Observe that k mismatches distributed among $x_1 \ldots x_n$ and $y_1 \ldots y_n$ may affect at most k of these $k + 1$ groups and therefore at least one of them will remain unaffected by these mismatches. $\square$

This theorem motivates the following *l-mer filtration* algorithm for query matching with k mismatches:

- *Potential match detection.* Find all matches of l-mers in both the query and the text for $l = \lfloor \frac{n}{k+1} \rfloor$.

- *Potential match verification.* Verify each potential match by extending it to the left and to the right until the first $k + 1$ mismatches are found (or the beginning or end of the query or the text is found).

Potential match detection in this algorithm can be implemented either by hashing or by the suffix tree approach. For most practical values of n and k, the number of potential matches between the text and the query is small, yielding a fast algorithm.

9.8   **BLAST**: Comparing a Sequence against a Database

Using shared l-mers for finding similarities, as **FASTA** does, has some disadvantages. For example, two proteins can have different amino acid sequences but still be biologically similar. This frequently occurs when the genes that code for the two proteins evolve, but continue to produce proteins with a particular function. A common construct in many heuristic similarity search algorithms, including **BLAST**, is that of a scoring matrix similar to the scoring matrices introduced in chapter 6. These scoring matrices reveal similarities between proteins even if they do not share a single l-mer.

BLAST, the dominant database search tool in molecular biology, uses scoring matrices to improve the efficiency of filtration and to introduce more accurate rules for locating potential matches. Another powerful feature of **BLAST** is the use of Altschul-Dembo-Karlin statistics for estimating the statistical significance of found matches. For any two l-mers $x_1 \ldots x_l$ and $y_1 \ldots y_l$, **BLAST** defines the *segment score* as $\sum_{i=1}^{l} \delta(x_i, y_i)$, where $\delta(x, y)$ is the similarity score between amino acids x and y. A *segment pair* is just a pair of l-mers, one from each sequence. The *maximal segment pair* is a segment pair with the best score over all segment pairs in the two sequences. A segment pair is *locally maximal* if its score cannot be improved either by extending or by shortening both segments. A researcher is typically interested in all statistically significant locally maximal segments, rather than only the highest scoring segment pair.

BLAST attempts to find all locally maximal segment pairs in the query and database sequences with scores above some set threshold. It finds all l-mers in the text that have scores above a threshold when scored against some l-mer in the query sequence. A fast algorithm for finding such l-mers is the key ingredient of **BLAST**. An important observation is that if the threshold is high enough, then the set of all l-mers that have scores above a threshold is not too large. In this case the database can be searched for exact occurrences of the strings from this set. This is a Multiple Pattern Matching problem, and the Aho-Corasick algorithm finds the location of any of these strings in the database.[12] After the potential matches are located, **BLAST** attempts to extend them to see whether the resulting score is above the threshold. In recent years **BLAST** has been further improved by allowing insertions and deletions and combining matches on the same and close diagonals.

---

12. **BLAST** has evolved in the last fifteen years and has become significantly more involved than this simple description.

The choice of the threshold in **BLAST** is guided by the Altschul-Dembo-Karlin statistics, which allow one to identify the smallest value of the segment score that is unlikely to happen by chance. **BLAST** reports matches to sequences that either have one segment score above the threshold or that have several closely located segment pairs that are statistically significant when combined. According to the Altschul-Dembo-Karlin statistics, the number of matches with scores above θ is approximately Poisson-distributed, with mean

$$E(\theta) = Kmne^{-\lambda\theta},$$

where m and n are the lengths of compared sequences, and K is a constant. Parameter λ is a positive root of the equation

$$\sum_{x,y \in \mathcal{A}} p_x p_y e^{\lambda \delta(x,y)} = 1$$

where p_x and p_y are frequencies of amino acids x and y from the twenty-letter alphabet $\mathcal{A}$ and δ is the scoring matrix. The probability that there is a match of score greater that θ between two "random" sequences of length n and m is computed as $1 - e^{E(\theta)}$. This probability guides the choice of **BLAST** parameters and allows one to evaluate the statistical significance of found matches.

9.9   **Notes**

Linear-time algorithms for exact pattern matching were discovered in the late 1970s (58; 16). The linear-time algorithm for approximate pattern matching with k mismatches was discovered by Gadi Landau and Udi Vishkin (61) in 1985. Although this algorithm yields the best worst-case performance, it is not the best in practice. Consequently, several filtration-based approaches have emphasized the expected running time, rather than the worst-case running time. The idea of filtration was first described by Richard Karp and Michael Rabin in 1987 (54) for exact pattern matching. The filtration approach presented in this chapter was described in (86).

Keyword trees were used to develop a multiple pattern matching algorithm by Alfred Aho and Margaret Corasick in 1975 (2). Suffix trees were invented by Peter Weiner (111), who also proposed the linear-time algorithm for their construction. The REPUTER algorithm (60) uses an efficient implementation of suffix tree to find all repeats in a text.

FASTA was developed by David Lipman and William Pearson in 1985 (67). **BLAST**, the dominant database search tool in molecular biology, was developed by Steven Altschul, Warren Gish, Webb Miller, Gene Myers, and David Lipman in 1990 (4). **BLAST** uses Altschul-Dembo-Karlin statistics (52; 27) to estimate the statistical significance of found similarities. Currently there exist many variations and improvements of the BLAST tool, each tuned to a particular application (5).

Gene Myers (born 1953 in Idaho) is currently a professor at the University of California at Berkeley and a member of the National Academy of Engineering. He has contributed fundamental algorithmic methods for pattern matching and computational biology. He made key contributions to the development of BLAST, the most widely used biosequence search engine, and both proposed to shotgun-sequence the human genome and developed an assembler that did it. Myers spent his youth in the Far East— Pakistan, India, Indonesia, Japan, and Hong Kong—following his father from post to post for Exxon. He believes this early exposure to diverse cultures and mindsets contributed to his interest in interdisciplinary work. He was fascinated by science and recalls studying a *Gray's Anatomy*, which he still has, and reading everything by Asimov, Heinlein, and Bradbury. As an undergraduate at Caltech, Myers was indoctrinated in a "can do" attitude toward science that has stayed with him. As a computer science graduate student at the University of Colorado, he became fascinated with algorithm design. By chance he fell in with an eclectic discussion group led by Andrzej Ehrenfeucht, that included future bioinformaticians David Haussler and Gary Stormo. Myers was attracted to Ehrenfeucht because of his broad range of interests that included psychology, artificial intelligence, formal language theory, and computational molecular biology, even though it had no name back then.

In 1981 Myers took a position at the University of Arizona where he completely forgot about his early foray into computational biology and worked on traditional computer science problems for the next few years. In 1984, David Mount needed a computer science partner for a proposal for a computational biology center he envisioned at Arizona. Myers was working on comparing files at the time (the algorithm at the heart of GNU *diff* is his). The work was the closest topic in the computer science department to DNA sequence comparison, so he joined the project. He recalls picking the problem of DNA fragment assembly and also of finding complex patterns in sequences as topics for the proposal. A stellar review panel, including David Sankoff, Temple Smith, and Rich Roberts, visited Arizona and were so impressed with Myers' preliminary work that he was invited to a meeting at Waterville Valley, the first ever big bioinformatics meeting. Myers says:

> I loved interacting with the biologists and the mathematicians and the sense of excitement that something big was in the making. So in 1984 I became a computational biologist. Well, truthfully, a computer scientist interested in problems motivated by computational biology. Over the next twenty years that changed.

In the 1980s Myers worked a great deal with Webb Miller, another bioinformatics pioneer, and they developed some fundamental ideas for sequence comparison, for example, linear-space alignment. He says:

> It was Miller that mentored me and really helped me to develop as a young researcher. I had been struggling with my ability to write and with my confidence in the quality of my own work. Webb, ten years senior to me, encouraged me and introduced me to the craft of technical writing.

In 1989, Myers was working on how to find approximate matches to strings in less than quadratic time in response to a challenge from Zvi Galil. He was a smoker at the time and recalls talking with David Lipman outside the NIH while on a cigarette break about his ideas. David, the coinventor of FASTA, felt that a heuristic reformulation of the concepts would make for a fast and practical protein search tool. Over the next few months, Miller and Myers took turns coding different versions of the idea, with Lipman always driving and challenging. As the work developed Steven Altschul contributed new statistical work he had been doing with Sam Karlin, and Warren Gish, a real code ace, built the final product. BLAST was published in 1990 and put on the NCBI web server. It quickly became the most used tool in bioinformatics and the most cited paper in science for several years after.

Myers did finally achieve his subquadratic algorithm in 1994 and continued to work on similarity search and fragment assembly. In 1994, at the same meeting, Waterman and Idury, and Myers both presented fundamentally new ideas for fragment assembly, one based on the idea of an Euler string graph and the other on collapsing maximal uncontested interval subgraphs. But despite being significant algorithmic advances, Phil Green's beautifully engineered and tuned PHRAP assembler became the mainstay of the ever

growing sequencing centers as they prepared for the assault on the human
genome. Myers thought he was out of the race. In 1995, Craig Venter and his
team shotgun sequenced *Haemophilus influenzae*, a 1.8 Mbp bacterium and
assembled it with a program written by the then very young Granger Sutton. This was a rather startling success as most bioinformaticians viewed a
large clone, such as a BAC at 150 kbp, as the limit of what could be successfully shotgun-sequenced. Indeed the entire human genome program was
gearing up around the production of a collection of such BACs that spanned
the genome, and then shotgun-sequencing each of these in a hierarchical approach. Much of this impression was based on a misinterpretation of the
implication of a statistical theorem by Lander and Waterman about DNA
mapping. Myers says:

> I recall beginning to rethink what this theorem was really saying about
> shotgun-sequencing. At the same time, geneticist Jim Weber was think-
> ing about shotgunning as a way to accelerate the Human Genome
> Project, as he wanted to get to human genetic variation as quickly as
> possible for his research. Weber called Lipman, looking for a bioinfor-
> matician to help him put together a proposal. Lipman referred Weber
> to me. I said yes, let's think about it and in 1996 we began pushing the
> idea and trying to publish simulation results showing that it was, in
> principle, possible. The reaction to the proposal was incredibly nega-
> tive; we were basically accused of being fools.

After several rejections, a very condensed version of their proposal and
Myers' simulation was finally published in 1997 under the condition that a
rebuttal, by Phil Green, be published with it. Myers figured the proposal was
dead, but he continued to work on it at a simulation level with his students
because he found it intellectually interesting. He says:

> I kept giving talks about it, hoping someone might get excited enough
> to fund a $10 million pilot project. In 1998, Applied Biosystems, the
> prime manufacturer of sequencing machines, and Craig Venter announced
> that they would be forming a company to sequence the human genome
> using a shotgun approach. To me it was as if someone had decided to
> spend $300 million to try "my" project.

Myers had known Venter for ten years and quickly came to the decision
to join the company, Celera, at its inception in August 1998. With more re-
sources than he imagined, his problem became to deliver on the promise. He
says:

It was an incredible time of my life. I thought I could do it, but I didn't yet know I could do it. The pressure was incredible, but so was the excitement.

A year and a half-million lines of code later a team of ten, including Granger Sutton, achieved their first assembly of the *Drosophila* genome and presented the results in October 1999. Myers had his 120 Mbp pilot project and it was a success. The race for the human genome was then on. The Celera team's main obstacle was in how to scale up another 30-fold for the human genome—computer memory and time were real problems. In 20,000 CPU hours and a month's time, a first assembly of the human genome was accomplished in May 2000. Subsequent refinements took place and the first reconstructions of the genome were published by Celera and the public consortium in February 2001. Today, whole-genome shotgun sequence is not really debatable: it is the way that every genome is being sequenced. Myers says:

Now at Berkeley, I have become fascinated with how cells work from the level of particle systems and in particular how multicellular organisms develop. I figure the next great challenge is to "decode the cell." I am driven by the intellectual curiosity to have some sense of understanding the cell as a system before the end of my career. Biology drives what I do today; my skills as an algorithmist are tools to this end as opposed to my primary focus, although I still get pretty excited when I come up with a tricky new way to solve a particular computational problem.

9.10   Problems

Problem 9.1

Derive the expectation and the variance of the number of occurrences of patterns AA and AT in a random text of length n. Are the expectations and variances for AA and AT the same?

Problem 9.2

Derive the expectation and the variance of the number of occurrences of a given pattern of length l in a random text of length n.

Let $\mathcal{X}$ be the set of all l-mers. Given an l-mer x and a text $\mathbf{t}$, define $x(\mathbf{t})$ as the number of occurrences of x in $\mathbf{t}$. *Statistical distance* between texts $\mathbf{t}$ and $\mathbf{u}$ is defined as

$$d(\mathbf{t}, \mathbf{u}) = \sqrt{\sum_{x \in \mathcal{X}} (x(\mathbf{t}) - x(\mathbf{u}))^2}.$$

Although statistical distance can be computed very fast, it can miss weak similarities that do not preserve shared l-mers.

Problem 9.3

Evaluate an expected statistical distance between two random n-letter strings.

Problem 9.4

Prove that the average time that PATTERNMATCHING takes to check whether a pattern of length n appears at a given position in a random text is $2 - \frac{n+1}{A^n} + \frac{n}{A^{n+1}}$ (A is the size of alphabet).

Problem 9.5

Write an efficient algorithm that will construct a keyword tree given a list of patterns $\mathbf{p}^1, \mathbf{p}^2, \ldots, \mathbf{p}^k$.

Problem 9.6

Design an algorithm for a generalization of the Approximate Pattern Matching problem in the case where the pattern can match a text with up to k mismatches, insertions, or deletions (rather than the mismatch-only model as before).

Problem 9.7

A string $\mathbf{p} = p_1 \ldots p_n$ is a *palindrome* if it spells the same string when read backward, that is, $p_i = p_{n+1-i}$ for $1 \leq i \leq n$. Design an efficient algorithm for finding all palindromes (of all lengths) in a text.

Problem 9.8

Design an efficient algorithm for finding the longest exact repeat within a text.

Problem 9.9

Design an efficient algorithm for finding the longest exact tandem repeat within a text.

Problem 9.10

Design an efficient algorithm for finding the longest exact repeat with at most one mismatch in a text.

Problem 9.11

Design an efficient algorithm for finding the longest string shared by two given texts.

Problem 9.12

Design an efficient algorithm for finding a shortest nonrepeated string in a text, that is, a shortest string that appears in the text only once.

Problem 9.13

Design an efficient algorithm that finds a shortest string in text t_1 that does not appear in text t_2.

Problem 9.14

Implement an l-tuple filtration algorithm for the Query Matching problem.

10 *Clustering and Trees*

A common problem in biology is to partition a set of experimental data into groups (clusters) in such a way that the data points within the same cluster are highly similar while data points in different clusters are very different. This problem is far from simple, and this chapter covers several algorithms that perform different types of clustering. There is no simple recipe for choosing one particular approach over another for a particular clustering problem, just as there is no universal notion of what constitutes a "good cluster." Nonetheless, these algorithms can yield significant insight into data and allow one, for example, to identify clusters of genes with similar functions even when it is not clear what particular role these genes play. We conclude this chapter with studies of evolutionary tree reconstruction, which is closely related to clustering.

10.1 Gene Expression Analysis

Sequence comparison often helps to discover the function of a newly sequenced gene by finding similarities between the new gene and previously sequenced genes with known functions. However, for many genes, the sequence similarity of genes in a functional family is so weak that one cannot reliably derive the function of the newly sequenced gene based on sequence alone. Moreover, genes with the same function sometimes have no sequence similarity at all. As a result, the functions of more than 40% of the genes in sequenced genomes are still unknown.

In the last decade, a new approach to analyzing gene functions has emerged. DNA arrays allow one to analyze the *expression levels* (amount of mRNA produced in the cell) of many genes under many time points and conditions and to reveal which genes are switched on and switched off in the cell.[1]   The outcome of this type of study is an $n \times m$ *expression matrix* $\mathbf{I}$, with the n rows corresponding to genes, and the m columns corresponding to different time points and different conditions. The expression matrix $\mathbf{I}$ represents *intensities* of hybridization signals as provided by a DNA array. In reality, expression matrices usually represent transformed and normalized intensities rather than the raw intensities obtained as a result of a DNA array experiment, but we will not discuss this transformation.

The element $I_{i,j}$ of the expression matrix represents the expression level of gene i in experiment j; the entire ith row of the expression matrix is called the *expression pattern* of gene i. One can look for pairs of genes in an expression matrix with similar expression patterns, which would be manifested as two similar rows. Therefore, if the expression patterns of two genes are similar, there is a good chance that these genes are somehow related, that is, they either perform similar functions or are involved in the same biological process.[2] Accordingly, if the expression pattern of a newly sequenced gene is similar to the expression pattern of a gene with known function, a biologist may have reason to suspect that these genes perform similar or related functions. Another important application of expression analysis is in the deciphering of regulatory pathways; similar expression patterns usually imply coregulation.   However, expression analysis should be done with caution since DNA arrays typically produce noisy data with high error rates.

Clustering algorithms group genes with similar expression patterns into *clusters* with the hope that these clusters correspond to groups of functionally related genes. To cluster the expression data, the $n \times m$ expression matrix is often transformed into an $n \times n$ *distance matrix* $\mathbf{d} = (d_{i,j})$ where $d_{i,j}$ reflects how similar the expression patterns of genes i and j are (see figure 10.1).[3] The goal of clustering is to group genes into clusters satisfying the following two conditions:

• *Homogeneity.* Genes (rather, their expression patterns) within a cluster

---

1. Expression analysis studies implicitly assume that the amount of mRNA (as measured by a DNA array) is correlated with the amount of its protein produced by the cell. We emphasize that a number of processes affect the production of proteins in the cell (transcription, splicing, translation, post-translational modifications, protein degradation, etc.) and therefore this correlation may not be straightforward, but it is still significant.
2. Of course, we do not exclude the possibility that expression patterns of two genes may look similar simply by chance, but the probability of such a chance similarity decreases as we increase the number of time points.
3. The distance matrix is typically larger than the expression matrix since $n \gg m$ in most expression studies.

Time	1 hr	2 hr	3 hr
g_1	10.0	8.0	10.0
g_2	10.0	0.0	9.0
g_3	4.0	8.5	3.0
g_4	9.5	0.5	8.5
g_5	4.5	8.5	2.5
g_6	10.5	9.0	12.0
g_7	5.0	8.5	11.0
g_8	2.7	8.7	2.0
g_9	9.7	2.0	9.0
g_{10}	10.2	1.0	9.2

(a) Intensity matrix, **I**

	g_1	g_2	g_3	g_4	g_5	g_6	g_7	g_8	g_9	g_{10}
g_1	0.0	8.1	9.2	7.7	9.3	2.3	5.1	10.2	6.1	7.0
g_2	8.1	0.0	12.0	0.9	12.0	9.5	10.1	12.8	2.0	1.0
g_3	9.2	12.0	0.0	11.2	0.7	11.1	8.1	1.1	10.5	11.5
g_4	7.7	0.9	11.2	0.0	11.2	9.2	9.5	12.0	1.6	1.1
g_5	9.3	12.0	0.7	11.2	0.0	11.2	8.5	1.0	10.6	11.6
g_6	2.3	9.5	11.1	9.2	11.2	0.0	5.6	12.1	7.7	8.5
g_7	5.1	10.1	8.1	9.5	8.5	5.6	0.0	9.1	8.3	9.3
g_8	10.2	12.8	1.1	12.0	1.0	12.1	9.1	0.0	11.4	12.4
g_9	6.1	2.0	10.5	1.6	10.6	7.7	8.3	11.4	0.0	1.1
g_{10}	7.0	1.0	11.5	1.1	11.6	8.5	9.3	12.4	1.1	0.0

(b) Distance matrix, **d**

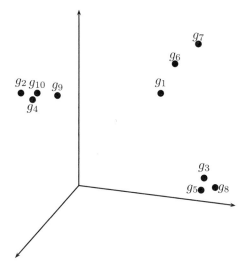

(c) Expression patterns as points in three-dimentsional space.

Figure 10.1   An "expression" matrix of ten genes measured at three time points, and the corresponding distance matrix. Distances are calculated as the Euclidean distance in three-dimensional space.

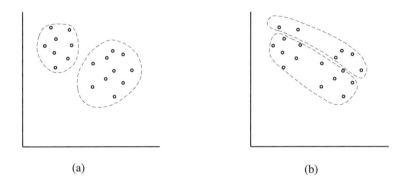

(a)   (b)

Figure 10.2   Data can be grouped into clusters. Some clusters are better than others: the two clusters in a) exhibit good homogeneity and separation, while the clusters in b) do not.

should be highly similar to each other. That is, $d_{i,j}$ should be small if i and j belong to the same cluster.

- *Separation.* Genes from different clusters should be very different. That is, $d_{i,j}$ should be large if i and j belong to different clusters.

An example of clustering is shown in figure 10.2. Figure 10.2 (a) shows a good partition according to the above two properties, while (b) shows a bad one. Clustering algorithms try to find a good partition.

A "good" clustering of data is one that adheres to these goals. While we hope that a better clustering of genes gives rise to a better grouping of genes on a functional level, the final analysis of resulting clusters is left to biologists.

Different tissues express different genes, and there are typically over 10,000 genes expressed in any one tissue. Since there are about 100 different tissue types, and since expression levels are often measured over many time points, gene expression experiments can generate vast amounts of data which can be hard to interpret. Compounding these difficulties, expression levels of related genes may vary by several orders of magnitude, thus creating the problem of achieving accurate measurements over a large range of expression levels; genes with low expression levels may be related to genes with high expression levels.

Figure 10.3   A schematic of hierarchical clustering.

10.2   Hierarchical Clustering

In many cases clusters have subclusters, these have subsubclusters, and so on. For example, mammals can be broken down into primates, carnivora, bats, marsupials, and many other orders. The order carnivora can be further broken down into cats, hyenas, bears, seals, and many others. Finally, cats can be broken into thirty seven species.[4]

---

4. Lions, tigers, leopards, jaguars, lynx, cheetahs, pumas, golden cats, domestic cats, small wild-cats, ocelots, and many others.

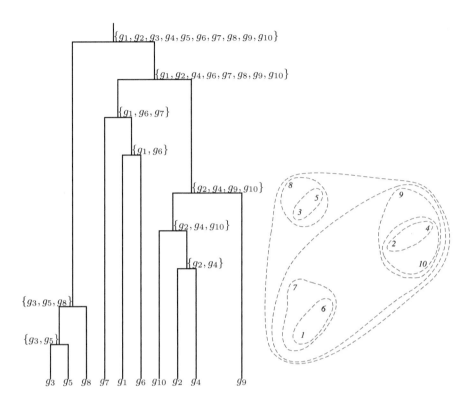

Figure 10.4   A hierarchical clustering of the data in figure 10.1

Hierarchical clustering (fig. 10.3) is a technique that organizes elements into a *tree*, rather than forming an explicit partitioning of the elements into clusters. In this case, the genes are represented as the leaves of a tree. The edges of the trees are assigned *lengths* and the distances between leaves—that is, the length of the path in the tree that connects two leaves—correlate with entries in the distance matrix. Such trees are used in both the analysis of expression data and in studies of molecular evolution which we will discuss below.

Figure 10.4 shows a tree that represents clustering of the data in figure 10.1. This tree actually describes a family of different partitions into clusters, each with a different number of clusters (one for each value from 1 to n). You can see what these partitions by drawing a horizontal line through the tree. Each line crosses the tree at i points ($1 \leq i \leq k$) and correspond to i clusters.

The HIERARCHICALCLUSTERING algorithm below takes an $n \times n$ distance matrix **d** as an input, and progressively generates n different partitions of the data as the tree it outputs. The largest partition has n single-element clusters, with every element forming its own cluster. The second-largest partition combines the two closest clusters from the largest partition, and thus has $n - 1$ clusters. In general, the ith partition combines the two closest clusters from the $(i - 1)$th partition and has $n - i + 1$ clusters.

HIERARCHICALCLUSTERING(**d**, n)
  1 Form n clusters, each with 1 element
  2 Construct a graph T by assigning an isolated vertex to each cluster
  3 **while** there is more than 1 cluster
  4   Find the two closest clusters C_1 and C_2
  5   Merge C_1 and C_2 into new cluster C with $|C_1| + |C_2|$ elements
  6   Compute distance from C to all other clusters
  7   Add a new vertex C to T and connect to vertices C_1 and C_2
  8   Remove rows and columns of **d** corresponding to C_1 and C_2
  9   Add a row and column to **d** for the new cluster C
 10 **return** T

Line 6 in the algorithm is (intentionally) left ambiguous; clustering algorithms vary in how they compute the distance between the newly formed cluster and any other cluster. Different formulas for recomputing distances yield different answers from the same hierarchical clustering algorithm. For example, one can define the distance between two clusters as the smallest distance between any pair of their elements

$$d_{min}(C^*, C) = \min_{x \in C^*, y \in C} d(x, y)$$

or the average distance between their elements

$$d_{avg}(C^*, C) = \frac{1}{|C^*||C|} \sum_{x \in C^*, y \in C} d(x, y).$$

Another distance function estimates distance based on the separation of C_1 and C_2 in HIERARCHICALCLUSTERING:

$$d(C^*, C) = \frac{d(C^*, C_1) + d(C^*, C_2) - d(C_1, C_2)}{2}$$

In one of the first expression analysis studies, Michael Eisen and colleagues used hierarchical clustering to analyze the expression profiles of 8600 genes over thirteen time points to find the genes responsible for the growth response of starved human cells. The HIERARCHICALCLUSTERING resulted in a tree consisting of five main subtrees and many smaller subtrees. The genes within these five clusters had similar functions, thus confirming that the resulting clusters are biologically sensible.

10.3   *k*-Means Clustering

One can view the n rows of the $n \times m$ expression matrix as a set of n points in m-dimensional space and partition them into k subsets, pretending that k—the number of clusters—is known in advance.

One of the most popular clustering methods for points in multidimensional spaces is called *k-Means clustering*. Given a set of n data points in m-dimensional space and an integer k, the problem is to determine a set of k points, or *centers*, in m-dimensional space that minimize the *squared error distortion* defined below. Given a data point v and a set of k centers $\mathcal{X} = \{x_1, \ldots x_k\}$, define the distance from v to the centers $\mathcal{X}$ as the distance from v to the closest point in $\mathcal{X}$, that is, $d(v, \mathcal{X}) = \min_{1 \leq i \leq k} d(v, x_i)$. We will assume for now that $d(v, x_i)$ is just the Euclidean[5] distance in m dimensions. The squared error distortion for a set of n points $\mathcal{V} = \{v_1, \ldots v_n\}$, and a set of k centers $\mathcal{X} = \{x_1, \ldots x_k\}$, is defined as the mean squared distance from each data point to its nearest center:

$$d(\mathcal{V}, \mathcal{X}) = \frac{\sum_{i=1}^{n} d(v_i, \mathcal{X})^2}{n}$$

---

k-**Means Clustering Problem**:

Given n data points, find k center points minimizing the squared error distortion.

   Input: A set, $\mathcal{V}$, consisting of n points and a parameter k.

   Output: A set $\mathcal{X}$ consisting of k points (called centers) that minimizes $d(\mathcal{V}, \mathcal{X})$ over all possible choices of $\mathcal{X}$.

---

5. Chapter 2 contains a sample algorithm to calculate this when m is 2.

While the above formulation does not explicitly address *clustering* n points into k clusters, a clustering can be obtained by simply assigning each point to its closest center. Although the k-Means Clustering problem looks relatively simple, there are no efficient (polynomial) algorithms known for it. The *Lloyd* k-Means clustering algorithm is one of the most popular clustering heuristics that often generates good solutions in gene expression analysis. The Lloyd algorithm randomly selects an arbitrary partition of points into k clusters and tries to improve this partition by moving some points between clusters. In the beginning one can choose arbitrary k points as "cluster representatives." The algorithm iteratively performs the following two steps until either it converges or until the fluctuations become very small:

- Assign each data point to the cluster C_i corresponding to the closest cluster representative x_i ($1 \leq i \leq k$)

- After the assignments of all n data points, compute new cluster representatives according to the center of gravity of each cluster, that is, the new cluster representative is $\frac{\sum_{v \in C} v}{|C|}$ for every cluster C.

The Lloyd algorithm often converges to a local minimum of the squared error distortion function rather than the global minimum. Unfortunately, interesting objective functions other than the squared error distortion lead to similarly difficult problems. For example, finding a good clustering can be quite difficult if, instead of the squared error distortion ($\sum_{i=1}^{n} d(v_i, \mathcal{X})^2$), one tries to minimize $\sum_{i=1}^{n} d(v_i, \mathcal{X})$ (*k-Median problem*) or $\max_{1 \leq i \leq n} d(v_i, \mathcal{X})$ (*k-Center problem*). We remark that all of these definitions of clustering cost emphasize the homogeneity condition and more or less ignore the other important goal of clustering, the separation condition. Moreover, in some unlucky instances of the k-Means Clustering problem, the algorithm may converge to a local minimum that is *arbitrarily* bad compared to an optimal solution (a problem at the end of this chapter).

While the Lloyd algorithm is very fast, it can significantly rearrange every cluster in every iteration. A more conservative approach is to move only one element between clusters in each iteration. We assume that every partition P of the n-element set into k clusters has an associated *clustering cost*, denoted $cost(P)$, that measures the quality of the partition P: the smaller the clustering cost of a partition, the better that clustering is.[6] The squared error distortion is one particular choice of $cost(P)$ and assumes that each center

---

6. "Better" here is better clustering, not a better biological grouping of genes.

point is the center of gravity of its cluster. The pseudocode below implicitly assumes that $cost(P)$ can be efficiently computed based either on the distance matrix or on the expression matrix. Given a partition P, a cluster C within this partition, and an element i outside C, $P_{i \to C}$ denotes the partition obtained from P by moving the element i from its cluster to C. This move improves the clustering cost only if $\Delta(i \to C) = cost(P) - cost(P_{i \to C}) > 0$, and the PROGRESSIVEGREEDYK-MEANS algorithm searches for the "best" move in each step (i.e., a move that maximizes $\Delta(i \to C)$ for all C and for all $i \notin C$).

PROGRESSIVEGREEDYK-MEANS(k)
1   Select an arbitrary partition P into k clusters.
2   **while** forever
3       $bestChange \leftarrow 0$
4   **for** every cluster C
5   **for** every element $i \notin C$
6   **if** moving i to cluster C reduces the clustering cost
7   **if** $\Delta(i \to C) > bestChange$
8                       $bestChange \leftarrow \Delta(i \to C)$
9                       $i^* \leftarrow i$
10                      $C^* \leftarrow C$
11   **if** $bestChange > 0$
12   change partition P by moving i^* to C^*
13   **else**
14   **return** P

Even though line 2 makes an impression that this algorithm may loop endlessly, the return statement on line 14 saves us from an infinitely long wait. We stop iterating when no move allows for an improvement in the score; this eventually has to happen.[7]

10.4   Clustering and Corrupted Cliques

A complete graph, written K_n, is an (undirected) graph on n vertices with every two vertices connected by an edge. A *clique graph* is a graph in which every connected component is a complete graph. Figure 10.5 (a) shows a

---

7. What would be the natural implication if there could always be an improvement in the score?

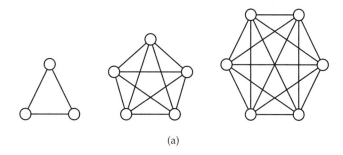

(a)

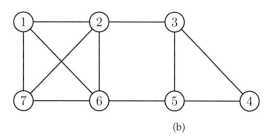

(b)

Figure 10.5   a) A clique graph consisting of the three connected components K_3, K_5, and K_6. b) A graph with 7 vertices that has 4 cliques formed by vertices $\{1, 2, 6, 7\}$, $\{2, 3\}$, $\{5, 6\}$, and $\{3, 4, 5\}$.

clique graph consisting of three connected components, K_3, K_5, and K_6. Every partition of n elements into k clusters can be represented by a clique graph on n vertices with k cliques. A subset of vertices $V' \subset V$ in a graph $G(V, E)$ forms a *complete subgraph* if the *induced subgraph* on these vertices is complete, that is, every two vertices v and w in V' are connected by an edge in the graph. For example, vertices 1, 6, and 7 form a complete subgraph of the graph in figure 10.5 (b). A *clique* in the graph is a maximal complete subgraph, that is, a complete subgraph that is not contained inside any other complete subgraph. For example, in figure 10.5 (b), vertices 1, 6, and 7 form a complete subgraph but do not form a clique, but vertices 1, 2, 6, and 7 do.

In expression analysis studies, the distance matrix $(d_{i,j})$ is often further transformed into a *distance graph* $G = G(\theta)$, where the vertices are genes and there is an edge between genes i and j if and only if the distance between them is below the threshold θ, that is, if $d_{i,j} < \theta$. A clustering of genes that satisfies the homogeneity and separation principles for an appropriately chosen θ will correspond to a distance graph that is also a clique graph. However, errors in expression data, and the absence of a "universally good" threshold θ often results in distance graphs that do not quite form clique graphs (fig. 10.6). Some elements of the distance matrix may fall below the distance threshold for unrelated genes (adding edges between different clusters), while other elements of the distance matrix exceed the distance threshold for related genes (removing edges within clusters). Such erroneous edges corrupt the clique graph, raising the question of how to transform the distance graph into a clique graph using the smallest number of edge removals and edge additions.

Corrupted Cliques Problem:

Determine the smallest number of edges that need to be added or removed to transform a graph into a clique graph.

Input: A graph G.

Output: The smallest number of additions and removals of edges that will transform G into a clique graph.

It turns out that the Corrupted Cliques problem is $\mathcal{NP}$-hard, so some heuristics have been proposed to approximately solve it. Below we describe the time-consuming PCC (*Parallel Classification with Cores*) algorithm, and the less theoretically sound, but practical, CAST (*Cluster Affinity Search Technique*) algorithm inspired by PCC.

Suppose we attempt to cluster a set of genes S, and suppose further that S' is a subset of S. If we are somehow magically given the correct clustering[8] $\{C_1, \ldots, C_k\}$ of S', could we extend this clustering of S' into a clustering of the entire gene set S? Let $S \setminus S'$ be the set of unclustered genes, and $N(j, C_i)$ be the number of edges between gene $j \in S \setminus S'$ and genes from the cluster C_i in the distance graph. We define the *affinity* of gene j to cluster C_i as $\frac{N(j,C_i)}{|C_i|}$.

---

8. By the "correct clustering of S'," we mean the classification of elements of S' (which is a subset of the entire gene set) into the same clusters as they would be in the clustering of S that has the optimal clustering score.

	g_1	g_2	g_3	g_4	g_5	g_6	g_7	g_8	g_9	g_{10}
g_1	0.0	8.1	9.2	7.7	9.3	**2.3**	**5.1**	10.2	**6.1**	**7.0**
g_2	8.1	0.0	12.0	**0.9**	12.0	9.5	10.1	12.8	**2.0**	**1.0**
g_3	9.2	12.0	0.0	11.2	**0.7**	11.1	8.1	**1.1**	10.5	11.5
g_4	7.7	**0.9**	11.2	0.0	11.2	9.2	9.5	12.0	**1.6**	**1.1**
g_5	9.3	12.0	**0.7**	11.2	0.0	11.2	8.5	**1.0**	10.6	11.6
g_6	**2.3**	9.5	11.1	9.2	11.2	0.0	**5.6**	12.1	7.7	8.5
g_7	**5.1**	10.1	8.1	9.5	8.5	**5.6**	0.0	9.1	8.3	9.3
g_8	10.2	12.8	**1.1**	12.0	**1.0**	12.1	9.1	0.0	11.4	12.4
g_9	**6.1**	**2.0**	10.5	**1.6**	10.6	7.7	8.3	11.4	0.0	**1.1**
g_{10}	**7.0**	**1.0**	11.5	**1.1**	11.6	8.5	9.3	12.4	**1.1**	0.0

(a) Distance matrix, **d** (distances shorter than 7 are shown in bold).

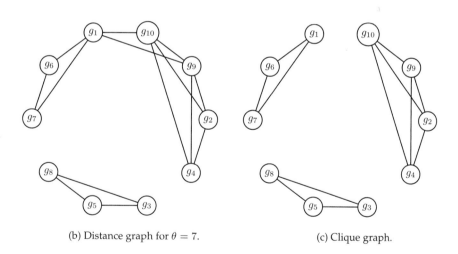

(b) Distance graph for $\theta = 7$.   (c) Clique graph.

Figure 10.6   The distance graph (b) for $\theta = 7$ is not quite a clique graph. However, it can be transformed into a clique graph (c) by removing edges (g_1, g_{10}) and (g_1, g_9).

A natural *maximal affinity* approach would be to put every unclustered gene j into the cluster C_i with the highest affinity to j, that is, the cluster that maximizes $\frac{N(j,C_i)}{|C_i|}$. In this way, the clustering of S' can be extended into clustering of the entire gene set S. In 1999, Amir Ben-Dor and colleagues developed the PCC clustering algorithm, which relies on the assumption that if S' is sufficiently large and the clustering of S' is correct, then the clustering of the entire gene set is likely to to be correct.

The only problem is that the correct clustering of S' is unknown! The way around this is to generate *all* possible clusterings of S', extend them into a clustering of the entire gene set S, and select the resulting clustering with the best clustering score. As attractive as it may sound, this is not practical (unless S' is very small) since the number of possible partitions of a set S' into k clusters is $k^{|S'|}$. The PCC algorithm gets around this problem by making S' extremely small and generating *all* partitions of this set into k clusters. It then extends each of these $k^{|S'|}$ partitions into a partition of the entire n-element gene set by the two-stage procedure described below. The distance graph G guides these extensions based on the maximal affinity approach. The function $score(P)$ is defined to be the number of edges necessary to add or remove to turn G into a clique graph according to the partition P.[9] The PCC algorithm below clusters the set of elements S into k clusters according to the graph G by extending partitions of subsets of S using the maximal affinity approach:

PCC(G, k)
   1   $S \leftarrow$ set of vertices in the distance graph G
   2   $n \leftarrow$ number of elements in S
   3   $bestScore \leftarrow \infty$
   4   Randomly select a "very small" set $S' \subset S$, where $|S'| = \log \log n$
   5   Randomly select a "small" set $S'' \subset (S \setminus S')$, where $|S''| = \log n$.
   6   **for** every partition P' of S' into k clusters
   7   Extend P' into a partition P'' of S''
   8   Extend P'' into a partition P of S
   9   **if** $score(P) < bestScore$
   10               $bestScore \leftarrow score(P)$
   11               $bestPartition \leftarrow P$
   12   **return** $bestPartition$

---

9. Computing this number is an easy (rather than $\mathcal{NP}$-complete) problem, since we are given P. To search for the minimum *score* without P, we would need to search over all partitions.

The number of iterations that PCC requires is given by the number of partitions of set S', which is $k^{|S'|} = k^{\log \log n} = (\log n)^{\log_2 k} = (\log n)^c$. The amount of work done in each iteration is $O(n^2)$, resulting in a running time of $O\left(n^2 (\log n)^c\right)$. Since this is too slow for most applications, a more practical heuristic called CAST is often used.

Define the distance between gene i and cluster C as the average distance between i and all genes in the cluster C: $d(i, C) = \frac{\sum_{j \in C} d(i,j)}{|C|}$. Given a threshold θ, a gene i is *close* to cluster C if $d(i, C) < \theta$ and *distant* otherwise. The CAST algorithm below clusters set S according to the distance graph G and the threshold θ. CAST iteratively builds the partition P of the set S by finding a cluster C such that no gene $i \notin C$ is close to C, and no gene $i \in C$ is distant from C. In the beginning of the routine, P is initialized to an empty set.

CAST(G, θ)
1   $S \leftarrow$ set of vertices in the distance graph G
2   $P \leftarrow \emptyset$
3   **while** $S \neq \emptyset$
4        $v \leftarrow$ vertex of maximal degree in the distance graph G.
5        $C \leftarrow \{v\}$
6   **while** there exists a close gene $i \notin C$ or distant gene $i \in C$
7   Find the nearest close gene $i \notin C$ and add it to C.
8   Find the farthest distant gene $i \in C$ and remove it from C.
9   Add cluster C to the partition P
10      $S \leftarrow S \setminus C$
11   Remove vertices of cluster C from the distance graph G
12   **return** P

Although CAST is a heuristic with no performance guarantee[10] it performs remarkably well with gene expression data.

10.5   Evolutionary Trees

In the past, biologists relied on morphological features, like beak shapes or the presence or absence of fins to construct evolutionary trees. Today biologists rely on DNA sequences for the reconstruction of evolutionary trees. Figure 10.7 represents a DNA-based evolutionary tree of bears and raccoons that helped biologists to decide whether the giant panda belongs to the bear family or the raccoon family. This question is not as obvious as it may at first sound, since bears and raccoons diverged just 35 million years ago and they share many morphological features.

For over a hundred years biologists could not agree on whether the giant panda should be classified in the bear family or in the raccoon family. In 1870 an amateur naturalist and missionary, Père Armand David, returned to Paris from China with the bones of the mysterious creature which he called simply "black and white bear." Biologists examined the bones and concluded that they more closely resembled the bones of a red panda than those of bears. Since red pandas were, beyond doubt, part of the raccoon family, giant pandas were also classified as raccoons (albeit big ones).

Although giant pandas look like bears, they have features that are unusual for bears and typical of raccoons: they do not hibernate in the winter like

---

10. In fact, CAST may not even converge; see the problems at the end of the chapter.

other bears do, their male genitalia are tiny and backward-pointing (like raccoons' genitalia), and they do not roar like bears but bleat like raccoons. As a result, Edwin Colbert wrote in 1938:

> So the quest has stood for many years with the bear proponents and the raccoon adherents and the middle-of-the-road group advancing their several arguments with the clearest of logic, while in the meantime the giant panda lives serenely in the mountains of Szechuan with never a thought about the zoological controversies he is causing by just being himself.

The giant panda classification was finally resolved in 1985 by Steven O'Brien and colleagues who used DNA sequences and algorithms, rather than behavioral and anatomical features, to resolve the giant panda controversy (fig. 10.7). The final analysis demonstrated that DNA sequences provide an important source of information to test evolutionary hypotheses. O'Brien's study used about 500,000 nucleotides to construct the evolutionary tree of bears and raccoons.

Roughly at the same time that Steven O'Brien resolved the giant panda controversy, Rebecca Cann, Mark Stoneking and Allan Wilson constructed an evolutionary tree of humans and instantly created a new controversy. This tree led to the *Out of Africa* hypothesis, which claims that humans have a common ancestor who lived in Africa 200,000 years ago. This study turned the question of human origins into an algorithmic puzzle.

The tree was constructed from mitochondrial DNA (mtDNA) sequences of people of different races and nationalities.[11] Wilson and his colleagues compared sequences of mtDNA from people representing African, Asian, Australian, Caucasian, and New Guinean ethnic groups and found 133 variants of mtDNA. Next, they constructed the evolutionary tree for these DNA sequences that showed a trunk splitting into two major branches. One branch consisted only of Africans, the other included some modern Africans and some people from everywhere else. They concluded that a population of Africans, the first modern humans, forms the trunk and the first branch of the tree while the second branch represents a subgroup that left Africa and later spread out to the rest of the world. All of the mtDNA, even samples from regions of the world far away from Africa, were strikingly similar. This

---

11. Unlike the bulk of the genome, mitochondrial DNA is passed solely from a mother to her children without recombining with the father's DNA. Thus it is well-suited for studies of recent human evolution. In addition, it quickly accumulates mutations and thus offers a quick-ticking molecular clock.

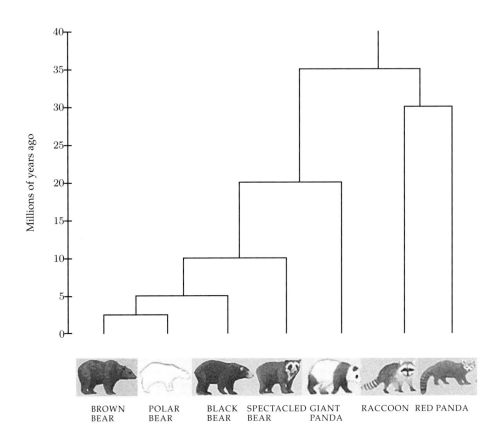

Figure 10.7   An evolutionary tree showing the divergence of raccoons and bears. Despite their difference in size and shape, these two families are closely related.

suggested that our species is relatively young. But the African samples had the most mutations, thus implying that the African lineage is the oldest and that all modern humans trace their roots back to Africa. They further estimated that modern man emerged from Africa 200,000 years ago with racial differences arising only 50,000 years ago.

Shortly after Allan Wilson and colleagues constructed the human mtDNA evolutionary tree supporting the Out of Africa hypothesis, Alan Templeton constructed 100 distinct trees that were also consistent with data that provide evidence *against* the African origin hypothesis! This is a cautionary tale suggesting that one should proceed carefully when constructing large evolutionary trees[12] and below we describe some algorithms for evolutionary tree reconstruction.

Biologists use either *unrooted* or *rooted* evolutionary trees;[13] the difference between them is shown in figure 10.8. In a rooted evolutionary tree, the root corresponds to the most ancient ancestor in the tree, and the path from the root to a leaf in the rooted tree is called an *evolutionary path*. Leaves of evolutionary trees correspond to the existing species while internal vertices correspond to hypothetical ancestral species.[14] In the unrooted case, we do not make any assumption about the position of an evolutionary ancestor (root) in the tree. We also remark that rooted trees (defined formally as undirected graphs) can be viewed as directed graphs if one directs the edges of the rooted tree from the root to the leaves.

Biologists often work with *binary weighted* trees where every internal vertex has degree equal to 3 and every edge has an assigned positive weight (sometimes referred to as the *length*). The *weight* of an edge (v, w) may reflect the number of mutations on the evolutionary path from v to w or a time estimate for the evolution of species v into species w. We sometimes assume the existence of a *molecular clock* that assigns a time $t(v)$ to every internal vertex v in the tree and a length of $t(w) - t(v)$ to an edge (v, w). Here, time corresponds to the "moment" when the species v produced its descendants; every leaf species corresponds to time 0 and every internal vertex presumably corresponds to some negative time.

---

12. Following advances in tree reconstruction algorithms, the critique of the Out of Africa hypothesis has diminished in recent years and the consensus today is that this hypothesis is probably correct.
13. We remind the reader that trees are undirected connected graphs that have no cycles. Vertices of degree 1 in the tree are called leaves. All other vertices are called internal vertices.
14. In rare cases like quickly evolving viruses or bacteria, the DNA of ancestral species is available (e.g., as a ten- to twenty-year-old sample stored in refrigerator) thus making sequences of some internal vertices real rather than hypothetical.

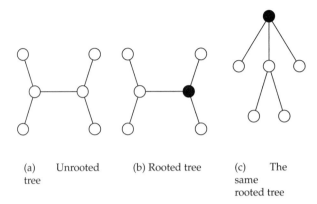

(a)   Unrooted   (b) Rooted tree   (c)   The
tree   same
                                         rooted tree

Figure 10.8 The difference between unrooted (a) and rooted (b) trees. These both describe the same tree, but the unrooted tree makes no assumption about the origin of species. Rooted trees are often represented with the root vertex on the top (c), emphasizing that the root corresponds to the ancestral species.

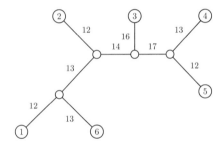

Figure 10.9 A weighted unrooted tree. The length of the path between any two vertices can be calculated as the sum of the weights of the edges in the path between them. For example, $d(1,5) = 12 + 13 + 14 + 17 + 12 = 68$.

10.6   Distance-Based Tree Reconstruction

If we are given a weighted tree T with n leaves, we can compute the length of the path between any two leaves i and j, $d_{i,j}(T)$ (fig. 10.9). Evolutionary biologists often face the opposite problem: they measure the $n \times n$ *distance matrix* $(D_{i,j})$, and then must search for a tree T that has n leaves and fits

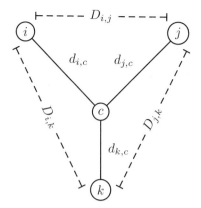

Figure 10.10   A tree with three leaves.

the data,[15] that is, $d_{i,j}(T) = D_{i,j}$ for every two leaves i and j. There are many ways to generate distance matrices: for example, one can sequence a particular gene in n species and define $D_{i,j}$ as the edit distance between this gene in species i and species j.

It is not difficult to construct a tree that fits any given 3×3 (symmetric non-negative) matrix D. This binary unrooted tree has four vertices i, j, k as leaves and vertex c as the center. The lengths of each edge in the tree are defined by the following three equations with three variables $d_{i,c}$, $d_{j,c}$, and $d_{k,c}$ (fig. 10.10):

$$d_{i,c} + d_{j,c} = D_{i,j} \qquad d_{i,c} + d_{k,c} = D_{i,k} \qquad d_{j,c} + d_{k,c} = D_{j,k}.$$

The solution is given by

$$d_{i,c} = \frac{D_{i,j} + D_{i,k} - D_{j,k}}{2} \quad d_{j,c} = \frac{D_{j,i} + D_{j,k} - D_{i,k}}{2} \quad d_{k,c} = \frac{D_{k,i} + D_{k,j} - D_{i,j}}{2}.$$

An unrooted binary tree with n leaves has $2n - 3$ edges, so fitting a *given* tree to an $n \times n$ distance matrix D leads to solving a system of $\binom{n}{2}$ equations with $2n - 3$ variables. For $n = 4$ this amounts to solving six equations with only five variables. Of course, it is not always possible to solve this system,

---

15. We assume that all matrices in this chapter are symmetric, that is, that they satisfy the conditions $D_{i,j} = D_{j,i}$ and $D_{i,j} \geq 0$ for all i and j. We also assume that the distance matrices satisfy the triangle inequality, that is, $D_{i,j} + D_{j,k} \geq D_{i,k}$ for all i, j, and k.

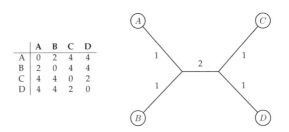

	A	B	C	D
A	0	2	4	4
B	2	0	4	4
C	4	4	0	2
D	4	4	2	0

(a) Additive matrix and the corresponding tree

	A	B	C	D
A	0	2	2	2
B	2	0	3	2
C	2	3	0	2
D	2	2	2	0

?

(b) Non-additive matrix

Figure 10.11   Additive and nonadditive matrices.

making it hard or impossible to construct a tree from D. A matrix $(D_{i,j})$ is called *additive* if there exists a tree T with $d_{i,j}(T) = D_{i,j}$, and *nonadditive* otherwise (fig. 10.11).

Distance-Based Phylogeny Problem:
Reconstruct an evolutionary tree from a distance matrix.

Input: An $n \times n$ distance matrix $(D_{i,j})$.

Output: A weighted unrooted tree T with n leaves fitting D, that is, a tree such that $d_{i,j}(T) = D_{i,j}$ for all $1 \leq i < j \leq n$ if $(D_{i,j})$ is additive.

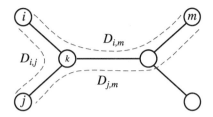

Figure 10.12   If i and j are neighboring leaves and k is their parent, then $D_{k,m} = \frac{D_{i,m}+D_{j,m}-D_{i,j}}{2}$ for every other vertex m in the tree.

The Distance-Based Phylogeny problem may not have a solution, but if it does—that is, if D is additive—there exists a simple algorithm to solve it. We emphasize the fact that we are somehow given the matrix of evolutionary distances between each pair of species, and we are searching for both the shape of the tree that fits this distance matrix and the weights for each edge in the tree.

10.7   Reconstructing Trees from Additive Matrices

A "simple" way to solve the Distance-Based Phylogeny problem for additive trees[16] is to find a pair of *neighboring* leaves, that is, leaves that have the same parent vertex.[17] Figure 10.12 illustrates that for a pair of neighboring leaves i and j and their parent vertex k, the following equality holds for every other leaf m in the tree:

$$D_{k,m} = \frac{D_{i,m} + D_{j,m} - D_{i,j}}{2}$$

Therefore, as soon as a pair of neighboring leaves i and j is found, one can remove the corresponding rows and columns i and j from the distance matrix and add a new row and column corresponding to their parent k. Since the distance matrix is additive, the distances from k to other leaves are recomputed as $D_{k,m} = \frac{D_{i,m}+D_{j,m}-D_{i,j}}{2}$. This transformation leads to a simple algorithm for the Distance-Based Phylogeny problem that finds a pair of neighboring leaves and reduces the size of the tree at every step.

One problem with the described approach is that it is not very easy to find neighboring leaves! One might be tempted to think that a pair of closest

---

16. To be more precise, we mean an "additive matrix," rather than an "additive tree"; the term "additive" applies to matrices. We use the term "additive trees" only because it dominates the

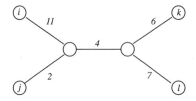

Figure 10.13   The two closest leaves (j and k) are not neighbors in this tree.

leaves (i.e., the leaves i and j with minimum $D_{i,j}$) would represent a pair of neighboring leaves, but a glance at figure 10.13 will show that this is not true. Since finding neighboring leaves using only the distance matrix is a nontrivial problem, we postpone exploring this until the next section and turn to another approach.

Figure 10.14 illustrates the process of shortening *all* "hanging" edges of a tree T, that is, edges leading to leaves. If we reduce the length of every hanging edge by the same small amount δ, then the distance matrix of the resulting tree is $(d_{i,j} - 2\delta)$ since the distance between any two leaves is reduced by 2δ. Sooner or later this process will lead to "collapsing" one of the leaves when the length of the corresponding hanging edge becomes equal to 0 (when δ is equal to the length of the shortest hanging edge). At this point, the original tree $T = T_n$ with n leaves will be transformed into a tree T_{n-1} with $n - 1$ or fewer leaves.

Although the distance matrix D does not explicitly contain information about δ, it is easy to derive both δ and the location of the collapsed leaf in T_{n-1} by the method described below. Thus, one can perform a series of tree transformations $T_n \rightarrow T_{n-1} \rightarrow \ldots \rightarrow T_3 \rightarrow T_2$, then construct the tree T_2 (which is easy, since it consists of only a single edge), and then perform a series of reverse transformations $T_2 \rightarrow T_3 \rightarrow \ldots \rightarrow T_{n-1} \rightarrow T_n$ recovering information about the collapsed edges at every step (fig. 10.14)

A triple of distinct elements $1 \leq i, j, k \leq n$ is called *degenerate* if $D_{i,j} + D_{j,k} = D_{i,k}$, which is essentially just an indication that j is located on the path from i to k in the tree. If D is additive, $D_{i,j} + D_{j,k} \geq D_{i,k}$ for every triple i, j, k. We call the entire matrix D degenerate if it has at least one degenerate triple. If (i, j, k) is a degenerate triple, and some tree T fits matrix D, then the

---

literature on evolutionary tree reconstruction.
17. Showing that every binary tree has neighboring leaves is left as a problem at the end of this chapter.

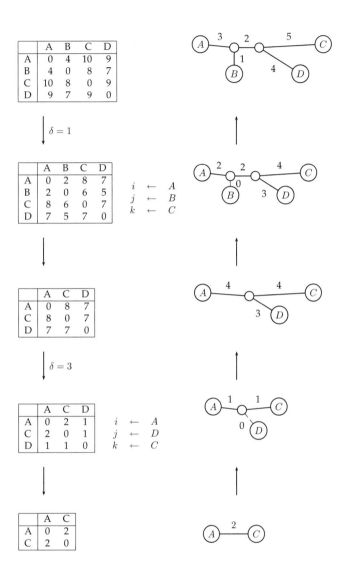

Figure 10.14   The iterative process of shortening the hanging edges of a tree.

vertex j should lie somewhere on the path between i and k in T.[18] Another way to state this is that j is attached to this path by an edge of weight 0, and the attachment point for j is located at distance $D_{i,j}$ from vertex i. Therefore, if an $n \times n$ additive matrix D has a degenerate triple, then it will be reduced to an $(n-1) \times (n-1)$ additive matrix by simply excluding vertex j from consideration; the position of j will be recovered during the reverse transformations. If the matrix D does not have a degenerate triple, one can start reducing the values of all elements in D by the same amount 2δ until the point at which the distance matrix becomes degenerate for the first time (i.e., δ is the minimum value for which $(D_{i,j} - 2\delta)$ has a degenerate triple for some i and j). Determining how to calculate the minimum value of δ (called the *trimming parameter*) is left as a problem at the end of this chapter. Though you do not have the tree T, this operation corresponds to shortening all of the hanging edges in T by δ until one of the leaves ends up on the evolutionary path between two other leaves for the first time. This intuition motivates the following recursive algorithm for finding the tree that fits the data.

ADDITIVEPHYLOGENY(D)
  1 **if** D is a 2×2 matrix
  2      $T \leftarrow$ the tree consisting of a single edge of length $D_{1,2}$.
  3   **return** T
  4 **if** D is non-degenerate
  5      $\delta \leftarrow$ trimming parameter of matrix D
  6   **for** all $1 \leq i \neq j \leq n$
  7          $D_{i,j} \leftarrow D_{i,j} - 2\delta$
  8 **else**
  9      $\delta \leftarrow 0$
10 Find a triple i, j, k in D such that $D_{ij} + D_{jk} = D_{ik}$
11 $x \leftarrow D_{i,j}$
12 Remove jth row and jth column from D.
13 $T \leftarrow$ ADDITIVEPHYLOGENY(D)
14 Add a new vertex v to T at distance x from i to k
15 Add j back to T by creating an edge (v, j) of length 0
16 **for** every leaf l in T
17   **if** distance from l to v in the tree T does not equal $D_{l,j}$
18   **output** "Matrix D is not additive"
19   **return**
20 Extend hanging edges leading to all leaves by δ
21 **return** T

---

18. To be more precise, vertex j partitions the path from i to k into paths of length $D_{i,j}$ and $D_{j,k}$.

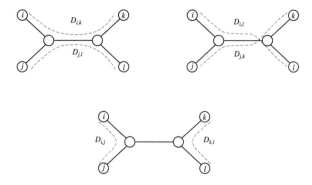

Figure 10.15 Representing three sums in a tree with 4 vertices.

The ADDITIVEPHYLOGENY algorithm above provides a way to check if the matrix D is additive. While this algorithm is intuitive and simple, it is not the most efficient way to construct additive trees. Another way to check additivity is by using the following "four-point condition". Let $1 \leq i, j, k, l \leq n$ be four distinct indices. Compute 3 sums: $D_{i,j} + D_{k,l}$, $D_{i,k} + D_{j,l}$, and $D_{i,l} + D_{j,k}$. If D is an additive matrix then these three sums can be represented by a tree with four leaves (fig. 10.15). Moreover, two of these sums represent the same number (the sum of lengths of all edges in the tree plus the length of the middle edge) while the third sum represents another smaller number (the sum of lengths of all edges in the tree minus the length of the middle edge). We say that elements $1 \leq i, j, k, l \leq n$ satisfy the *four-point condition* if two of the sums $D_{i,j} + D_{k,l}$, $D_{i,k} + D_{j,l}$, and $D_{i,l} + D_{j,k}$ are the same, and the third one is smaller than these two.

Theorem 10.1 *An $n \times n$ matrix D is additive if and only if the four point condition holds for every 4 distinct elements $1 \leq i, j, k, l \leq n$.*

If the distance matrix D is not additive, one might want instead to find a tree that approximates D using the sum of squared errors $\sum_{i,j}(d_{i,j}(T) - D_{i,j})^2$ as a measure of the quality of the approximation. This leads to the ($\mathcal{NP}$-hard) Least Squares Distance-Based Phylogeny problem:

Least Squares Distance-Based Phylogeny Problem:
Given a distance matrix, find the evolutionary tree that minimizes squared error.

> **Input:** An $n \times n$ distance matrix $(D_{i,j})$
>
> **Output:** A weighted tree T with n leaves minimizing $\sum_{i,j}(d_{i,j}(T) - D_{i,j})^2$ over all weighted trees with n leaves.

10.8   Evolutionary Trees and Hierarchical Clustering

Biologists often use variants of hierarchical clustering to construct evolutionary trees. UPGMA (*Unweighted Pair Group Method with Arithmetic Mean*) is a particularly simple clustering algorithm. The UPGMA algorithm is a variant of HIERARCHICALCLUSTERING that uses a different approach to compute the distance between clusters, and assigns *heights* to vertices of the constructed tree. Thus, the length of an edge (u, v) is defined to be the difference in heights of the vertices v and u. The height plays the role of the molecular clock, and allows one to "date" the divergence point for every vertex in the evolutionary tree.

Given clusters C_1 and C_2, UPGMA defines the distance between them to be the average pairwise distance: $D(C_1, C_2) = \frac{1}{|C_1||C_2|} \sum_{i \in C_1} \sum_{j \in C_2} D(i, j)$. At heart, UPGMA is simply another hierarchical clustering algorithm that "dates" vertices of the constructed tree.

UPGMA(D, n)
1   Form n clusters, each with a single element
2   Construct a graph T by assigning an isolated vertex to each cluster
3   Assign height $h(v) = 0$ to every vertex v in this graph
4   **while** there is more than one cluster
5   Find the two closest clusters C_1 and C_2
6   Merge C_1 and C_2 into a new cluster C with $|C_1| + |C_2|$ elements
7   **for** every cluster $C^* \neq C$
8             $D(C, C^*) = \frac{1}{|C| \cdot |C^*|} \sum_{i \in C} \sum_{j \in C^*} D(i, j)$
9   Add a new vertex C to T and connect to vertices C_1 and C_2
10       $h(C) \leftarrow \frac{D(C_1, C_2)}{2}$
11   Assign length $h(C) - h(C_1)$ to the edge (C_1, C)
12   Assign length $h(C) - h(C_2)$ to the edge (C_2, C)
13   Remove rows and columns of D corresponding to C_1 and C_2
14   Add a row and column to D for the new cluster C
15   **return** T

UPGMA produces a special type of rooted tree[19] that is known as *ultrametric*. In ultrametric trees the distance from the root to any leaf is the same.

We can now return to the "neighboring leaves" idea that we developed and then abandoned in the previous section. In 1987 Naruya Saitou and Masatoshi Nei developed an ingenious *neighbor joining* algorithm for phylogenetic tree reconstruction. In the case of additive trees, the neighbor joining algorithm somehow magically finds pairs of neighboring leaves and proceeds by substituting such pairs with the leaves' parent. However, neighbor joining works well not only for additive distance matrices but for many others as well: it does not assume the existence of a molecular clock and ensures that the clusters that are merged in the course of tree reconstruction are not only close to each other (as in UPGMA) but also are far apart from the rest.

For a cluster C, define $u(C) = \frac{1}{\text{number of clusters} - 2} \sum_{\text{all clusters } C'} D(C, C')$ as a measure of the separation of C from other clusters.[20] To choose which two clusters to merge, we look for the clusters C_1 and C_2 that are simultaneously close to each other and far from others. One may try to merge clusters that simultaneously minimize $D(C_1, C_2)$ and maximize $u(C_1) + u(C_2)$. However, it is unlikely that a pair of clusters C_1 and C_2 that simultaneously minimize $D(C_1, C_2)$ and maximize $u(C_1) + u(C_2)$ exists. As an alternative, one opts to minimize $D(C_1, C_2) - u(C_1) - u(C_2)$. This approach is used in the NEIGHBORJOINING algorithm below.

NEIGHBORJOINING(D, n)
1   Form n clusters, each with a single element
2   Construct a graph T by assigning an isolated vertex to each cluster
3   **while** there is more than one cluster
4   Find clusters C_1 and C_2 minimizing $D(C_1, C_2) - u(C_1) - u(C_2)$
5   Merge C_1 and C_2 into a new cluster C with $|C_1| + |C_2|$ elements
6   Compute $D(C, C^*) = \frac{D(C_1, C) + D(C_2, C)}{2}$ to every other cluster C^*
7   Add a new vertex C to T and connect it to vertices C_1 and C_2
8   Assign length $\frac{1}{2} D(C_1, C_2) + \frac{1}{2}(u(C_1) - u(C_2))$ to the edge (C_1, C)
9   Assign length $\frac{1}{2} D(C_1, C_2) + \frac{1}{2}(u(C_2) - u(C_1))$ to the edge (C_2, C)
10   Remove rows and columns of D corresponding to C_1 and C_2
11   Add a row and column to D for the new cluster C
12   **return** T

---

19. Here, the root corresponds to the cluster created last.
20. The explanation of that mysterious term "number of clusters − 2" in this formula is beyond the scope of this book.

10.9   Character-Based Tree Reconstruction

Evolutionary tree reconstruction often starts by sequencing a particular gene in each of n species. After aligning these genes, biologists end up with an $n \times m$ alignment matrix (n species, m nucleotides in each) that can be further transformed into an $n \times n$ distance matrix. Although the distance matrix could be analyzed by distance-based tree reconstruction algorithms, a certain amount of information gets lost in the transformation of the alignment matrix into the distance matrix, rendering the reverse transformation of distance matrix back into the alignment matrix impossible. A better technique is to use the alignment matrix directly for evolutionary tree reconstruction. *Character-based* tree reconstruction algorithms assume that the input data are described by an $n \times m$ matrix (perhaps an alignment matrix), where n is the number of species and m is the number of *characters*. Every row in the matrix describes an existing species and the goal is to construct a tree whose leaves correspond to the n existing species and whose internal vertices correspond to ancestral species. Each internal vertex in the tree is labeled with a character string that describes, for example, the hypothetical number of legs in that ancestral species. We want to determine what character strings at internal nodes would best explain the character strings for the n observed species.

The use of the word "character" to describe an attribute of a species is potentially confusing, since we often use the word to refer to letters from an alphabet. We are not at liberty to change the terminology that biologists have been using for at least a century, so for the next section we will refer to nucleotides as *states* of a character. Another possible character might be "number of legs," which is not very informative for mammalian evolutionary studies, but could be somewhat informative for insect evolutionary studies.

An intuitive score for a character-based evolutionary tree is the total number of mutations required to explain all of the observed character sequences. The *parsimony* approach attempts to minimize this score, and follows the philosophy of Occam's razor: find the simplest explanation of the data (see figure 10.16).[21]

---

21. Occam was one of the most influential philosophers of the fourteenth century. Strong opposition to his ideas from theology professors prevented him from obtaining his masters degree (let alone his doctorate). Occam's razor states that "plurality should not be assumed without necessity," and is usually paraphrased as "keep it simple, stupid." Occam used this principle to eliminate many pseudoexplanatory theological arguments. Though the parsimony principle is attributed to Occam, sixteen centuries earlier Aristotle wrote simply that *Nature operates in the shortest way possible.*

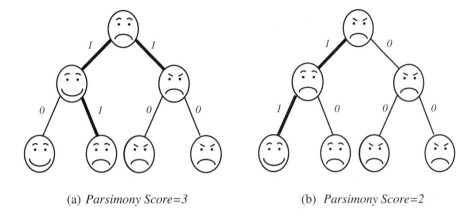

(a) *Parsimony Score=3*   (b) *Parsimony Score=2*

Figure 10.16   If we label a tree's leaves with characters (in this case, eyebrows and mouth, each with two states), and choose labels for each internal vertex, we implicitly create a *parsimony* score for the tree. By changing the labels in (a) we are able to create a tree with a better parsimony score in (b).

Given a tree T with every vertex labeled by an m-long string of characters, one can set the length of an edge (v, w) to the Hamming distance $d_H(v, w)$ between the character strings for v and w. The *parsimony score* of a tree T is simply the sum of lengths of its edges $\sum_{\text{All edges } (v,w) \text{ of the tree}} d_H(v, w)$. In reality, the strings of characters assigned to internal vertices are unknown and the problem is to find strings that minimize the parsimony score.

Two particular incarnations of character-based tree reconstruction are the Small Parsimony problem and the Large Parsimony problem. The Small Parsimony problem assumes that the tree is given but the labels of its internal vertices are unknown, while the vastly more difficult Large Parsimony problem assumes that neither the tree structure nor the labels of its internal vertices are known.

10.10   Small Parsimony Problem

Small Parsimony Problem:
Find the most parsimonious labeling of the internal vertices in an evolutionary tree.

Input: Tree T with each leaf labeled by an m-character string.

Output: Labeling of internal vertices of the tree T minimizing the parsimony score.

An attentive reader should immediately notice that, because the characters in the string are independent, the Small Parsimony problem can be solved independently for each character. Therefore, to devise an algorithm, we can assume that every leaf is labeled by a single character rather than by a string of m characters.

As we have seen in previous chapters, sometimes solving a more general—and seemingly more difficult—problem may reveal the solution to the more specific one. In the case of the Small Parsimony Problem we will first solve the more general *Weighted Small Parsimony* problem, which generalizes the notion of parsimony by introducing a scoring matrix. The length of an edge connecting vertices v and w in the Small Parsimony problem is defined as the Hamming distance, $d_H(v, w)$, between the character strings for v and w. In the case when every leaf is labeled by a single character in a k-letter alphabet, $d_H(v, w) = 0$ if the characters corresponding to v and w are the same, and $d_H(v, w) = 1$ otherwise. One can view such a scoring scheme as a $k \times k$ scoring matrix $(\delta_{i,j})$ with diagonal elements equal to 0 and all other elements equal to 1. The Weighted Small Parsimony problem simply assumes that the scoring matrix $(\delta_{i,j})$ is an arbitrary $k \times k$ matrix and minimizes the weighted parsimony score $\sum_{\text{all edges } (v, w) \text{ in the tree}} \delta_{v,w}$.

Weighted Small Parsimony Problem:
Find the minimal weighted parsimony score labeling of the internal vertices in an evolutionary tree.

Input: Tree T with each leaf labeled by elements of a k-letter alphabet and a $k \times k$ scoring matrix (δ_{ij}).

Output: Labeling of internal vertices of the tree T minimizing the weighted parsimony score.

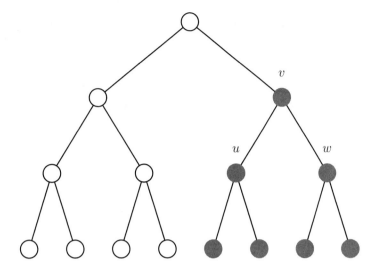

Figure 10.17   A subtree of a larger tree. The shaded vertices form a tree rooted at the topmost shaded node.

In 1975 David Sankoff came up with the following dynamic programming algorithm for the Weighted Small Parsimony problem. As usual in dynamic programming, the Weighted Small Parsimony problem for T is reduced to solving the Weighted Small Parsimony Problem for smaller subtrees of T. As we mentioned earlier, a rooted tree can be viewed as a directed tree with all of its edges directed away from the root toward the leaves. Every vertex v in the tree T defines a *subtree* formed by the vertices beneath v (fig. 10.17), which are all of the vertices that can be reached from v. Let $s_t(v)$ be the minimum parsimony score of the subtree of v under the assumption that vertex v has character t. For an internal vertex v with children u and w, the score $s_t(v)$ can be computed by analyzing k scores $s_i(u)$ and k scores $s_i(w)$ for $1 \leq i \leq k$ (below, i and j are characters):

$$s_t(v) = \min_i \left\{ s_i(u) + \delta_{i,t} \right\} + \min_j \left\{ s_j(w) + \delta_{j,t} \right\}$$

The initial conditions simply amount to an assignment of the scores $s_t(v)$ at the leaves according to the rule: $s_t(v) = 0$ if v is labeled by letter t and $s_t(v) = \infty$ otherwise. The minimum weighted parsimony score is given by the smallest score at the root, $s_t(r)$ (fig. 10.18). Given the computed values

$s_t(v)$ at all of the vertices in the tree, one can reconstruct an optimal assignment of labels using a backtracking approach that is similar to that used in chapter 6. The running time of the algorithm is $O(nk)$.

In 1971, even before David Sankoff solved the Weighted Small Parsimony problem, Walter Fitch derived a solution of the (unweighted) Small Parsimony problem. The Fitch algorithm below is essentially dynamic programming in disguise. The algorithm assigns a set of letters S_v to every vertex in the tree in the following manner. For each leaf v, S_v consists of single letter that is the label of this leaf. For any internal vertex v with children u and w, S_v is computed from the sets S_u and S_w according to the following rule:

$$S_v = \begin{cases} S_u \cap S_w, & \text{if } S_u \text{ and } S_w \text{ overlap} \\ S_u \cup S_w, & \text{otherwise} \end{cases}$$

To compute S_v, we traverse the tree in *post-order* as in figure 10.19, starting from the leaves and working toward the root. After computing S_v for all vertices in the tree, we need to decide on how to assign letters to the internal vertices of the tree. This time we traverse the tree using *preorder* traversal from the root toward the leaves. We can assign root r any letter from S_r. To assign a letter to an internal vertex v we check if the (already assigned) label of its parent belongs to S_v. If yes, we choose the same label for v; otherwise we label v by an arbitrary letter from S_v (fig. 10.20). The running time of this algorithm is also $O(nk)$.

At first glance, Fitch's labeling procedure and Sankoff's dynamic programming algorithm appear to have little in common. Even though Fitch probably did not know about application of dynamic programming for evolutionary tree reconstruction in 1971, the two algorithms are almost identical. To reveal the similarity between these two algorithms let us return to Sankoff's recurrence. We say that character t is optimal for vertex v if it yields the smallest score, that is, if $s_t(v) = \min_{1 \le i \le k} s_i(v)$. The set of optimal letters for a vertex v forms a set $S(v)$. If u and w are children of v and if $S(u)$ and $S(w)$ overlap, then it is easy to see that $S(v) = S(u) \cap S(w)$. If $S(u)$ and $S(w)$ do not overlap, then it is easy to see that $S(v) = S(u) \cup S(w)$. Fitch's algorithm uses exactly the same recurrence, thus revealing that these two approaches are algorithmic twins.

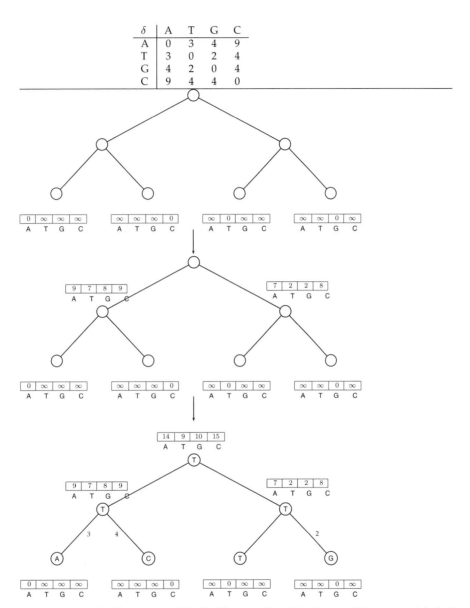

Figure 10.18 An illustration of Sankoff's algorithm. The leaves of the tree are labeled by A, C, T, G in order. The minimum weighted parsimony score is given by $s_T(root) = 0 + 0 + 3 + 4 + 0 + 2 = 9$.

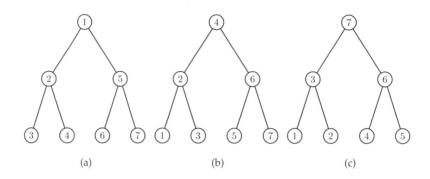

Figure 10.19   Three methods of traversing a tree. (a) Pre-order: SELF, LEFT, RIGHT. (b) In-order: LEFT, SELF, RIGHT. (c) Post-order: LEFT, RIGHT, SELF.

10.11   Large Parsimony Problem

Large Parsimony Problem:
Find a tree with n leaves having the minimal parsimony score.

Input: An $n \times m$ matrix M describing n species, each represented by an m-character string.

Output: A tree T with n leaves labeled by the n rows of matrix M, and a labeling of the internal vertices of this tree such that the parsimony score is minimized over all possible trees and over all possible labelings of internal vertices.

Not surprisingly, the Large Parsimony problem is $\mathcal{NP}$-complete. In the case n is small, one can explore all tree topologies with n leaves, solve the Small Parsimony problem for each topology, and select the best one. However, the number of topologies grows very fast with respect to n, so biologists often use local search heuristics (e.g., greedy algorithms) to navigate in the space of all topologies. *Nearest neighbor interchange* is a local search heuristic that defines neighbors in the space of all trees.[22] Every *internal* edge in a tree defines four subtrees A, B, C, and D (fig. 10.21) that can be combined into a

---

22. "Nearest neighbor" has nothing to do with the two closest leaves in the tree.

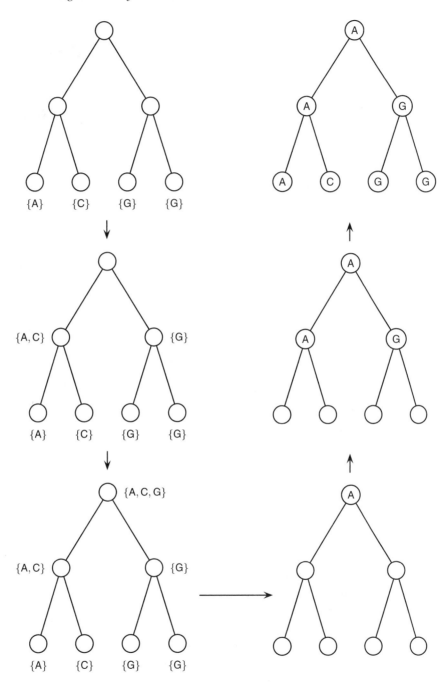

Figure 10.20 An illustration of Fitch's algorithm.

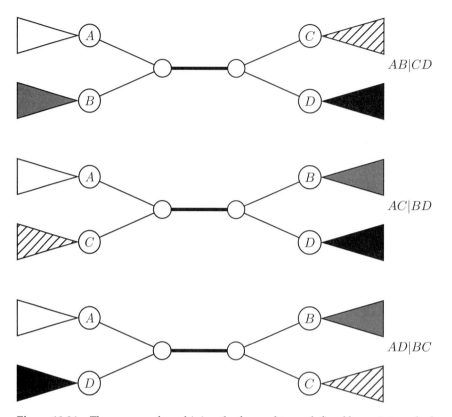

Figure 10.21   Three ways of combining the four subtrees defined by an internal edge.

tree in three different ways that we denote $AB|CD$, $AC|BD$, and $AD|BC$. These three trees are called *neighbors* under the nearest neighbor interchange transformation. Figure 10.22 shows all trees with five leaves and connects two trees if they are neighbors. Figure 10.23 shows two nearest neighbor interchanges that transform one tree into another. A greedy approach to the Large Parsimony problem is to start from an arbitrary tree and to move (by nearest neighbor interchange) from one tree to another if such a move provides the best improvement in the parsimony score among all neighbors of the tree T.

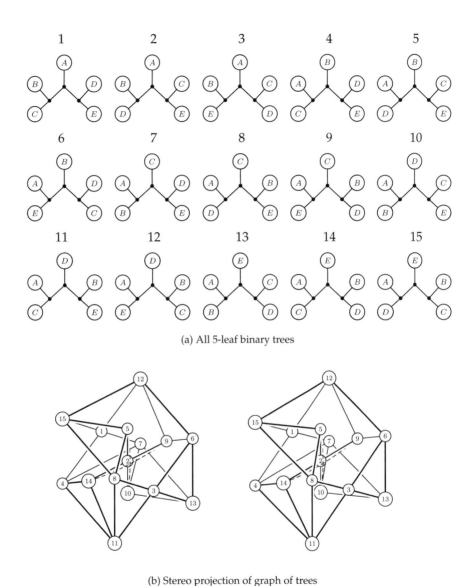

(a) All 5-leaf binary trees

(b) Stereo projection of graph of trees

Figure 10.22   (a) All unrooted binary trees with five leaves. (b) These can also be considered to be vertices in a graph; two vertices are connected if and only if their respective trees are interchangeable by a single nearest neighbor interchange operation. Shown is a three dimensional view of the graph as a stereo representation.

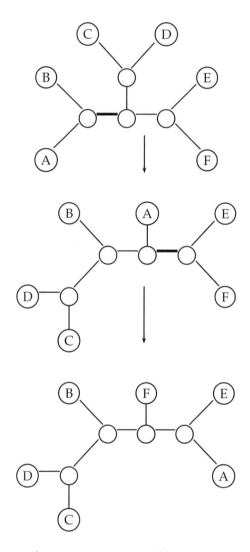

Figure 10.23   Two trees that are two nearest neighbor interchanges apart.

10.12 Notes

The literature on clustering can be traced back to the nineteenth century. The k-Means clustering algorithm was introduced by Stuart Lloyd in 1957 (68), popularized by MacQueen in 1965 (70), and has since inspired dozens of variants. Applications of hierarchical clustering for gene expression analyses were pioneered by Michael Eisen, Paul Spellman, Patrick Brown and David Botstein in 1998 (33). The PCC and CAST algorithms were developed by Amir Ben-Dor, Ron Shamir, and Zohar Yakhini in 1999 (11).

The molecular solution of the giant panda riddle was proposed by Stephen O'Brien and colleagues in 1985 (81) and described in more details in O'Brien's book (80). The Out of Africa hypothesis was proposed by Rebecca Cann, Mark Stoneking, and Allan Wilson in 1987. Even today it continues to be debated by Alan Templeton and others (103).

The simple and intuitive UPGMA hierarchical clustering approach was developed in 1958 by Charles Michener and Robert Sokal (98). The neighbor joining algorithm was developed by Naruya Saitou and Masatoshi Nei in 1987 (90).

The algorithm to solve the Small Parsimony problem was developed by Walter Fitch in 1971 (37), four years before David Sankoff developed a dynamic programming algorithm for the more general Weighted Small Parsimony problem (92). The four-point condition was first formulated in 1965 (113), promptly forgotten, and rediscovered by Peter Buneman six years later (19). The nearest neighbor interchange approach to exploring trees was proposed in 1971 by David Robinson (88).

Ron Shamir (born 1953 in Jerusalem) is a professor at the School of Computer Science at Tel Aviv University. He holds an undergraduate degree in mathematics and physics from Hebrew University, Jerusalem and a PhD in operations research from the University of California at Berkeley. He was a pioneer in the application of sophisticated graph-theoretical techniques in bioinformatics. Shamir's interests have always revolved around the design and analysis of algorithms and his PhD dissertation was on studies of linear programming algorithms. Later he developed a keen interest in graph algorithms that eventually brought him (in a roundabout way) to computational biology. Back in 1990 Shamir was collaborating with Martin Golumbic on problems related to temporal reasoning. In these problems, one has a set of events, each modeled by an unknown interval on the timeline, and a collection of constraints on the relations between each two events, for example, whether two intervals overlap or not. One has to determine if there exists a realization of the events as intervals on the line, satisfying all the constraints. This is a generalization of the problem that Seymour Benzer faced trying to establish the linearity of genes, but at the time Shamir did not know about Benzer's work. He says:

I presented my work on temporal reasoning at a workshop in 1990 and the late Gene Lawler told me this model fits perfectly the DNA mapping problem. I knew nothing about DNA at the time but started reading, and I also sent a note to Eric Lander describing the temporal reasoning result with the mapping consequences. A few weeks later Rutgers University organized a kickoff day on computational biology and brought Mike Waterman and Eric Lander who gave fantastic talks on the genome project and the computational challenges. Eric even mentioned our result as an important example of computer science contribution to the field. I think he was exaggerating quite a bit on this point, but the two talks sent me running to read biology textbooks and papers. I got a lot of help during the first years from my wife

Michal, who happens to be a biologist and patiently answered all my trivial questions.

This chance meeting with Gene Lawler and Eric Lander converted Shamir into a bioinformatician and since 1991 he has been devoting more and more time to algorithmic problems in computational biology. He still keeps an interest in graph algorithms and often applies graph-theoretical techniques to biological problems. In recent years, he has tended to combine discrete mathematics techniques and probabilistic reasoning. Ron says:

I still worry about the complexity of the problems and the algorithms, and would care about (and enjoy) the $\mathcal{NP}$-hardness proof of a new problem, but biology has trained me to be "tolerant" to heuristics when the theoretical computer science methodologies only take you so far, and the data require more.

Shamir was an early advocate of rigorous clustering algorithms in bioinformatics. Although clustering theory has been under development since the 1960s, his interest in clustering started in the mid-1990s as part of a collaboration with Hans Lehrach of the Max Planck Institute in Berlin. Lehrach is one of the fathers of the oligonucleotide fingerprinting technology that predates the DNA chips techniques. Lehrach had a "factory" in Berlin that would array thousands of unknown cDNA clones, and then hybridize them to k-mer probes (typically 8-mers or 9-mers). This was repeated with dozens of different probes, giving a fingerprint for each cDNA clone. To find which clones correspond to the same gene, one has to cluster them based on their fingerprints, and the noise level in the data makes accurate clustering quite a challenge.

It was this challenge that brought Shamir to think about clustering. As clustering theory is very heterogeneous and scattered over many different fields, he set out to develop his algorithm first, before learning the literature. Naturally, Shamir resorted to graph algorithms, modeled the clustered elements (clones in the fingerprinting application) as vertices, and connected clones with highly similar fingerprints by edges. It was a natural next step to repeatedly partition the graph based on minimum cuts (sets of edges whose removal makes the graph disconnected) and hence his first clustering algorithm was born. Actually the idea looked so natural to Shamir that he and his student Erez Hartuv searched the literature for a long time just to make sure it had not been published before. They did not find the same approach but found similar lines of thinking in works a decade earlier. According to Shamir,

> Given the ubiquity of clustering, I would not be surprised to unearth in the future an old paper with the same idea in archeology, zoology, or some other field.

Later, after considering the pros and cons of the approach, together with his student Roded Sharan, Shamir improved the algorithm by adding a rigorous probabilistic analysis and created the popular clustering algorithm CLICK. The publication of vast data sets of microarray expression profiles was a great opportunity and triggered Shamir to further develop PCC and CAST together with Amir Ben-Dor and Zohar Yakhini.

Shamir is one of the very few scientists who is regarded as a leader in both algorithms and bioinformatics and who was credited with an important breakthrough in algorithms even before he became a bioinformatician. He says:

> Probably the most exciting moment in my scientific life was my PhD result in 1983, showing that the average complexity of the Simplex algorithm is quadratic. The Simplex algorithm is one of the most commonly used algorithms, with thousands of applications in virtually all areas of science and engineering. Since the 1950s, the Simplex algorithm was known to be very efficient in practice and was (and is) used ubiquitously, but worst-case exponential-time examples were given for many variants of the Simplex, and none was proved to be efficient. This was a very embarrassing situation, and resolving it was very gratifying. Doing it together with my supervisors Ilan Adler (a leader in convex optimization, and a student of Dantzig, the father of the Simplex algorithm) and Richard Karp (one of the great leaders of theoretical computer science) was extremely exciting. Similar results were obtained independently at the same time by other research groups. There were some other exhilarating moments, particularly those between finding out an amazing new result and discovering the bug on the next day, but these are the rewards and frustrations of scientific research.

Shamir views himself as not committed to a particular problem. One of the advantages of scientific research in the university setting is the freedom to follow one's interests and nose, and not be obliged to work on the same problem for years. The exciting rapid developments in biology bring about a continuous flow of new problems, ideas, and data. Of course, one has to make sure breadth does not harm the research depth. In practice, in spite of this "uncommitted" ideology, Ron finds himself working on the same areas for several years, and following the literature in these fields a while later. He tends to work on several problem areas simultaneously but tries to choose topics that are overlapping to some extent, just to save on the effort of understanding and following several fields. Shamir says:

> In my opinion, an open mind, high antennae, a solid background in the relevant disciplines, and hard work are the most important elements of discovery. As a judge once said, "Inspiration only comes to the law library at 3 AM". This is even more true in science. Flexibility and sensitivity to the unexpected are crucial.

Shamir views the study of gene networks, that is, understanding how genes, proteins, and other molecules work together in concert, as a long-standing challenge.

10.13   Problems

Problem 10.1

Determine the number of different ways to partition a set of n elements into k clusters.

Problem 10.2

Construct an instance of the k-Means Clustering problem for which the Lloyd algorithm produces a particularly bad solution. Derive a performance guarantee of the Lloyd algorithm.

Problem 10.3

Find an optimal algorithm for solving the k-Means Clustering problem in the case of $k = 1$. Can you find an analytical solution in this case?

Problem 10.4

Estimate the number of iterations that the Lloyd algorithm will require when $k = 1$. Repeat for $k = 2$.

Problem 10.5

Construct an example for which the CAST algorithm does not converge.

Problem 10.6

How do you calculate the trimming parameter δ in ADDITIVEPHYLOGENY?

Problem 10.7

Prove that a connected graph in which the number of vertices exceeds the number of edges by 1 is a tree.

Problem 10.8

Some of 8 Hawaiian islands are connected by airlines. It is known one can reach every island from any other (probably with some intermediate stops). Prove that you can visit all islands making no more than 12 flights.

Problem 10.9

Show that any binary tree with n leaves has $2n - 3$ edges.

Problem 10.10

How many different unrooted binary trees on n vertices are there?

Problem 10.11

Prove that every binary tree has at least two neighboring leaves.

Problem 10.12

Design a backtracking procedure to reconstruct the optimal assignment of characters in the Sankoff algorithm for the Weighted Small Parsimony problem.

Problem 10.13

While we showed how to solve the Weighted Small Parsimony problem using dynamic programming, we did not show how to construct a Manhattan-like graph for this problem. Cast the Weighted Small Parsimony problem in terms of finding a path in an appropriate Manhattan-like directed acyclic graph.

Problem 10.14

The *nearest neighbor interchange distance* between two trees is defined as the minimum number of interchanges to transform one tree into another. Design an approximation algorithm for computing the nearest neighbor interchange distance.

Problem 10.15

Find two binary trees with six vertices that are the maximum possible nearest neighbor interchange distance apart from each other.

11 *Hidden Markov Models*

Hidden Markov Models are a popular *machine learning* approach in bioinformatics. Machine learning algorithms are presented with *training data*, which are used to derive important insights about the (often hidden) parameters. Once an algorithm has been suitably trained, it can apply these insights to the analysis of a *test sample*. As the amount of training data increases, the accuracy of the machine learning algorithm typically increases as well. The parameters that are learned during training represent knowledge; application of the algorithm with those parameters to new data (not used in the training phase) represents the algorithm's use of that knowledge. The Hidden Markov Model (HMM) approach, considered in this chapter, learns some unknown probabilistic parameters from training samples and uses these parameters in the framework of dynamic programming (and other algorithmic techniques) to find the best explanation for the experimental data.

11.1 *CG*-Islands and the "Fair Bet Casino"

The least frequent dinucleotide in many genomes is CG. The reason for this is that the C within CG is easily *methylated*, and the resulting methyl-C has a tendency to mutate into T.[1] However, the methylation is often suppressed around genes in areas called *CG-islands* in which CG appears relatively frequently. An important problem is to define and locate CG-islands in a long genomic text.

Finding CG-islands can be modeled after the following toy gambling problem. The "Fair Bet Casino" has a game in which a dealer flips a coin and

---

1. Cells often biochemically modify DNA and proteins. Methylation is the most common DNA modification and results in the addition of a methyl (CH_3) group to a nucleotide position in DNA.

the player bets on the outcome (heads or tails). The dealer in this (crooked) casino uses either a fair coin (heads or tails are equally likely) or a biased coin that will give heads with a probability of $\frac{3}{4}$. For security reasons, the dealer does not like to change coins, so this happens relatively rarely, with a probability of 0.1. Given a sequence of coin tosses, the problem is to find out when the dealer used the biased coin and when he used the fair coin, since this will help you, the player, learn the dealer's psychology and enable you to win money. Obviously, if you observe a long line of heads, it is likely that the dealer used the biased coin, whereas if you see an even distribution of heads and tails, he likely used the fair one. Though you can never be certain that a long string of heads is not just a fluke, you are primarily interested in the most probable explanation of the data. Based on this sensible intuition, we might formulate the problem as follows:

Fair Bet Casino Problem:
Given a sequence of coin tosses, determine when the dealer used a fair coin and when he used a biased coin.

Input: A sequence $x = x_1\, x_2\, x_3 \ldots x_n$ of coin tosses (either H or T) made by two possible coins (F or B).

Output: A sequence $\pi = \pi_1\, \pi_2\, \pi_3 \cdots \pi_n$, with each π_i being either F or B indicating that x_i is the result of tossing the fair or biased coin, respectively.

Unfortunately, this problem formulation simply makes no sense. The ambiguity is that *any* sequence of coins could *possibly* have generated the observed outcomes, so technically $\pi = FFF \ldots FF$ is a valid answer to this problem for every observed sequence of coin flips, as is $\pi = BBB \ldots BB$. We need to incorporate a way to grade different coin sequences as being better answers than others. Below we explain how to turn this ill-defined problem into the Decoding problem based on HMM paradigm.

First, we consider the problem under the assumption that the dealer never changes coins. In this case, letting 0 denote tails and 1 heads, the question is which of the two coins he used, fair ($p^+(0) = p^+(1) = \frac{1}{2}$) or biased ($p^-(0) = \frac{1}{4}$, $p^-(1) = \frac{3}{4}$). If the resulting sequence of tosses is $x = x_1 \ldots x_n$, then the probability that x was generated by a fair coin is[2]

$$P(x|\text{fair coin}) = \prod_{i=1}^{n} p^+(x_i) = \frac{1}{2^n}.$$

On the other hand, the probability that x was generated by a biased coin is

$$P(x|\text{biased coin}) = \prod_{i=1}^{n} p^-(x_i) = \left(\frac{1}{4^{n-k}}\right)\left(\frac{3^k}{4^k}\right) = \frac{3^k}{4^n}.$$

Here k is the number of heads in x. If $P(x|\text{fair coin}) > P(x|\text{biased coin})$, then the dealer most likely used a fair coin; on the other hand, we can see that if $P(x|\text{fair coin}) < P(x|\text{biased coin})$, then the dealer most likely used a biased coin. The probabilities $P(x|\text{fair coin}) = \frac{1}{2^n}$ and $P(x|\text{biased coin}) = \frac{3^k}{4^n}$ become equal at $k = \frac{n}{\log_2 3}$. As a result, when $k < \frac{n}{\log_2 3}$, the dealer most likely used a fair coin, and when $k > \frac{n}{\log_2 3}$, he most likely used a biased coin. We can define the *log-odds ratio* as follows:

$$\log_2 \frac{P(x|\text{fair coin})}{P(x|\text{biased coin})} = \sum_{i=1}^{k} \log_2 \frac{p^+(x_i)}{p^-(x_i)} = n - k \log_2 3$$

However, we know that the dealer *does* change coins, albeit rarely. One approach to making an educated guess as to which coin the dealer used at each point would be to slide a window of some width along the sequence of coin flips and calculate the log-odds ratio of the sequence under each window. In effect, this is considering the log-odds ratio of short regions of the sequence. If the log-odds ratio of the short sequence falls below 0, then the dealer most likely used a biased coin while generating this window of sequence; otherwise the dealer most likely used a fair coin.

Similarly, a naive approach to finding **CG**-islands in long DNA sequences is to calculate log-odds ratios for a sliding window of some particular length, and to declare windows that receive positive scores to be potential **CG**-islands. Of course, the disadvantage of this approach is that we do not know the length of **CG**-islands in advance and that some overlapping windows may classify the same nucleotide differently. HMMs represent a different probabilistic approach to this problem.

---

2. The notation $P(x|y)$ is shorthand for the "probability of x occurring under the assumption that (some condition) y is true." The notation $\prod_{i=1}^{n} a_i$ means $a_1 \cdot a_2 \cdot a_3 \cdots a_n$.

11.2   The Fair Bet Casino and Hidden Markov Models

An HMM can be viewed as an abstract machine that has an ability to produce some output using coin tossing. The operation of the machine proceeds in discrete steps: at the beginning of each step, the machine is in a *hidden state* of which there are k. During the step, the HMM makes two decisions: (1) "What state will I move to next?" and (2) "What symbol—from an alphabet Σ—will I emit?" The HMM decides on the former by choosing randomly among the k states; it decides on the latter by choosing randomly among the $|\Sigma|$ symbols. The choices that the HMM makes are typically biased, and may follow arbitrary probabilities. Moreover, the *probability distributions*[3] that govern which state to move to and which symbols to emit change from state to state. In essence, if there are k states, then there are k different "next state" distributions and k different "symbol emission" distributions. An important feature of HMMs is that an observer can see the emitted symbols but has no ability to see what state HMM is in at any step, hence the name *Hidden* Markov Models. The goal of the observer is to infer the most likely states of the HMM by analyzing the sequences of emitted symbols. Since an HMM effectively uses dice to emit symbols, the sequence of symbols it produces does not form any readily recognizable pattern.

Formally, an HMM $\mathcal{M}$ is defined by an alphabet of emitted symbols Σ, a set of (hidden) states Q, a matrix of state transition probabilities A, and a matrix of emission probabilities E, where

- Σ is an alphabet of symbols;

- Q is a set of states, each of which will emit symbols from the alphabet Σ;

- $A = (a_{kl})$ is a $|Q| \times |Q|$ matrix describing the probability of changing to state l after the HMM is in state k; and

- $E = (e_k(b))$ is a $|Q| \times |\Sigma|$ matrix describing the probability of emitting the symbol b during a step in which the HMM is in state k.

Each row of the matrix A describes a "state die"[4] with $|Q|$ sides, while each row of the matrix E describes a "symbol die" with $|\Sigma|$ sides. The Fair

---

3. A probability distribution is simply an assignment of probabilities to outcomes; in this case, the outcomes are either symbols to emit or states to move to. We have seen probability distributions, in a disguised form, in the context of motif finding. Every column of a profile, when each element is divided by the number of sequences in the sample, forms probability distributions.
4. Singular of "dice."

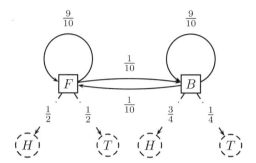

Figure 11.1 The HMM designed for the Fair Bet Casino problem. There are two states: F (fair) and B (biased). From each state, the HMM can emit either heads (H) or tails (T), with the probabilities shown. The HMM will switch between F and B with probability $1/10$.

Bet Casino process corresponds to the following HMM $\mathcal{M}(\Sigma, Q, A, E)$ shown in figure 11.1:

- $\Sigma = \{0, 1\}$, corresponding to tails (0) or heads (1)

- $Q = \{F, B\}$, corresponding to a fair (F) or biased (B) coin

- $a_{FF} = a_{BB} = 0.9$, $a_{FB} = a_{BF} = 0.1$

- $e_F(0) = \frac{1}{2}$, $e_F(1) = \frac{1}{2}$, $e_B(0) = \frac{1}{4}$, $e_B(1) = \frac{3}{4}$

A *path* $\pi = \pi_1 \ldots \pi_n$ in the HMM $\mathcal{M}$ is a sequence of states. For example, if a dealer used the fair coin for the first three and the last three tosses and the biased coin for five tosses in between, the corresponding path π would be $\pi = $ FFFBBBBBFFF. If the resulting sequence of tosses is 01011101001, then the following shows the matching of x to π and the probability of x_i being generated by π_i at each flip:

$$
\begin{array}{c}
x \\
\pi \\
P(x_i | \pi_i)
\end{array}
=
\begin{pmatrix}
0 & 1 & 0 & 1 & 1 & 1 & 0 & 1 & 0 & 0 & 1 \\
F & F & F & B & B & B & B & B & F & F & F \\
\frac{1}{2} & \frac{1}{2} & \frac{1}{2} & \frac{3}{4} & \frac{3}{4} & \frac{3}{4} & \frac{1}{4} & \frac{3}{4} & \frac{1}{2} & \frac{1}{2} & \frac{1}{2}
\end{pmatrix}
$$

We write $P(x_i | \pi_i)$ to denote the probability that symbol x_i was emitted from state π_i—these values are given by the matrix E. We write $P(\pi_i \to \pi_{i+1})$

to denote the probability of the transition from state π_i to π_{i+1}—these values are given by the matrix A.

The path $\pi = $ FFFBBBBBFFF includes only two switches of coins, first from F to B (after the third step), and second from B to F (after the eighth step). The probability of these two switches, $\pi_3 \rightarrow \pi_4$ and $\pi_8 \rightarrow \pi_9$, is $\frac{1}{10}$, while the probability of all other transitions, $\pi_{i-1} \rightarrow \pi_i$, is $\frac{9}{10}$ as shown below:[5]

$$
\begin{array}{c}
x \\
\pi \\
P(x_i|\pi_i) \\
P(\pi_{i-1} \rightarrow \pi_i)
\end{array}
=
\left(
\begin{array}{ccccccccccc}
0 & 1 & 0 & 1 & 1 & 1 & 0 & 1 & 0 & 0 & 1 \\
F & F & F & B & B & B & B & B & F & F & F \\
\frac{1}{2} & \frac{1}{2} & \frac{1}{2} & \frac{3}{4} & \frac{3}{4} & \frac{3}{4} & \frac{1}{4} & \frac{3}{4} & \frac{1}{2} & \frac{1}{2} & \frac{1}{2} \\
\frac{1}{2} & \frac{9}{10} & \frac{9}{10} & \frac{1}{10} & \frac{9}{10} & \frac{9}{10} & \frac{9}{10} & \frac{9}{10} & \frac{1}{10} & \frac{9}{10} & \frac{9}{10}
\end{array}
\right)
$$

The probability of generating x through the path π (assuming for simplicity that in the first moment the dealer is equally likely to have a fair or a biased coin) is roughly 2.66×10^{-6} and is computed as:

$$
\left(\frac{1}{2} \cdot \frac{1}{2}\right)\left(\frac{1}{2} \cdot \frac{9}{10}\right)\left(\frac{1}{2} \cdot \frac{9}{10}\right)\left(\frac{3}{4} \cdot \frac{1}{10}\right)\left(\frac{3}{4} \cdot \frac{9}{10}\right)\left(\frac{3}{4} \cdot \frac{9}{10}\right)\left(\frac{1}{4} \cdot \frac{9}{10}\right)\left(\frac{3}{4} \cdot \frac{9}{10}\right)\left(\frac{1}{2} \cdot \frac{1}{10}\right)\left(\frac{1}{2} \cdot \frac{9}{10}\right)\left(\frac{1}{2} \cdot \frac{9}{10}\right)
$$

In the above example, we assumed that we knew π and observed x. However, in reality we do not have access to π. If you only observe that $x = 01011101001$, then you might ask yourself whether or not $\pi =$FFFBBBBBFFF is the "best" explanation for x. Furthermore, if it is not the best explanation, is it possible to reconstruct the best one? It turns out that FFFBBBBBFFF is not the most probable path for $x = 01011101001$: FFFBBBFFFFF is slightly better, with probability 3.54×10^{-6}.

$$
\begin{array}{c}
x \\
\pi \\
P(x_i|\pi_i) \\
P(\pi_{i-1} \rightarrow \pi_i)
\end{array}
=
\left(
\begin{array}{ccccccccccc}
0 & 1 & 0 & 1 & 1 & 1 & 0 & 1 & 0 & 0 & 1 \\
F & F & F & B & B & B & F & F & F & F & F \\
\frac{1}{2} & \frac{1}{2} & \frac{1}{2} & \frac{3}{4} & \frac{3}{4} & \frac{3}{4} & \frac{1}{2} & \frac{1}{2} & \frac{1}{2} & \frac{1}{2} & \frac{1}{2} \\
\frac{1}{2} & \frac{9}{10} & \frac{9}{10} & \frac{1}{10} & \frac{9}{10} & \frac{9}{10} & \frac{1}{10} & \frac{9}{10} & \frac{9}{10} & \frac{9}{10} & \frac{9}{10}
\end{array}
\right)
$$

The probability that sequence x was generated by the path π, given the model $\mathcal{M}$, is

$$
P(x|\pi) = P(\pi_0 \rightarrow \pi_1) \cdot \prod_{i=1}^{n} P(x_i|\pi_i)P(\pi_i \rightarrow \pi_{i+1}) = a_{\pi_0,\pi_1} \cdot \prod_{i=1}^{n} e_{\pi_i}(x_i) \cdot a_{\pi_i,\pi_{i+1}}.
$$

---

5. We have added a fictitious term, $P(\pi_0 \rightarrow \pi_1) = \frac{1}{2}$ to model the initial condition: the dealer is equally likely to have either a fair or a biased coin before the first flip.

For convenience, we have introduced π_0 and π_{n+1} as the fictitious initial and terminal states *begin* and *end*.

This model defines the probability $P(x|\pi)$ for a given sequence x and a given path π. Since only the dealer knows the real sequence of states π that emitted x, we say that π is *hidden* and attempt to solve the following Decoding problem:

Decoding Problem:
Find an optimal hidden path of states given observations.

    Input: Sequence of observations $x = x_1 \ldots x_n$ generated by an HMM $\mathcal{M}(\sum, Q, A, E)$.

    Output: A path that maximizes $P(x|\pi)$ over all possible paths π.

The Decoding problem is an improved formulation of the ill-defined Fair Bet Casino problem.

11.3 Decoding Algorithm

In 1967 Andrew Viterbi used an HMM-inspired analog of the Manhattan grid for the Decoding problem, and described an efficient dynamic programming algorithm for its solution. Viterbi's Manhattan is shown in figure 11.2 with every choice of $\pi_1, \ldots, \pi_n$ corresponding to a path in this graph. One can set the edge weights in this graph so that the product of the edge weights for path $\pi = \pi_1 \ldots \pi_n$ equals $P(x|\pi)$. There are $|Q|^2(n-1)$ edges in this graph with the weight of an edge from (k, i) to $(l, i + 1)$ given by $e_l(x_{i+1}) \cdot a_{kl}$. Unlike the alignment approaches covered in chapter 6 where the set of valid directions was restricted to south, east, and southeast edges, the Manhattan built to solve the decoding problem only forces the tourists to move in any eastward direction (e.g., northeast, east, southeast, etc.), and places no additional restrictions (fig. 11.3). To see why the length of the edge between the vertices (k, i) and $(l, i + 1)$ in the corresponding graph is given by $e_l(x_{i+1}) \cdot a_{kl}$, one should compare $p_{k,i}$ [the probability of a path ending in vertex (k, i)] with the probability

$$p_{l,i+1} = \prod_{j=1}^{i+1} e_{\pi_j}(x_j) \cdot a_{\pi_{j-1},\pi_j}$$

$$= \left(\prod_{j=1}^{i} e_{\pi_j}(x_j) \cdot a_{\pi_{j-1},\pi_j} \right) \cdot (e_{\pi_{i+1}}(x_{i+1}) \cdot a_{\pi_i,\pi_{i+1}})$$

$$= p_{k,i} \cdot e_l(x_{i+1}) \cdot a_{kl}$$

$$= p_{k,i} \cdot \text{weight of edge from } (k,i) \text{ to } (l,i+1)$$

Therefore, the decoding problem is reduced to finding a longest path in the directed acyclic graph (DAG) shown in figure 11.2, which poses no problems to the algorithm presented in chapter 6. We remark that, in this case, the length of the path is defined as the product of its edges' weights, rather than the sum of weights used in previous examples of dynamic programming algorithms, but the application of logarithms makes the problems the same.

The idea behind the Viterbi algorithm is that the optimal path for the $(i+1)$-prefix $x_1 \ldots x_{i+1}$ of x uses a path for $x_1 \, x_2 \cdots x_i$ that is optimal among the paths ending in some unknown state π_i. Let k be some state from Q, and let i be between 1 and n. Define $s_{k,i}$ to be the probability of the most likely path for the prefix $x_1 \ldots x_i$ that ends at state k. Then, for any state l,

$$s_{l,i+1} = \max_{k \in Q} \{ s_{k,i} \cdot \text{weight of edge between } (k,i) \text{ and } (l,i+1) \}$$

$$= \max_{k \in Q} \{ s_{k,i} \cdot a_{kl} \cdot e_l(x_{i+1}) \}$$

$$= e_l(x_{i+1}) \cdot \max_{k \in Q} \{ s_{k,i} \cdot a_{kl} \}$$

We initialize $s_{begin,0} = 1$ and $s_{k,0} = 0$ for $k \neq begin$. If π^* is an optimal path, then the value of $P(x|\pi^*)$ is

$$P(x|\pi^*) = \max_{k \in Q} \{ s_{k,n} \cdot a_{k,end} \}$$

As in chapter 6, these recurrence relations and the initial conditions determine the entire dynamic programming algorithm, so we do not provide pseudocode here.

The Viterbi algorithm runs in $O(n|Q|^2)$ time. The computations in the Viterbi algorithm are usually done using logarithmic scores $S_{k,i} = \log s_{k,i}$

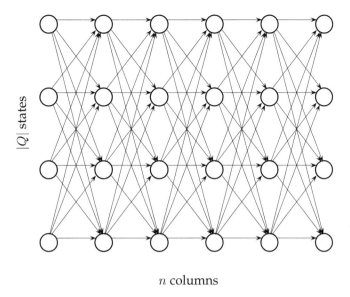

Figure 11.2   Manhattan, according to Viterbi, consists of $|Q|$ rows, n columns, and $|Q|^2$ edges per layer ($|Q| = 4$ and $n = 6$ in the example above).

to avoid overflow[6]:

$$S_{l,i+1} = \log e_l(x_{i+1}) + \max_{k \in Q}\{S_{k,i} + \log(a_{kl})\}.$$

As we showed above, every path π through the graph in figure 11.2 has probability $P(x|\pi)$. The Viterbi algorithm is essentially a search through the space of all possible paths in that graph for the one that maximizes the value of $P(x|\pi)$.

We can also ask a slightly different question: given x and the HMM, what is the probability $P(\pi_i = k|x)$ that the HMM was in state k at time i? In the casino analogy, we are given a sequence of coin tosses and are interested in the probability that the dealer was using a biased coin at a particular time.

We define $P(x) = \sum_{\pi} P(x|\pi)$ as the sum of probabilities of all paths and $P(x, \pi_i = k) = \sum_{\text{all } \pi \text{ with } \pi_i = k} P(x|\pi)$ as the sum of probabilities of all

---

6. Overflow occurs in real computers because there are only a finite number of bits (binary digits) in which to hold a number.

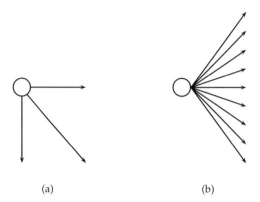

(a)   (b)

Figure 11.3 The set of valid directions in the alignment problem (a) is usually limited to south, east, and southeast edges, while the set of valid directions in the decoding problem (b) includes any eastbound edge.

paths with $\pi_i = k$. The ratio $\frac{P(x, \pi_i = k)}{P(x)}$ defines the probability $P(\pi_i = k|x)$ that we are trying to compute.

A simple variation of the Viterbi algorithm allows us to compute the probability $P(x, \pi_i = k)$. Let $f_{k,i}$ be the probability of emitting the prefix $x_1 \ldots x_i$ and reaching the state $\pi_i = k$. It can be expressed as follows.

$$f_{k,i} = e_k(x_i) \cdot \sum_{l \in Q} f_{l,i-1} \cdot a_{lk}$$

The only difference between the *forward algorithm* that calculates $f_{k,i}$ and the Viterbi algorithm is that the "max" sign in the Viterbi algorithm changes into a "$\sum$" sign in the forward algorithm.

However, forward probability $f_{k,i}$ is not the only factor affecting $P(\pi_i = k|x)$. The sequence of transitions and emissions that the HMM undergoes between π_{i+1} and π_n also affects $P(\pi_i = k|x)$. The backward probability $b_{k,i}$ is defined as the probability of being at state $\pi_i = k$ and emitting the *suffix*

$x_{i+1} \ldots x_n$. The *backward algorithm* uses a similar recurrence:

$$b_{k,i} = \sum_{l \in Q} e_l(x_{i+1}) \cdot b_{l,i+1} \cdot a_{kl}$$

Finally, the probability that the dealer had a biased coin at moment i is given by

$$P(\pi_i = k | x) = \frac{P(x, \pi_i = k)}{P(x)} = \frac{f_k(i) \cdot b_k(i)}{P(x)}.$$

11.4 HMM Parameter Estimation

The preceding analysis assumed that we know the state transition and emission probabilities of the HMM. Given these parameters, it is easy for an intelligent gambler to figure out that the dealer in the Fair Bet Casino is using a biased coin, simply by noticing that 0 and 1 have different expected frequencies ($\frac{3}{8}$ vs $\frac{5}{8}$). If the ratio of zeros to ones in a daylong sequence of tosses is suspiciously low, then it is likely that the dealer is using a biased coin. Unfortunately, the most difficult problem in the application of HMMs is that the HMM parameters are usually unknown and therefore need to be estimated from data. It is more difficult to estimate the transition and emission probabilities of an HMM than it is to reconstruct the most probable sequence of states it went through when you do know the probabilities. In this case, we are given the set of states, Q, but we do not know with what probability the HMM moves from one state to another, or with what probability it emits any particular symbol.

Let Θ be a vector combining the unknown transition and emission probabilities of the HMM $\mathcal{M}$. Given an observed symbol string x that the HMM emitted, define $P(x|\Theta)$ as the maximum probability of x given the assignment of parameters Θ. Our goal is to find

$$\max_{\Theta} P(x|\Theta).$$

Usually, instead of a single string x, we can obtain a sample of *training sequences* $x^1, \ldots, x^m$, so a natural goal is to find

$$\max_{\Theta} \prod_{j=1}^{m} P(x^j|\Theta).$$

This results in a difficult optimization problem in the multidimensional parameter space Θ. Commonly used algorithms for this type of parameter optimization are heuristics that use local improvement strategies. If we know the path $\pi_1 \ldots \pi_n$ corresponding to the observed states $x_1 \ldots x_n$, then we can scan the sequences and compute empirical estimates for transition and emission probabilities. If A_{kl} is the number of transitions from state k to l and $E_k(b)$ is the number of times b is emitted from state k, then the reasonable estimators are

$$a_{kl} = \frac{A_{kl}}{\sum_{q \in Q} A_{kq}} \qquad e_k(b) = \frac{E_k(b)}{\sum_{\sigma \in \sum} E_k(\sigma)}.$$

However, we do not usually know the state sequence $\pi = \pi_1 \ldots \pi_n$, and in this case we can start from a wild guess for $\pi_1 \ldots \pi_n$, compute empirical estimates for transition and emission probabilities using this guess, and solve the decoding problem to find a new, hopefully less wild, estimate for π. The commonly used iterative local improvement strategy, called the *Baum-Welch* algorithm, uses a similar approach to estimate HMM parameters.

11.5   Profile HMM Alignment

Given a family of functionally related biological sequences, one can search for new members of the family from a database using pairwise alignments between family members and sequences from the database. However, this approach may fail to identify distantly related sequences because distant cousins may have weak similarities that do not pass the statistical significance test. However, if the sequence has weak similarities with many family members, it is likely to belong to the family. The problem then is to somehow align a sequence to *all* members of the family at once, using the whole set of functionally related sequences in the search.

The simplest representation of a family of related proteins is given by their multiple alignment and the corresponding profile.[7] As with sequences, profiles can also be compared and aligned against each other since the dynamic programming algorithm for aligning two sequences works if both of the input sequences are profiles.

---

7. While in chapter 4 we defined profile element p_{ij} as a *count* of the nucleotide i in the jth column of alignment matrix, biologists usually define p_{ij} as *frequency* of the nucleotide i in the jth column of the alignment matrix, that is, they divide all of the counts by t (see figure 11.4). In order to avoid columns that contain one or more letters with probabilities of 0, small numbers

A	.72	.14	0	0	.72	.72	0	0
T	.14	.72	0	0	0	.14	.14	.86
G	.14	.14	.86	.44	0	.14	0	0
C	0	0	.14	.56	.28	0	.86	.14

Figure 11.4   A profile represented in terms of frequencies of nucleotides.

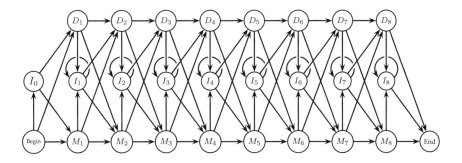

Figure 11.5   A profile HMM.

HMMs can also be used for sequence comparison, in particular for aligning a sequence against a profile. The simplest HMM for a profile P contains n sequentially linked *match* states $M_1, \ldots, M_n$ with emission probabilities $e_i(a)$ taken from the profile P (fig. 11.5). The probability of a string $x_1 \ldots x_n$ given the profile P is $\prod_{i=1}^{n} e_i(x_i)$. To model insertions and deletions we add *insertion* states $I_0, \ldots, I_n$ and *deletion* states $D_1, \ldots, D_n$ to the HMM and assume that

$$e_{I_j}(a) = p(a),$$

where $p(a)$ is the frequency of the occurrence of the symbol a in all the sequences. The transition probabilities between matching and insertion states can be defined in the affine gap penalty model by assigning a_{MI}, a_{IM}, and a_{II} in such a way that $\log(a_{MI}) + \log(a_{IM})$ equals the gap initiation penalty

called *pseudocounts* can be added. We will not do so here, however.

and $\log(a_{II})$ equals the gap extension penalty. The (silent) deletion states do not emit any symbols.

Define $v_j^M(i)$ as the logarithmic likelihood score of the best path for matching $x_1 \ldots x_i$ to profile HMM ending with x_i emitted by the state M_j. Define $v_j^I(i)$ and $v_j^D(i)$ similarly. The resulting dynamic programming recurrence for the decoding problem is very similar to the standard alignment recurrence:

$$v_j^M(i) = \log \frac{e_{M_j}(x_i)}{p(x_i)} + \max \left\{ \begin{array}{l} v_{j-1}^M(i-1) + \log(a_{M_{j-1},M_j}) \\ v_{j-1}^I(i-1) + \log(a_{I_{j-1},M_j}) \\ v_{j-1}^D(i-1) + \log(a_{D_{j-1},M_j}) \end{array} \right.$$

The values $v_j^I(i)$ and $v_j^D(i)$ are defined similarly:

$$v_j^I(i) = \log \frac{e_{I_j}(x_i)}{p(x_i)} + \max \left\{ \begin{array}{l} v_j^M(i-1) + \log(a_{M_j,I_j}) \\ v_j^I(i-1) + \log(a_{I_j,I_j}) \\ v_j^D(i-1) + \log(a_{D_j,I_j}) \end{array} \right.$$

$$v_j^D(i) = \max \left\{ \begin{array}{l} v_{j-1}^M(i) + \log(a_{M_{j-1},D_j}) \\ v_{j-1}^I(i) + \log(a_{I_{j-1},D_j}) \\ v_{j-1}^D(i) + \log(a_{D_{j-1},D_j}) \end{array} \right.$$

Figure 11.6 shows how a path in the edit graph gives instructions on how to traverse a path in the profile HMM. The path in the edit graph can be coded in a three-letter alphabet, namely, DDDVDDHDHV, where D, H, and V denote the diagonal, horizontal, and vertical edges respectively. The sequence DDDHDDVDVH serves as an instruction for moving in a three-layer profile HMM graph, where the HMM moves between two match states when it encounters a D, and switches to either an insertion or deletion state when it encounters a V or an M, respectively.

11.6   Notes

Although the roots of HMM theory can be traced back to the 1950s, the first practical applications of HMMs had to wait until Andrew Viterbi and Leonard Baum and colleagues developed algorithms for HMM decoding and parameter estimation in the late 1970s (106; 9). Pierre Baldi, Gary Churchill, David Haussler and their colleagues pioneered the application of HMMs in computational biology (8; 22; 45; 59). Profile HMM alignments were later

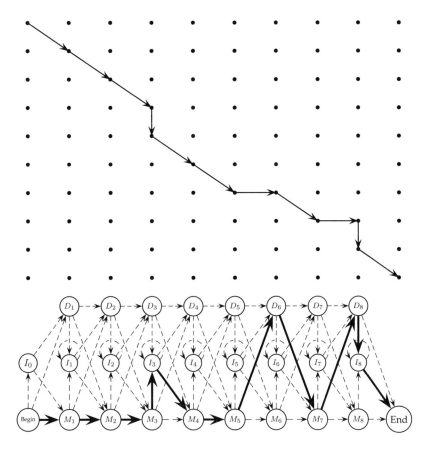

Figure 11.6 A path through an edit graph, and the corresponding path through a profile HMM.

used for developing *Pfam*, a database of protein domain families (99) and in many other applications.

David Haussler (born October 1953 in California) currently holds the University of California Presidential Chair in Computer Science at the University of California at Santa Cruz (UCSC). He is also an investigator with the Howard Hughes Medical Institute. He was a pioneer in the application of machine learning techniques to bioinformatics and he has played a key role in sequencing the human genome.

While an undergraduate studying mathematics, Haussler worked for his brother Mark in a molecular biology laboratory at the University of Arizona. David has fond memories of this time:

> We extracted vitamin D hormone receptors from the intestines of chicks that were deprived of vitamin D and used the extract to study the level of vitamin D in human blood samples. My jobs were to sacrifice the chicks, extract the vitamin D receptors from their guts, perform the assay with them, and finally do the mathematical analysis on the results of these experiments. The work was quite successful, and led to a publication in *Science*. But it was there that I decided that I was more fond of mathematics than I was of molecular biology.

After sacrificing many chicks, David decided to pursue his doctorate at the University of Colorado to study with Professor Andrzej Ehrenfeucht in the Department of Computer Science. Excited about the interaction between computation and logic, he recognized that Ehrenfeucht was one of the leaders in that area and sought him out as an advisor. While his early papers were in various areas of mathematics he participated in a discussion group organized by Ehrenfeucht that was dominated by discussions about DNA. This was at a time in the early 1980s when the first complete sequences from a few viruses had become available. Two other students in this group went directly on to careers in bioinformatics: Gene Myers, who put together the human genome assembly for Celera Genomics, and Gary Stormo who did pioneering work on motif finding. While Haussler did produce a few papers in bioinformatics in this period, at the time he felt there were not enough data to sustain an entire field of bioinformatics. So he remained rather aloof from the field, waiting for technological advances that would allow it to take off. Haussler instead followed another interest—the study of artificial intelligence—because he wanted to try to understand how the brain works. He became involved with building artificial neural networks and designed adaptive computer algorithms that can improve their performance as they encounter more data. The study of adaptation and learning theory led him into HMMs.

By the early 1990s, molecular biologists had begun to churn out data much more rapidly. Haussler's interest in molecular biology was rekindled, and he switched his scientific goal from understanding how the brain works to understanding how cells work. He began to apply the same types of models used for speech and formal grammar analysis to the biological sequences, providing the foundation for further work along these lines by many other scientists in the field. In particular, his group developed HMM approaches to gene prediction and protein classification. The HMM vocabulary is rich enough to allow a bioinformaticist to build a model that captures much of the quirkiness of actual DNA. So these models caught on.

The HMM aproach did not appear from nowhere. Foundational work by David Sankoff, Michael Waterman, and Temple Smith had formalized the dynamic programming methods to align biosequences. David Searls had made the analogy between biosequences and the strings produced by a formal grammar. Gary Stormo and Chip Lawrence had introduced key probabilistic ideas to the problem of sequence classification and motif finding. HMMs were just begging to be introduced into the field since they combined all these things in a simple and natural framework.

Haussler began applying HMMs to biosequence analysis when Anders Krogh joined his group as a postdoc. Both were familiar with HMMs, and Haussler still maintained an interest in DNA and protein sequences. One day they began to talk about the crazy idea that HMMs might make good models for protein sequences, and after a quick examination of the literature, they were surprised to find that the young field of bioinformatics was then ripe with the beginnings of this idea, but no one had really put all the pieces together and exploited it. So they dove in and built HMMs to recognize different protein families, demonstrating how this methodology could unify the dynamic programming, grammatical, and maximum-likelihood methods of statistical inference that were then becoming popular in bioinformatics. David Haussler says:

> The epiphany that Anders and I had about HMMs for protein sequences

was certainly one of the most exciting moments in my career. However, I would have to say that the most exciting moment came later, when my group participated in efforts to sequence the human genome, and Jim Kent, then a graduate student at UCSC, was the first person able to computationally assemble the public sequencing project's data to form a working draft. We raced with Gene Myers' team at Celera as they assembled their draft genome. On July 7, 2000, shortly after Francis Collins and Craig Venter jointly announced in a White House ceremony that their two teams had successfully assembled the human genome, we released the public working draft onto the World Wide Web. That moment on July 7, when the flood of A, C, T, and G of the human genome sequence came across my computer screen, passing across the net as they were to thousands of other people all over the world, was the most exciting moment in my scientific life. For me, it was symbolic of the entire Human Genome Project, both public and private, and the boundless determination of the many scientists involved. It seems unthinkable that out of a primordial soup of organic molecules, life forms would eventually evolve whose cells would carry their vital messages forward in DNA from generation to generation, ever diversifying and expanding their abilities as this message became more complex, until one day one of these species would accumulate the facility to decode and share its own DNA message. Yet that day had arrived.

Haussler's approach to research has always been at the extreme interdisciplinary end of the spectrum. Most of his insights have come from the application of perspectives of one field to problems encountered in another.

I've tried to stay focused on the big scientific questions, and never limit my approach to conform to the norms of any well-established and narrow method of inquiry. I try to always listen carefully to the few best scientists in all fields, rather than all the scientists in one field.

Haussler thinks that one key to discovery is picking the right problem. Important scientific problems become ripe at particular times. Before that they are unapproachable because the foundation required for their solution has not been laid. After that they are no longer as important because the heart of the problem has already been solved. Knowing when a scientific problem is ripe for solution is a difficult art, however. Breadth of focus helps. Luck doesn't hurt either. Haussler says:

We have not even really begun to understand how a cell works, so there are plenty of interesting problems left for bioinformaticians. Computational models for this problem probably won't look anything like what we have today. Problems associated with understanding how cells form organs, bodies, and minds that function as they do is also likely to keep anyone from being bored in this field for quite some time. Applying what we learn to the practice of medicine will pose further challenging problems. However, much foundational works remains to be done before we can do justice to the loftiest of these problems. One important piece of foundational work where bioinformatics will play the key role is in understanding the evolutionary history of entire genomes. With the full genome sequences of many different species, we can begin to try to reconstruct the key events in the evolution of our own genome using new comparative genomics approaches. This will require more than raw genome data. It will demand the development of new mathematics and algorithms appropriate to the task. People have always been fascinated with origins, and the mystery of our own origins has been paramount among these fascinations. Thus, I anticipate no lack of interest in such investigations.

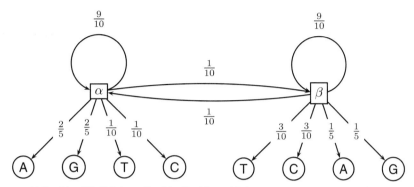

Figure 11.7   The HMM described in Problem 11.4.

11.7   Problems

Problem 11.1

The Decoding problem can be formulated as a longest path problem in a DAG. This motivates a question about a space-efficient version of the decoding algorithm. Does there exist a linear-space algorithm for the decoding problem?

Problem 11.2

To avoid suspicion, the dealer in the Fair Bet Casino keeps every coin for at least ten tosses, regardless of whether it is fair or biased. Describe both the corresponding HMM and how to modify the decoding algorithm for this case.

Problem 11.3

Suppose you are given the dinucleotide frequences in CG-islands and the dinucleotide frequencies outside CG-islands. Design an HMM for finding CG-islands in genomic sequences.

Problem 11.4

Figure 11.7 shows an HMM with two states α and β. When in the α state, it is more likely to emit purines (A and G). When in the β state, it is more likely to emit pyrimidines (C and T). Decode the most likely sequence of states (α/β) for sequence GGCT. Use log-scores, rather than straight probability scores.

Problem 11.5

Suppose a dishonest dealer has two coins, one fair and one biased; the biased coin has heads probability $1/4$. Assume that the dealer never switches the coins. Which coin is more likely to have generated the sequence HTTTHHHTTTTHTHHTT? It may be useful to know that $\log_2(3) = 1.585$

Problem 11.6

Consider a different game where the dealer is not flipping a coin, but instead rolling a three-sided die with labels 1, 2, and 3. (Try not to think about what a three-sided die might look like.) The dealer has two loaded dice D_1 and D_2. For each die D_i, the probability of rolling the number i is $1/2$, and the probability of each of the other two outcomes is $1/4$. At each turn, the dealer must decide whether to (1) keep the same die, (2) switch to the other die, or (3) end the game. He chooses (1) with probability $1/2$ and each of the others with probability $1/4$. At the beginning the dealer chooses one of the two dice with equal probability.

- Give an HMM for this situation. Specify the alphabet, the states, the transition probabilities, and the emission probabilities. Include a start state *start*, and assume that the HMM begins in state *start* with probability 1. Also include an end state *end*.

- Suppose that you observe the following sequence of die rolls: 1 1 2 1 2 2. Find a sequence of states which best explains the sequence of rolls. What is the probability of this sequence? Find the answer by completing the Viterbi table. Include backtrack arrows in the cells so you can trace back the sequence of states. Some of the following facts may be useful:

  $\log_2(0) = -\infty$
  $\log_2(1/4) = -2$
  $\log_2(1/2) = -1$
  $\log_2(1) = 0$

- There are actually two optimal sequences of states for this sequence of die rolls. What is the other sequence of states?

12 *Randomized Algorithms*

Randomized algorithms make random decisions throughout their operation. At first glance, making random decisions does not seem particularly helpful. Basing an algorithm on random decisions sounds like a recipe for disaster, but an eighteenth-century French naturalist, Comte de Buffon, proved the opposite by developing an algorithm to accurately compute π by randomly dropping needles on a sheet of paper with parallel lines. The fact that a randomized algorithm undertakes a nondeterministic sequence of operations often means that, unlike deterministic algorithms, no input can reliably produce worst-case results. Randomized algorithms are often used in hard problems where an exact, polynomial-time algorithm is not known. In this chapter we will see how randomized algorithms solve the Motif Finding problem.

12.1 The Sorting Problem Revisited

The QUICKSORT algorithm below is another fast and simple sorting technique. It selects an element m (typically, the first) from an array $\mathbf{c}$ and simply partitions the array into two subarrays: $\mathbf{c}_{small}$, containing all elements from $\mathbf{c}$ that are smaller than m; and $\mathbf{c}_{large}$ containing all elements larger than m. This partitioning can be done in linear time, and by following a divide-and-conquer strategy, QUICKSORT recursively sorts each subarray in the same way. The sorted list is easily created by simply concatenating the sorted $\mathbf{c}_{small}$, element m, and the sorted $\mathbf{c}_{large}$.

QuickSort($\mathbf{c}$)
1   **if** $\mathbf{c}$ consists of a single element
2   **return c**
3   $m \leftarrow c_1$
4   Determine the set of elements $\mathbf{c}_{small}$ smaller than m
5   Determine the set of elements $\mathbf{c}_{large}$ larger than m
6   QuickSort($\mathbf{c}_{small}$)
7   QuickSort($\mathbf{c}_{large}$)
8   Combine $\mathbf{c}_{small}$, m, and $\mathbf{c}_{large}$ into a single sorted array $\mathbf{c}_{sorted}$
9   **return $\mathbf{c}_{sorted}$**

It turns out that the running time of QuickSort depends on how lucky we are with our selection of the element m. If we happen to choose m in such a way that $\mathbf{c}$ is split into even halves (i.e., $|\mathbf{c}_{small}| = |\mathbf{c}_{large}|$), then

$$T(n) = 2T(n/2) + an,$$

where $T(n)$ represents the time taken by QuickSort to sort an array of n numbers, and an represents the time required to split the array of size n into two parts; a is a positive constant. This is exactly the same recurrence we saw in MergeSort and we already know that it leads to $O(n \log n)$ running time. However, if we choose m in such a way that it splits $\mathbf{c}$ unevenly (e.g., an extreme case occurs when $\mathbf{c}_{small}$ is empty and $\mathbf{c}_{large}$ has $n - 1$ elements), then the recurrence looks like

$$T(n) = T(n - 1) + an$$

This is exactly the recurrence we saw for SelectionSort and we already know that it leads to $O(n^2)$ running time, something we want to avoid. Indeed, QuickSort takes quadratic time to sort the array $(n, n - 1, \ldots, 2, 1)$. Worse yet, it requires $O(n^2)$ time to process $(1, 2, \ldots, n - 1, n)$, which seems unnecessary since the array is already sorted.

The QuickSort algorithm so far seems like a bad imitation of MergeSort. However, if we can choose a good "splitter" m that breaks an array into two equal parts, we might improve the running time. To achieve $O(n \log n)$ running time, it is not actually necessary to find a perfectly equal (50/50) split. For example, a split into approximately equal parts of size, say, 51/49 will also work. In fact, one can prove that the algorithm will achieve $O(n \log n)$ running time as long as the sets $\mathbf{c}_{small}$ and $\mathbf{c}_{large}$ are both larger in size than $\frac{n}{4}$, which implies that, of n possible choices for m, at least $\frac{3}{4}n - \frac{1}{4}n = \frac{1}{2}n$ of them make good splitters! In other words, if we choose m uniformly at random (i.e., every element of **c** has the same probability to be chosen), there is at least a 50% chance that it will be a good splitter. This observation motivates the following randomized algorithm:

RANDOMIZEDQUICKSORT(**c**)
1   **if** **c** consists of a single element
2   **return c**
3   Choose element m uniformly at random from **c**
4   Determine the set of elements $\mathbf{c}_{small}$ smaller than m
5   Determine the set of elements $\mathbf{c}_{large}$ larger than m
6   RANDOMIZEDQUICKSORT($\mathbf{c}_{small}$)
7   RANDOMIZEDQUICKSORT($\mathbf{c}_{large}$)
8   Combine $\mathbf{c}_{small}$, m, and $\mathbf{c}_{large}$ into a single sorted array $\mathbf{c}_{sorted}$
9   **return** $\mathbf{c}_{sorted}$

RANDOMIZEDQUICKSORT is a very fast algorithm in practice but its worst-case running time remains $O(n^2)$ since there is still a possibility that it selects bad splitters. Although the behavior of a randomized algorithm varies on the same input from one execution to the next, one can prove that its *expected* running time is $O(n \log n)$.[1]

The key advantage of randomized algorithms is performance: for many practical problems randomized algorithms are faster (in the sense of expected running time) than the best known deterministic algorithms. Another attractive feature of randomized algorithms, as illustrated by RANDOMIZEDQUICK-SORT and other algorithms in this chapter, is their simplicity.

We emphasize that RANDOMIZEDQUICKSORT, despite making random decisions, always returns the correct solution of the sorting problem. The only variable from one run to another is its running time, not the result. In contrast, other randomized algorithms we consider in this chapter usually produce incorrect (or, more gently, approximate) solutions. Randomized algorithms that always return correct answers are called *Las Vegas* algorithms, while algorithms that do not are called *Monte Carlo* algorithms. Of course, computer scientists prefer Las Vegas algorithms to Monte Carlo algorithms but the former are often difficult to come by. Although for some applications

---

1. The running time of a randomized algorithm is a random variable, and computer scientists are often interested in the mean value of this random variable. This is referred to as the *expected running time*.

Monte Carlo algorithms are not appropriate (when approximate solutions are of no value), they have been popular in different applications for over a hundred years and often provide good approximations to optimal solutions.

12.2   Gibbs Sampling

In 1993, Chip Lawrence and colleagues suggested using *Gibbs sampling* to find motifs in DNA sequences. Given a set of t sequences that are each n nucleotides long, and an integer l, the Gibbs sampler attempts to solve the Motif Finding problem that was presented in chapter 4, that is, to find an l-mer in each of t sequences such that the similarity between these l-mers is maximized.[2] Let $\mathbf{s} = (s_1, \ldots, s_t)$ be the starting positions of the chosen l-mers in t sequences. These substrings form a $t \times l$ alignment matrix and the corresponding $4 \times l$ profile $\mathbf{P}(\mathbf{s}) = (p_{ij})$.[3]

Given a profile $\mathbf{P}$ and an arbitrary l-mer $\mathbf{a} = a_1 a_2 \cdots a_l$, consider the quantity $P(\mathbf{a}|\mathbf{P}) = \prod_{i=1}^{l} p_{a_i,i}$, the probability that $\mathbf{a}$ was generated by $\mathbf{P}$. l-mers similar to the consensus string of the profile will have higher probabilities while dissimilar l-mers will have lower probabilities. For example, for the profile $\mathbf{P}$ in figure 11.4, which has consensus string ATGCAACT, $P(\mathsf{ATGCAACT}|\mathbf{P}) = 9.6 \times 10^{-2}$, while $P(\mathsf{TACGCGTC}|\mathbf{P}) = 9.3 \times 10^{-7}$. Given a profile $\mathbf{P}$, one can thus evaluate the probability of every l-mer in sequence i and to find the l-mer that was most likely to have been generated by $\mathbf{P}$—this l-mer will be called the $\mathbf{P}$-most probable l-mer in the sequence. This motivates the following GREEDYPROFILEMOTIFSEARCH algorithm for the Motif Finding problem:

GREEDYPROFILEMOTIFSEARCH(DNA, t, n, l)
1   Randomly select starting positions $\mathbf{s} = (s_1, \ldots, s_t)$ in DNA
2   Form profile $\mathbf{P}$ from $\mathbf{s}$
3    $bestScore \leftarrow 0$
4   **while** $Score(\mathbf{s}, DNA) > bestScore$
5         $bestScore \leftarrow Score(\mathbf{s}, DNA)$
6   **for** $i \leftarrow 1$ **to** t
7   Find a $\mathbf{P}$-most probable l-mer $\mathbf{a}$ from the ith sequence
8              $s_i \leftarrow$ starting position of $\mathbf{a}$
9   **return** $bestScore$

---

2. We deliberately do not specify here which particular objective function the Gibbs sampling algorithm optimizes.
3. As in chapter 11, we view a profile as the frequency of letters in an alignment, rather than the count of letters. Thus, the each column in the profile matrix forms a *probability distribution*.

GREEDYPROFILEMOTIFSEARCH starts from a *random seed* by selecting starting positions s uniformly at random, and attempts to improve on it using a greedy strategy. Since the computational space of starting positions is huge, randomly selected seeds will rarely come close to an optimal motif, and there is little chance that random seeds will be able to guide us to the optimal solution via the greedy strategy. Thus, GREEDYPROFILEMOTIFSEARCH is typically run a large number of times with the hope that one of these thousands of runs will generate a seed close to the optimum simply by chance. Needless to say, GREEDYPROFILEMOTIFSEARCH is unlikely to stumble on the optimal motif.

GREEDYPROFILEMOTIFSEARCH changes starting positions $(s_1, s_2, \ldots, s_t)$ between every iteration, and may change as many as all t positions in a single iteration. Gibbs sampling is an iterative procedure that at each iteration discards one l-mer from the alignment and replaces it with a new one. In other words, it changes at most one position in s in each iteration and thus moves with more caution in the space of all starting positions. Like GREEDYPRO-FILEMOTIFSEARCH, Gibbs sampling starts with randomly choosing l-mers in each of t DNA sequences but makes a random, rather than a greedy, choice at every iteration.

1. Randomly select starting positions $s = (s_1, \ldots, s_t)$ in DNA and form the set of l-tuples starting at these positions.

2. Randomly choose one sequence out of t DNA sequences.

3. Create a profile **P** from the l-mers in the remaining $t - 1$ sequences.

4. For each position i in the chosen DNA sequence, calculate the probability p_i that the l-mer starting at this position is generated by profile **P** ($1 \leq i \leq n - l + 1$).

5. Choose the new starting position in the chosen DNA sequence randomly, according to the distribution proportional to $(p_1, p_2, \ldots, p_{n-l+1})$.

6. Repeat until convergence.[4]

Although Gibbs sampling works well in many cases, it may converge to a local optimum rather than a global one, particularly for difficult search problems with subtle motifs. Motif finding becomes particularly difficult if the nucleotide distribution in the sample is skewed, that is, if some nucleotides in

---

4. We deliberately do not specify when the algorithm is considered to have converged.

the sample are more frequent than others. In this case, searching for a signal with the maximum number of occurrences may lead to patterns composed from the most frequent nucleotides that may not be biologically significant. For example, if A has a frequency of 70% and T, G, and C have frequencies of 10%, then poly(A) may be the most frequent motif, perhaps disguising the biologically relevant motif.

To find motifs in biased samples, some algorithms use *relative entropy* to highlight the motif among the patterns composed from frequent nucleotides. Given a profile of length l, the relative entropy is defined as

$$\sum_{j=1}^{l} \sum_{r \in \{A,T,G,C\}} p_{rj} \log_2 \frac{p_{rj}}{b_r},$$

where p_{rj} is the frequency of nucleotide r in position j of the alignment and b_r is the background frequency of r. Gibbs sampling can be adjusted to work with relative entropies.

12.3   Random Projections

The RANDOMPROJECTIONS algorithm is another randomized approach to motif finding. If an l-long pattern is "implanted" in DNA sequences without mutations, then motif finding simply reduces to counting the number of occurrences of l-mers from the sample (the most common one reveals the implanted pattern). However, motif finding becomes very difficult when the pattern is implanted with mutations. Could we possibly reveal mutated patterns in the same way that we deal with nonmutated patterns, that is, by considering the positions that were not mutated? For example, in any instance of a pattern of length 8 with two mutated positions, six positions remain unaffected by mutations and it is tempting to use these six positions as a basis for motif finding.

There are two complications with this approach. First, the six constant positions do not necessarily form a contiguous string: the mutated nucleotides may be in positions 3 and 7, leaving the other six conserved positions to form a *gapped* pattern. Second, different instances of the pattern can mutate in different positions. For example, three different instances of the mutated pattern may differ in positions 3 and 7, 3 and 6, and 2 and 6 respectively. The key observation is that although the three mutated patterns[5]

5. Here we use the notation "*" to represent a position that is conserved, and "X" to represent a

X*X*, **X**X**, *X***X** are all different, their "consensus" gapped pattern *XX**XX* remains unaffected by mutations. If we knew which gapped pattern was unaffected by mutations we could use it for motif search as if it were an implanted gap pattern without mutations.[6]

The problem, however is that the locations of the four unaffected positions in *XX**XX* are unknown. To bypass this problem, RANDOMPROJECTION tries different randomly selected sets of k (out of l) positions to reveal the original implanted pattern. These sets of positions are called *projections*.

We will define a (k, l)-*template* to be any set of k distinct integers $1 \leq t_1 < \cdots < t_k \leq l$. For a (k, l)-template $\mathbf{t} = (t_1, \ldots, t_k)$ and and an l-mer $\mathbf{a} = a_1, \ldots, a_l$, define $Projection(\mathbf{a}, \mathbf{t}) = a_{t_1}, a_{t_2} \ldots a_{t_k}$ to be the concatenation of nucleotides from $\mathbf{a}$ as defined by the template $\mathbf{t}$. For example, if $\mathbf{a}$=ATGCATT and $\mathbf{t} = (2, 5, 7)$, then $Projection(\mathbf{a}, \mathbf{t})$=TAT. The RANDOMPROJECTIONS algorithm, below, chooses a random (k, l)-template and projects every l-mer in the sample onto it; the resulting k-mers are recorded via a hash table. We expect that k-mers that correspond to projections of the implanted pattern appear more frequently than other k-mers. Therefore, k-mers that appear many times (e.g., whose count in the hash table is higher than a predefined threshold θ) as projections of l-mers from the sample are likely to represent projections of the implanted pattern. Of course, this is subject to noise and a single (k, l)-template does not necessarily reveal the implanted pattern. The RANDOMPROJECTIONS algorithm repeatedly selects a given number, m, of random (k, l)-template and aggregates the data obtained for all m iterations. As RANDOMPROJECTIONS chooses different random templates, the locations of the implanted pattern become clearer.

As compared to other motif finding algorithms, RANDOMPROJECTIONS requires additional parameters: k (the number of positions in the template), θ (the threshold that determines which bins in the hash table to consider after projecting all l-mers), and m (the number of chosen random templates). The RANDOMPROJECTIONS algorithm creates a table, **Bins**, of size 4^k such that every possible projection (k-mer) corresponds to a unique address in this table. For a given (k, l)-template $\mathbf{r}$, **Bins**($\mathbf{x}$) holds the count of l-mers $\mathbf{a}$ in DNA such that $Projection(\mathbf{a}, \mathbf{r}) = \mathbf{x}$.

---

position that can be mutated.

6. In real life, all l positions may be affected by a mutation in some instances of pattern. However, it is very likely that there is a relatively large set of instances that share the same nonmutated positions and the RANDOMPROJECTION algorithm below uses them to find motifs.

RANDOMPROJECTIONS($DNA, t, n, l, k, \theta, m$)
1   create a $t \times n$ array **motifs** and fill it with zeros
2   **for** m iterations
3   create a table **Bins** of size 4^k and fill it with zeros
4          $\mathbf{r} \leftarrow$ a random (k, l)-template.
5   **for** $i \leftarrow 1$ **to** t
6   **for** $j \leftarrow 1$ **to** $n - l + 1$
7                  $\mathbf{a} \leftarrow j$th l-mer in ith DNA sequence
8                  $\mathbf{Bins}(Projection(\mathbf{a}, \mathbf{r})) = \mathbf{Bins}(Projection(\mathbf{a}, \mathbf{r})) + 1$
9   **for** $i \leftarrow 1$ **to** t
10   **for** $j \leftarrow 1$ **to** $n - l + 1$
11                 $\mathbf{a} \leftarrow j$th l-mer in ith DNA sequence
12   **if** $\mathbf{Bins}(Projection(\mathbf{a}, \mathbf{r})) > \theta$
13                     $motifs_{i,j} \leftarrow motifs_{i,j} + 1$
14 **for** $i \leftarrow 1$ **to** t
15          $s_i \leftarrow$ Index of the largest element in row i of **motifs**.
16 **return** s

RANDOMPROJECTIONS provides no guarantee of returning the correct pattern, but one can prove that RANDOMPROJECTIONS returns the correct motif with *high probability*, assuming that the parameters are chosen in a sensible way. The main difference between this toy algorithm and the practical **Projection** algorithm developed by Jeremy Buhler and Martin Tompa is the way in which this algorithm evaluates the results from hashing all the l-mer projections. The method that we have presented here to choose $(s_1, s_2, \ldots, s_t)$ is crude, while the **Projection** algorithm uses a heuristic method that is harder to trick by accidentally large counts in the array **motifs**.[7]

12.4   Notes

QUICKSORT was discovered by Tony Hoare in 1962 (50). Gibbs sampling is a variation of Markov chain Monte Carlo methods that can be traced to the classic paper by Nicholas Metropolis and colleagues in 1953 (75). Charles Lawrence and colleagues pioneered applications of Gibbs sampling for motif finding in 1993 (63). The random projections algorithm was developed by Jeremy Buhler and Martin Tompa in 2001 (18).

---

7. Specifically, they use an Expectation Maximization algorithm, which is a local search technique used in many bioinformatics algorithms.

12.5   Problems

Problem 12.1

Show how to partition an array $\mathbf{c}$ into arrays $\mathbf{c}_{small}$ and $\mathbf{c}_{large}$ in QUICKSORT without requesting additional memory.

Problem 12.2

Prove that QUICKSORT takes linear time to sort an array if all splits are chosen in such a way that $\frac{1}{3} < \frac{|c_{large}|}{|c_{small}|} < 3$.

Problem 12.3

The Viterbi algorithm is a deterministic algorithm for solving the Decoding problem. Design a randomized algorithm for solving the Decoding problem that starts from a randomly chosen assignment of states and tries to improve it using coin tossing.

Problem 12.4

The k-Means clustering algorithm randomly selects an original partition into clusters and deterministically rearranges clusters afterward. Design a randomized version of the k-Means algorithm that uses coin tossing to rearrange clusters.

Problem 12.5

Design a randomized algorithm for solving the Large Parsimony problem that uses coin tossing to choose among the nearest neighbor interchanges available at every step.

Problem 12.6

The Gibbs sampler algorithm "moves slowly" in the space of all starting positions by changing starting position s_i in only one DNA sequence at every iteration. In contrast, GREEDYPROFILEMOTIFSEARCH "moves fast" and may change positions s_i in all DNA sequences. Describe a version of the Gibbs sampler that may change many positions at every iteration. Explain the advantages and disadvantages of your algorithm as compared to the Gibbs sampler described in the book.

Using Bioinformatics Tools (www.bioalgorithms.info)

No bioinformatics textbook would be complete without some application of the theory; we hope that you will actually use the algorithmic principles outlined in this text. To get you started, we have compiled a number of challenging and useful practical exercises. Each of these exercises involves using a computer and software to solve a biological question. Unfortunately, the tools and techniques that bioinformaticians use change so quickly that we could not hope to present these exercises in a static book form, so we have made them available through the book's website at

www.bioalgorithms.info

The exercises cover a range of bioinformatics problems including motif finding, sequence comparison, searching biological databases, applications of HMMs, gene expression analysis, among others.

A large amount of supporting information can be found from the same website. For example, we have made available many sets of lecture notes that professors and students alike can use during a bioinformatics course, a bulletin board to discuss sections of the book, a searchable index, and a glossary.

419

Bibliography

[1] A. Aho, J. Hopcroft, and J. Ullman. *Data Structures and Algorithms*. Addison-Wesley, Boston, 1983.

[2] A.V. Aho and M.J. Corasick. Efficient string matching: an aid to bibliographic search. *Communication of ACM*, 18:333–340, 1975.

[3] B. Alberts, D. Bray, J. Lewis, M. Raff, K. Roberts, and J. Watson. *Molecular Biology of the Cell*. Garland Publishing, New York, 1994.

[4] S. Altschul, W. Gish, W. Miller, E. Myers, and J. Lipman. Basic local alignment search tool. *Journal of Molecular Biology*, 215:403–410, 1990.

[5] S.F. Altschul, T.L. Madden, A.A. Schaffer, J. Zhang, Z. Zhang, W. Miller, and D.J. Lipman. Gapped BLAST and Psi-BLAST: A new generation of protein database search programs. *Nucleic Acids Research*, 25:3389–3402, 1997.

[6] V.L. Arlazarov, E. A. Dinic, M. A. Kronrod, and I. A. Faradzev. On economical construction of the transitive closure of an oriented graph. *Soviet Math. Dokl.*, 11:1209–1210, 1970.

[7] P. Baldi and S. Brunak. *Bioinformatics: The Machine Learning Approach*. MIT Press, Cambridge, MA, 1997.

[8] P. Baldi, Y. Chauvin, T. Hunkapiller, and M. McClure. Hidden Markov models of biological primary sequence information. *Proceedings of the National Academy of Sciences of the United States of America*, 91:1059–1063, 1994.

[9] L. Baum, T. Petrie, G. Soules, and N. Weiss. A maximization technique occurring in the statistical analysis of probabilistic functions of Markov Chains. *Annals of Mathematical Satistics*, 41:164–171, 1970.

[10] A. Baxevanis and B.F. Ouellette. *Bioinformatics: A Practical Guide to the Analysis of Genes and Proteins*. Wiley-Interscience, Hoboken, NJ, 1998.

[11] A. Ben-Dor, R. Shamir, and Z. Yakhini. Clustering gene expression patterns. *Journal of Computational Biology*, 6:281–297, 1999.

[12] S.M. Berget, C. Moore, and P.A. Sharp. Spliced segments at the 5′ terminus of adenovirus 2 late mRNA. *Proceedings of the National Academy of Sciences of the United States of America*, 74:3171–3175, 1977.

[13] P. Berman, S. Hannenhalli, and M. Karpinski. 1.375-approximation algorithm for sorting by reversals. In *European Symposium on Algorithms*, volume 2461 of *Lecture Notes in Computer Science*, pages 200–210, Rome, Italy, 2002. Springer-Verlag.

[14] M. Borodovsky and J. McIninch. Recognition of genes in DNA sequences with ambiguities. *BioSystems*, 30:161–171, 1993.

[15] P. Bourne and H. Weissig (eds). *Structural Bioinformatics*. Wiley–Liss, Hoboken, NJ, 2002.

[16] R.S. Boyer and J.S. Moore. A fast string searching algorithm. *Communication of ACM*, 20:762–772, 1977.

[17] T. Brown. *Genomes*. John Wiley and Sons, New York, 2002.

[18] J. Buhler and M. Tompa. Finding motifs using random projections. In *Proceedings of the Fifth Annual International Conference on Computational Molecular Biology (RECOMB-01)*, pages 69–76, Montreal, Canada, April, 2001.

[19] P. Buneman. *The Recovery of Trees from Measures of Dissimilarity*. Edinburgh University Press, Edinburgh, 1971.

[20] C. Burge and S. Karlin. Prediction of complete gene structures in human genomic DNA. *Journal of Molecular Biology*, 268:78–94, 1997.

[21] L.T. Chow, R.E. Gelinas, T.R. Broker, and R.J. Roberts. An amazing sequence arrangement at the 5′ ends of adenovirus 2 messenger RNA. *Cell*, 12:1–8, 1977.

[22] G. Churchill. Stochastic models for heterogeneous DNA sequences. *Bulletin of Mathematical Biology*, 51:79–94, 1989.

[23] S. A. Cook. The complexity of theorem-proving procedures. In *Proceedings of the 3rd annual ACM Symposium on Theory of Computing*, pages 151–158, Shaker Heights, OH, 1971. ACM Press.

[24] T. H. Cormen, C. L. Leieserson, R. L. Rivest, and C. Stein. *Introduction to Algorithms*. MIT Press, Cambridge MA, 2001.

[25] V. Dancik, T. A. Addona, K. R. Clauser, J. E. Vath, and P. A. Pevzner. De novo peptide sequencing via tandem mass spectrometry. *Journal of Computational Biology*, 6:327–342, 1999.

[26] K. J. Danna, G. H. Sack Jr., and D. Nathans. Studies of simian virus 40 DNA. VII. A cleavage map of the SV40 genome. *Journal of Molecular Biology*, 78:363–376, 1973.

[27] A. Dembo and S. Karlin. Strong limit theorem of empirical functions for large exceedances of partial sums of i.i.d. variables. *Annals of Probability*, 19:1737–1755, 1991.

[28] R. F. Doolittle, M. W. Hunkapiller, L. E. Hood, S. G. Devare, K. C. Robbins, S. A. Aaronson, and H. N. Antoniades. Simian sarcoma virus oncogene, v-sis, is derived from the gene (or genes) encoding a platelet-derived growth factor. *Science*, 221:275–277, 1983.

[29] R. Drmanac, I. Labat, I. Brukner, and R. Crkvenjakov. Sequencing of megabase plus DNA by hybridization: Theory of the method. *Genomics*, 4:114–128, 1989.

[30] A. Duarat, Y. Gerard, and M. Nivat. The chords problem. *Theoretical Computer Science*, 282:319–336, 2002.

[31] R. Durbin, S. Eddy, A. Krogh, and G. Mitchinson. *Biological Sequence Analysis*. Cambridge University Press, Cambridge, England, 1998.

[32] D. Dussoix and W. Arber. Host specificity of infectious DNA from bacteriophage lambda. *Journal of Molecular Biology*, 11:238–246, 1965.

[33] M. B. Eisen, P. T. Spellman, P. O. Brown, and D. Botstein. Cluster analysis and display of genome-wide expression patterns. *Proceedings of the National Academy of Sciences of the United States of America*, 95:14863–14868, 1998.

[34] J. K. Eng, A. L. McCormack, and J. R. Yates (III). An approach to correlate tandem mass spectral data of peptides with amino acid sequences in a protein database. *J Am. Soc. Mass Spectrom.*, 5:976–989, 1995.

[35] E. Eskin and P. A. Pevzner. Finding composite regulatory patterns in DNA sequences. *Bioinformatics*, 18:S354–363, 2002.

[36] D. Feng and R. F. Doolittle. Progressive sequence alignment as a prerequisite to correct phylogenetic trees. *J. Mol. Evol.*, 60:351–360, 1987.

[37] W. M. Fitch. Toward defining the course of evolution: Minimum change for a specific tree topology. *Systematic Zoology*, 20:406–416, 1971.

[38] S.P.A. Fodor, J.L. Read, M.S. Pirrung, L. Stryer, A.T. Lu, and D. Solas. Light-directed spatially addressable parallel chemical synthesis. *Science*, 251:767–773, 1991.

[39] M. R. Garey and D. S. Johnson. *Computers and Intractability: A Guide to the Theory of NP-completeness*. W. H. Freeman and Co., 1979.

[40] W. Gates and C. Papadimitriou. Bounds for sorting by prefix reversals. *Discrete Mathematics*, 27:45–57, 1979.

[41] M.S. Gelfand, A.A. Mironov, and P.A. Pevzner. Gene recognition via spliced sequence alignment. *Proceedings of the National Academy of Sciences of the United States of America*, 93:9061–9066, 1996.

[42] O. Gotoh. An improved algorithm for matching biological sequences. *Journal of Molecular Biology*, 162:705–708, 1982.

[43] M. Gribskov, M. McLachlan, and D. Eisenberg. Profile analysis: detection of distantly related proteins. *Proceedings of the National Academy of Sciences of the United States of America*, 84:4355–4358, 1987.

[44] D. Gusfield. *Algorithms on Strings, Trees, and Sequences. Computer Science and Computational Biology.* Cambridge University Press, Cambridge, England, 1997.

[45] D. Haussler, A. Krogh, I. S. Mian, and K. Sjölander. Protein modeling using hidden Markov models: Analysis of globins. In *Proceedings of the Hawaii International Conference on System Sciences*, pages 792–802, Los Alamitos, CA, 1993. IEEE Computer Society Press.

[46] G. Z. Hertz, G. W. Hartzell 3rd, and G. D. Stormo. Identification of consensus patterns in unaligned DNA sequences known to be functionally related. *Computer Applications in Bioscience*, 6:81–92, 1990.

[47] G.Z. Hertz and G.D. Stormo. Identifying DNA and protein patterns with statistically significant alignments of multiple sequences. *Bioinformatics*, 15:563–577, 1999.

[48] D.G. Higgins, J.D. Thompson, and T.J. Gibson. Using CLUSTAL for multiple sequence alignments. *Methods in Enzymology*, 266:383–402, 1996.

[49] D. Hirschberg. A linear space algorithm for computing maximal common subsequences. *Communication of ACM*, 18:341–343, 1975.

[50] C. A. R. Hoare. Quicksort. *Computer Journal*, 5:10–15, 1962.

[51] S. Hopper, R. S. Johnson, J. E. Vath, and K. Biemann. Glutaredoxin from rabbit bone marrow. purification, characterization, and amino acid sequence determined by tandem mass spectrometry. *Journal of Biological Chemistry*, 264:20438–20447, 1989.

[52] S. Karlin and S.F. Altschul. Methods for assessing the statistical significance of molecular sequence features by using general scoring schemes. *Proceedings of the National Academy of Sciences of the United States of America*, 87:2264–2268, 1990.

[53] R. M. Karp. Reducibility among combinatorial problems. In *Complexity of Computer Computations*, pages 85–103, Yorktown Heights, NY, 1972. Plenum Press.

[54] R.M. Karp and M.O. Rabin. Efficient randomized pattern-matching algorithms. *IBM Journal of Research and Development*, 31:249–260, 1987.

[55] J. Kececioglu and E.W. Myers. Combinatorial algorithms for DNA sequence assembly. *Algorithmica*, 13:7–51, 1995.

[56] J. Kececioglu and D. Sankoff. Exact and approximation algorithms for the inversion distance between two permutation. *Algorithmica*, 13:180–210, 1995.

[57] D. E. Knuth. *The Art of Computer Programming.* Addison-Wesley, 1998.

[58] D.E. Knuth, J.H. Morris, and V.R. Pratt. Fast pattern matching in strings. *SIAM Journal on Computing*, 6:323–350, 1977.

[59] A. Krogh, M. Brown, I.S. Mian, K. Sjölander, and D. Haussler. Hidden Markov models in computational biology: Applications to protein modeling. *Journal of Molecular Biology*, 235:1501–1531, 1994.

[60] S. Kurtz, J. V. Choudhuri, E. Ohlebusch, C. Schleiermacher, J. Stoye, and R. Giegerich. Reputer: the manifold applications of repeat analysis on a genomic scale. *Nucleic Acids Research*, 29:4633–4642, 2001.

[61] G.M. Landau and U. Vishkin. Efficient string matching in the presence of errors. In *26th Annual Symposium on Foundations of Computer Science*, pages 126–136, Los Angeles, October 1985.

[62] E. Lander et al. Initial sequencing and analysis of the human genome. *Nature*, 409:860–921, 2001.

[63] C.E. Lawrence, S.F. Altschul, M.S. Boguski, J.S. Liu, A.F. Neuwald, and J.C. Wootton. Detecting subtle sequence signals: A Gibbs sampling strategy for multiple alignment. *Science*, 262:208–214, 1993.

[64] V. I. Levenshtein. Binary codes capable of correcting deletions, insertions, and reversals. *Cybernetics and Control Theory*, 10:707–710, 1966.

[65] L. Levin. Universal sorting problems. *Problems of Information Transmission*, 9:265–266, 1973.

[66] B. Lewin. *Genes VII*. Oxford University Press, Oxford, UK, 1999.

[67] D.J. Lipman and W.R. Pearson. Rapid and sensitive protein similarity searches. *Science*, 227:1435–1441, 1985.

[68] S. P. Lloyd. Least squares quantization in PCM. *IEEE Transactions on Information Theory*, 28:129–137, 1982.

[69] Y. Lysov, V. Florent'ev, A. Khorlin, K. Khrapko, V. Shik, and A. Mirzabekov. DNA sequencing by hybridization with oligonucleotides. *Doklady Academy Nauk USSR*, 303:1508–1511, 1988.

[70] J. MacQueen. On convergence of k-means and partitions with minimum average variance. *Annals of Mathematical Statistics*, 36:1084, 1965.

[71] M. Mann and M. Wilm. Error-tolerant identification of peptides in sequence databases by peptide sequence tags. *Analytical Chemistry*, 66:4390–4399, 1994.

[72] L. Marsan and M. F. Sagot. Algorithms for extracting structured motifs using a suffix tree with an application to promoter and regulatory site consensus identification. *Journal of Computational Biology*, 7:345–362, 2000.

[73] W. J. Masek and M. S. Paterson. A faster algorithm computing string edit distances. *Journal of Computational System Science*, 20:18–31, 1980.

[74] A. M. Maxam and W. Gilbert. A new method for sequencing DNA. *Proceedings of the National Academy of Sciences of the United States of America*, 74:560–564, 1977.

[75] N. Metropolis, A. W. Rosenbluth, M. N. Rosenbluth, A. H. Teller, and E. Teller. Equation of state calculations by fast computing machines. *Journal of Chemical Physics*, 21:1087–1092, 1953.

[76] D. Mount. *Bioinformatics: Sequence and Genome Analysis*. Cold Spring Harbor Press, Cold Spring Harbor, NY, 2001.

[77] E.W. Myers and W. Miller. Optimal alignments in linear space. *Computer Applications in Biosciences*, 4:11–17, 1988.

[78] J. Nadeau and B. Taylor. Lengths of chromosome segments conserved since divergence of man and mouse. *Proceedings of the National Academy of Sciences of the United States of America*, 81:814–818, 1984.

[79] S.B. Needleman and C.D. Wunsch. A general method applicable to the search for similarities in the amino acid sequence of two proteins. *Journal of Molecular Biology*, 48:443–453, 1970.

[80] S. J. O'Brien. *Tears of the Cheetah*. Thomas Dunne Books, New York, 2003.

[81] S. J. O'Brien, W. G. Nash, D. E. Wildt, M. E. Bush, and R. E. Benveniste. A molecular solution to the riddle of the giant panda's phylogeny. *Nature*, 317:140–144, 1985.

[82] H. Peltola, H. Soderlund, and E. Ukkonen. SEQAID: A DNA sequence assembling program based on a mathematical model. *Nucleic Acids Research*, 12:307–321, 1984.

[83] P.A. Pevzner. *l*-Tuple DNA sequencing: computer analysis. *Journal of Biomolecular Structure and Dynamics*, 7:63–73, 1989.

[84] P.A. Pevzner, V. Dancik, and C.L. Tang. Mutation-tolerant protein identification by mass-spectrometry. *Journal of Computational Biology*, 7:777–787, 2000.

[85] P.A. Pevzner and G. Tesler. Genome rearrangements in mammalian evolution: Lessons from human and mouse genomes. *Genome Research*, 13:37–45, 2003.

[86] P.A. Pevzner and M.S. Waterman. Multiple filtration and approximate pattern matching. *Algorithmica*, 13:135–154, 1995.

[87] J.C. Roach, C. Boysen, K. Wang, and L. Hood. Pairwise end sequencing: A unified approach to genomic mapping and sequencing. *Genomics*, 26:345–353, 1995.

[88] D. F. Robinson. Comparison of labeled trees with valency three. *Journal of Combinatorial Theory (B)*, 11:105–119, 1971.

[89] J. R. Sadler, M. S. Waterman, and T. F. Smith. Regulatory pattern identification in nucleic acid sequences. *Nucleic Acids Research*, 11:2221–2231, 1983.

[90] N. Saitou and M. Nei. The neighbor-joining method: A new method for reconstructing phylogenetic trees. *Molecular Biological Evolution*, 4:406–425, 1987.

[91] F. Sanger, S. Nicklen, and A. R. Coulson. DNA sequencing with chain-terminating inhibitors. *Proceedings of the National Academy of Sciences of the United States of America*, 74:5463–5467, 1977.

[92] D. Sankoff. Minimal mutation trees of sequences. *SIAM Journal of Applied Mathematics*, 28:35–42, 1975.

[93] D. Sankoff. Edit distances for genome comparisons based on non-local operations. In *Third Annual Symposium on Combinatorial Pattern Matching*, volume 644 of *Lecture Notes in Computer Science*, pages 121–135, Tucson, AZ, 1992. Springer-Verlag.

[94] S.S. Skiena, W.D. Smith, and P. Lemke. Reconstructing sets from interpoint distances. In *Proceedings of Sixth Annual Symposium on Computational Geometry*, pages 332–339, Berkeley, CA, June, 1990.

[95] H.O. Smith and K.W. Wilcox. A restriction enzyme from Hemophilus influenzae. I. Purification and general properties. *Journal of Molecular Biology*, 51:379–391, 1970.

[96] T.F. Smith and M.S. Waterman. Identification of common molecular subsequences. *Journal of Molecular Biology*, 147:195–197, 1981.

[97] E.E. Snyder and G.D. Stormo. Identification of protein coding regions in genomic DNA. *Journal of Molecular Biology*, 248:1–18, 1995.

[98] R. R. Sokal and C. D. Michener. A statistical method for evaluating systematic relationships. *University of Kansas Science Bulletin*, 38:1409–1438, 1958.

[99] E.L. Sonnhammer, S.R. Eddy, and R. Durbin. Pfam: a comprehensive database of protein domain families based on seed alignments. *Proteins*, 28:405–420, 1997.

[100] E. Southern. United Kingdom patent application GB8810400. 1988.

[101] G. Stormo, T. Schneider, L. Gold, and A. Ehrenfeucht. Use of the perceptron algorithm to distinguish translational initiation sites in E. coli. *Nucleic Acids Research*, 10:2997–3011, 1982.

[102] A. H. Sturtevant and T. Dobzhansky. Inversions in the third chromosome of wild races of *Drosophila pseudoobscura*, and their use in the study of the history of the species. *Proceedings of the National Academy of Sciences of the United States of America*, 22:448–450, 1936.

[103] A. R. Templeton. Out of Africa again and again. *Nature*, 416:45–51, 2002.

[104] J.C. Venter et al. The sequence of the human genome. *Science*, 291:1304–1351, 2001.

[105] T.K. Vintsyuk. Speech discrimination by dynamic programming. *Kibernetika*, 4:52–57, 1968.

[106] A. Viterbi. Error bounds for convolutional codes and an asymptotically optimal decoding algorithm. *IEEE Transactions on Information Theory*, 13:260–269, 1967.

[107] M. S. Waterman. *Skiing the Sun*. (unpublished manuscript), 2004.

[108] M.S. Waterman. *Introduction to Computational Biology*. Chapman Hall, New York, 1995.

[109] M.S. Waterman and M. Eggert. A new algorithm for best subsequence alignments with application to tRNA–rRNA comparisons. *Journal of Molecular Biology*, 197:723–728, 1987.

[110] J. Weber and G. Myers. Whole genome shotgun sequencing. *Genome Research*, 7:401–409, 1997.

[111] P. Weiner. Linear pattern matching algorithms. In *Proceedings of the 14th IEEE Symposium on Switching and Automata Theory*, pages 1–11, University of Iowa, October 1973.

[112] J. R. Yates III, J. K. Eng, A. L. McCormack, and D. Schieltz. Method to correlate tandem mass spectra of modified peptides to amino acid sequences in the protein database. *Analytical Chemistry*, 67:1426–1436, 1995.

[113] K. Zaretskii. Constructing a tree on the basis of a set of distances between the hanging vertices. *Uspekhi Mat. Nauk.*, 20:90–92, 1965.

[114] Z. Zhang. An exponential example for a partial digest mapping algorithm. *Journal of Computational Biology*, 1:235–239, 1994.

Index